Cases on Collaborative Experiential Ecological Literacy for Education

Lizoon Nahar
Indiana University of Pennsylvania, USA

Nada Tayem
Indiana University of Pennsylvania, USA

A volume in the Advances in Early Childhood and K-12 Education (AECKE) Book Series

IGI Global
PUBLISHER of TIMELY KNOWLEDGE

Published in the United States of America by
IGI Global
Information Science Reference (an imprint of IGI Global)
701 E. Chocolate Avenue
Hershey PA, USA 17033
Tel: 717-533-8845
Fax: 717-533-8661
E-mail: cust@igi-global.com
Web site: http://www.igi-global.com

Library of Congress Cataloging-in-Publication Data

Names: Nahar, Lizoon, 1983- editor. | Tayem, Nada, 1979- editor.
Title: Cases on collaborative experiential ecological literacy for education / edited by Lizoon Nahar, Nada Tayem.
Description: Hershey, PA : Information Science Reference, 2024. | Includes bibliographical references and index. | Summary: "In this book, the goal is to convey the profound influence of environmental education and literacy rooted in hands-on experiences"-- Provided by publisher.
Identifiers: LCCN 2024011679 (print) | LCCN 2024011680 (ebook) | ISBN 9798369317105 (hardcover) | ISBN 9798369345245 (paperback) | ISBN 9798369317112 (ebook)
Subjects: LCSH: Environmental education--Case studies. | Environmental literacy--Case studies. | Experiential learning--Case studies.
Classification: LCC GE70 .C37 2024 (print) | LCC GE70 (ebook) | DDC 304.2071--dc23/eng/20240404
LC record available at https://lccn.loc.gov/2024011679
LC ebook record available at https://lccn.loc.gov/2024011680

British Cataloguing in Publication Data
A Cataloguing in Publication record for this book is available from the British Library.

All work contributed to this book is new, previously-unpublished material.
The views expressed in this book are those of the authors, but not necessarily of the publisher.

For electronic access to this publication, please contact: eresources@igi-global.com.

Table of Contents

Detailed Table of Contents

The educational system should cultivate a generation committed to environmental stewardship and equipped to combat climate change. While individuals of all ages can glean knowledge about climate issues from media sources, educators hold a unique capacity to instill environmental literacy in their students who need adequate preparation and support. This research utilized an unstructured direct observation method to explore the impact of an environmental initiative on empowering educators involving 15 high school teachers and 150 students from the USA, India, and Bangladesh. Data analysis was conducted through the lens of the Environmental Education Strategies Framework. The findings indicated that effective strategies for conveying information and building understanding supported teachers in empowering students with improved skills to conduct sustainable actions. This chapter also highlights the challenges faced by the coordination team for project execution and shares recommendations for all stakeholders for future planning.

This chapter introduces the exciting world of cutting-edge technologies in controlled environment agriculture (CEA) and their potential impact on high school education. The objective of this chapter is to establish the groundwork for transforming education through the creation of state-of-the-art augmented reality (AR) and virtual reality (VR) learning platforms, and research within the domain of CEA. These innovative technologies provide immersive learning experiences, allowing high school students and new entrepreneurs with diverse capabilities to explore new agricultural technologies in a virtual environment before implementing them in the world. While AR and VR technologies hold significant promise, there are notable challenges in effectively integrating AR and VR technologies within educational and CEA settings. These challenges highlight the need for thoughtful integration, aiming to enrich learning experiences. Through this chapter, the authors examine various ideas that have successfully harnessed the benefits of AR and VR in education and discuss a proposed framework.

The climate crisis poses a major threat to sustainability, highlighting the need for climate education that develops students' systems thinking skills regarding dynamic environmental issues. For this reason, this project, funded by the United Nations Development Programme (UNDP), was developed by a research university, a non-governmental organization, and a middle school partnership. As part of the project, this study investigated a ten-week after-school climate education program for 8th-grade students that aimed to improve their understanding of the climate system through simulation and policymaking. Semi-structured interviews were analyzed to assess the program's model alignment with students' explanations. The most frequently mentioned category was the economy, followed by the carbon cycle; population and the greenhouse effect were the least common. The study offers valuable insights into using a systems approach in climate education to comprehend climate complexity and dynamics by considering the importance of economy and population in understanding human-caused climate change.

Environmental awareness is crucial for ensuring a sustainable future. This chapter presents how fostering environmental awareness in children is important for securing a safe Mother Earth for future generations. This chapter showcases an initiative implemented through an art-integrated project for environmental awareness across three public schools in Lahore, Pakistan. Beyond strengthening academic and creative skills besides social-emotional learning (SEL), the program also cultivated a deep sense of environmental responsibility in students, teachers, and administrators. Olomopolo Media (Pakistan) launched the global "River of Hope" British Council program to raise awareness about rivers and their deep connection to the cities they flow through. This program used traditional printmaking to raise awareness about river Ravi's problems. It focused on inspiring action for its restoration by presenting the power of art for environmental causes resulting in the culmination of a student-created printmaking exhibition displayed during the Queen's Platinum Jubilee in England.

This study aims to develop competencies for sustainable development and create corresponding indicators designed to address the current conditions, to study desirable conditions in accordance with the concept of the whole-school approach, to develop innovations and assess the effectiveness of the integrative learning process to enhance competencies for sustainable development. It also aims to propose education management strategies for sustainable development. The study's findings were developing competencies for sustainable development and indicators and dividing them into eight competencies. The study on the students' current conditions revealed average competencies ($\bar{x}$ = 3.73, SD = 1.34) regarding the desirable conditions of competencies. Students' overall wishes were at the highest level ($\bar{x}$ = 4.68, SD = 0.67). The integrative learning process consisted of four steps. The impoverished students had at least 70% of the competencies for sustainable development in each cycle. Recommendations on education management strategies for sustainable development consisted of four strategies.

Teaching

Unconscious resource consumption and environmental problems pose a threat to humanity. Therefore, it is necessary for young individuals to gain ecological literacy knowledge and reflect it in their daily lives in order to know only the changing and developing world. The aim of this research is to adapt a scale to determine the ecological literacy levels of high school students. The sample of the study consists of approximately 530 volunteer high school students. In terms of construct validity, exploratory and confirmatory factor analyzes were performed on the data. Factor analysis of the scale was performed. Sixteen items with an item load of less than 0.45 and burdening more than one factor were removed from the scale, and the remaining 24 items were found to be grouped under five factors. A Cronbach's alpha (α=0.816) scale with high reliability, consisting of five factors and 24 items, including ecological behavioral literacy, information literacy, emotional literacy, ethical literacy, mindfulness literacy, was obtained. The reliability value of the scale is high.

For a sustainable future, it is crucial to cultivate individuals with ecological literacy skills. As in every field, the process that begins within the family to increase the number of ecologically literate individuals must be supported and sustained through education in schools. Consequently, curricula and courses should be structured to promote ecological literacy. This study examines the presence of ecological literacy in the high school biology curriculum using the document analysis method. The research findings indicate that topics and learning outcomes related to ecological literacy are limited in the overall program and are not included at all grade levels. Moreover, it has been determined that the outcomes related to the ethical, emotional, and behavioral dimensions of ecological literacy are insufficient. Recommendations have been made to ensure that the curriculum is structured to support ecological literacy.

The climate over the Earth's surface has remarkably been affected by various inconsiderate usages of natural resources and human-made disasters worldwide, such as substantial growth of fossil fuel consumption and coal-energy-dependent infrastructures, abundant use of synthetic materials, and the absence of environment-friendly technologies and solutions. In addition, in recent years, people seem to be more indulged in hyper-consumerism, quickly changeable products and fashion, and unsustainable urban design and living habits, which generates more waste and gives birth to numerous complex environmental problems. For the last few decades, the delicate balance of the Earth's ecosystems has continued to face constant threats, ranging from the profound impacts of climate change to the depletion of biodiversity and the severe consequences of pollution.

Foreword

A NEED OF THE HOUR TO EDUCATE PEOPLE ABOUT A CHANGING WORLD

Climate change, a complex, emerging, and wide-ranging threat, is an imminent crisis exacerbated by the lack of education from academia and institutional settings. This book, born from a similar void, is a significant collection of environmental education models, projects, and new media integrations. By teaching, educating, and raising awareness among teachers, students, and scholars about sustainable solutions, we foster mobilization and create a conscious community ready to engage and survive this unprecedented threat. This book is a collection of cutting-edge multimodal educational strategies that encourage critical, community-based, and collaborative initiatives to generate environmental awareness among teachers and students. It brings together experiential, experimental, creative, and artful approaches from various scholars. While teaching plays a key role in enriching climate change discussions, the book intersects social, political, cultural, and policy-related challenges to extend beyond conventional climate change discussions for public awareness. In addition to the key elements, steps, methods, and decisions needed to be integrated into environmental education, the book emphasizes the uniform dissemination of environmental education through 21st-century educational technologies including AR, VR, and many more.

This book takes a climate-action-oriented approach from academic and public perspectives by concerning a range of climate change topics situated in global discourse. By breaking away from the deliberation of widespread science-heavy climate change, the book brings art and humanities to the forefront. This book outlines the interrelations of climate, education, and teaching through diverse examples of engaging digital literacy models in various classrooms. The book rationalizes the need for collaborative and collective humanities-based information infrastructure to develop effective educational outcomes by expanding the intellectual matrix and integrating dialogues from students, teachers, community learners, etc. To do so, one of the primary goals this edited volume focuses on is to deliver potent examples of innovative environmental education models, approaches, and methods to disseminate action-based knowledge and expand resources.

This collection offers unique teaching models as deliverables concerning student engagement, education, and communication of environmental issues. The book's two sections, which have eleven chapters, explore and concern action on a range of ways to bring climate change to classrooms. The first section investigates a range of independent project-based initiatives by discussing case studies and experiential notions from diverse classrooms across the globe and assessing the summative and generative impacts of those projects and practices on students recognizing and educating others about environmental issues. Moreover, the section broadly interspersed effective strategies to install environmental literacy modules for students from various levels, classrooms, and ethnicities by fostering collaboration between teachers, researchers, and stakeholders. The second section has a broader scope by exploring interdisciplinary interconnections between environmental literacy and technological teaching to promote problem-solving, ecological stewardship, community engagement, experimental teaching, and assessment. So, the book is a unique collection that extends its horizon to facilitate the need for climate education, policies, and technological installation.

Eleven chapters of the book share unique parameters to generate importance, precision, and outcomes of environmental and climate education for sustainable development. The book begins with a reflective chapter that discusses the impact of an environmental initiative involving teachers and students in several countries. It highlights the effective strategies that help teachers empower students with the skills to undertake sustainable actions. The second chapter takes a giant leap by asserting an experimental approach to explore controlled ecological agriculture and its gamifying impact on high school students through VR and AR. Some chapters address UNDP goals as part of collaborative educational curricula installed in research universities, NGOs, and climate education programs to improve climate policy and action through simulation. When it comes to teaching climate change in the Humanities, art forms matter, for which the book rightly uses media ecologic structure in education by integrating the "River of Hope" program developed by the British Council and integration of art to harness creative climate problem-solving skills. The use of educational technologies and digital media shapes the next chapter, showcasing technology-assisted approaches to assess the risk of environmental hazards related to pollution, resource depletion, and biodiversity conservation and find possible solutions. The next chapter addresses the whole-school approach to developing innovative, integrative learning strategies through eight different competencies that would help meet the education mitigation for impoverished students. Some chapters develop aspiring teaching models by using multimodal texts to teach fiction, nonfiction, and poetry. Additionally, there are chapters that pull off various uses of digital technologies to facilitate environmental education beyond institutional settings, validity, and reliability to assess ecological literacy, and use of environmental literacy in disciplinary educational settings like biology, art, literature, and beyond.

While efforts to design an educational curriculum for climate change have not fully met the demands of the Paris Agreement, this book effectively addresses education, training, teaching, public awareness, and participation in climate change. By developing a robust compilation of climate-action-based teaching applications, and project-based integration, the book brings out the critical, practical, and creative potential of teachers, students, scholars, and professionals. It empowers them to push the conventions of climate actions. For 21st-century learners, educators, and innovators, this book is not only an important read for delving into interdisciplinary methods and techniques for environmental education but also a guide for enhancing climate education in classrooms, society, and beyond.

Zeeshan Faiez Siddique
Indiana University of Pennsylvania, USA

Preface

The earth is exhibiting its growing agony through a rising number of environmental catastrophes, such as floods, heatwaves, and forest fires, which are regularly making headlines nowadays. In order to tackle this predicament, it is crucial to have a strong dedication to ecological literacy in school and beyond. Having a deep comprehension of ecosystems and their activities, known as ecological literacy, is essential for attaining sustainability objectives and getting ready for forthcoming problems. It provides individuals with the necessary skills and knowledge to actively engage in society and make educated and responsible decisions that contribute to the well-being of the planet.

The significance of the educational system in fostering ecological consciousness among citizens is evident. Nevertheless, this raises the question of the extent to which our education system is prepared to integrate environmental education, particularly in light of the present imperative to equip students with the necessary competencies for an always changing labor market. While leading a high school-level environmental awareness project[1] in 2023 in Pennsylvania, USA, funded and supported by the U.S. Department of State, we (Lizoon Nahar and Nada Tayem) faced challenges in obtaining some schools' approval and involving teachers and students in the project. These challenges became a primary concern for us as educators with over a decade of experience in environmental education, both in the United States and internationally. Our shared concern led to a thought-provoking discussion that generated further inquiries.

Do teachers possess the curricular standards, pedagogical skills, technology tools, and other resources necessary for incorporating educational interventions on environmental issues into their teaching? Does the situation in the USA differ from other regions of the world? What are the latest trends in environmental programs designed to improve ecological literacy? Which instructional methodologies are more efficacious in fostering environmental literacy? And what obstacles do teachers encounter while fostering environmental consciousness in their students, and how do they address those obstacles?

As educators and academics, we acknowledge the importance of finding solutions to these inquiries. In addition, we acknowledge the significance of staying well-informed, maintaining connections, and continuously expanding our knowledge and development in order to effectively support our students and communities in becoming ecologically literate. We acknowledged the necessity of persistently innovating our learning and teaching methods to cater to the changing demands of the present circumstances. We have come to the realization that simply acquiring knowledge about these matters is insufficient. We should collect the responses and distribute them to a wider audience, including teachers, faculty members, curriculum coordinators, school administrators, independent researchers, and policymakers who are strongly committed to restoring ecological equilibrium in nature and fostering students' ecological literacy. We came up with the idea of publishing this book with the goal of empowering those who care to influence others to become future guardians of our world.

We decided to initiate this collaborative endeavor in June 2023 with the goal of publishing a book on environmental education that caters to students across all academic levels. This publication effort aimed to engage teacher practitioners, independent project directors, and environmental education scholars in sharing their practical experiences and research discoveries on environmental education through this book. Therefore, we have gathered researchers and practitioners from different regions of the globe in order to provide our readers with examples of cutting-edge educational programs and practices that aim to enhance ecological literacy.

This book offers a range of materials for environmental educators, catering to both classroom teachers seeking engaging environmental activities and college instructors aiming to enhance their students' environmental consciousness. This book also provides pragmatic advice and tactics for integrating environmental education into many disciplines and grade levels, rendering it a valuable asset for instructors from all backgrounds. In addition, the book incorporates case studies and success stories to motivate and direct educators in their pursuit of establishing a more sustainable future. The objective of this book is to address the lack of environmental education resources and enable educators to involve students in significant, practical learning opportunities that promote a deeper understanding of the environment. By furnishing a thorough manual replete with tactics, interactive exercises, and practical instances, instructors will be prepared to easily include environmental education into their curriculum. Our goal is to assist educators in their efforts to foster environmentally knowledgeable individuals.

This book is structured into two primary sections. The first section, named "Advancing Ecological Literacy through Independent Projects, "analyzes various case studies of independent projects in different educational contexts worldwide. These case studies serve to demonstrate the practical application of these practices and their influence on student involvement and comprehension of environmental matters. Furthermore, it offers diverse perspectives on successful approaches to incorporating environmental education into different academic disciplines and grade levels, highlighting the significance of cooperation among teachers and interested parties in fostering environmental knowledge among students. This section comprises five chapters that specifically address various educational aspects pertaining to independent environmental projects that receive funding from esteemed organizations such as the United Nations Development Program (UNDP) Accelerator Lab, the United States Department of State, the British Council, and the National Institute of Food and Agriculture (NIFA). The authors of this section reported projects that were carried out in both developed nations, such as the USA, and least developed nations, like Pakistan.

Chapter 1 emphasizes the significance of equipping high school educators with the necessary knowledge and skills to improve students' understanding of the environment. This case study investigates an environmental program that engages educators and students from the United States, India, and Bangladesh. This illustrates that effective teaching methods have the ability to empower educators to promote sustainable behaviors among their students. The study examines many scenarios to underscore the global reach of environmental education and the imperative of cross-cultural cooperation and support in effectively addressing climate change.

The second chapter explores the capacity of cutting-edge technology and the engaging encounters that teachers can create to cultivate a more profound comprehension and admiration for sustainable practices among students from various cultural backgrounds. The authors thoroughly review literature on immersive agricultural education and explore how technologies might facilitate immersive learning experiences. The study found how these experiences enable students to virtually examine agricultural advancements. The authors discuss the difficulties in incorporating augmented reality (AR) and virtual reality (VR) into educational environments, with the potential advantages for improving student engagement and comprehension.

Chapter 3 focuses on how middle school students can apply systems thinking in climate education. The study sought to enhance students' comprehension of the intricate nature of the climate system by incorporating a ten-week program that involved simulation and policymaking activities. The authors conducted semi-structured interviews to assess the program's alignment with students' explanations. They reported that the economy was the most frequently mentioned category, followed by the carbon cycle, while population and the greenhouse effect were the least common. The study provides valuable insights into how to use a systems approach in climate education to grasp climate complexity and dynamics.

Chapter 4 demonstrates the efficacy of art as a potent instrument for cultivating environmental consciousness in young kids. In Lahore, Pakistan, a project successfully incorporated traditional printmaking into an environmental education program, effectively connecting artistic expression with environmental stewardship. The authors find that the art integration technique not only enhanced kids' artistic and social-emotional abilities but also cultivated a stronger bond with local environmental concerns. This project showcases the efficacy of utilizing art to include students in environmental stewardship.

Chapter 5 emphasizes the cultivation of sustainable development competencies through a comprehensive learning process. The study establishes a framework for improving sustainable development education by developing and evaluating indicators for eight compentencies. The authors emphasized the necessity for inventive educational approaches that tackle present circumstances and foster sustainable behaviors among pupils.

The second section, titled "Exploring Pedagogical Aspects in Environmental Education Teaching," comprises six chapters that expand the scope of environmental education. It delves into various educational aspects of environmental awareness practices that have the potential to boost students' learning in all areas. The characteristics encompass interdisciplinary linkages, promote critical thinking and problem-solving abilities, cultivate environmental stewardship, create community engagement, and utilize innovative teaching methods and assessment tools.

Chapter 6 highlights the significance of literature in promoting environmental awareness among university students in Algeria and the United States. The instructors strive to actively involve students in the study of environmental issues by combining different teaching methods in English literature, such as analyzing Romantic and Modernist poetry and writing place-based reports and problem-solution essays. The multimodal strategy not only improves students' literary abilities but also encourages critical analysis of environmental concerns, demonstrating the capacity of humanities to tackle worldwide challenges.

Chapter 7 explores the worldwide advancement of environmental education (EE) using digital technologies and media. The author emphasizes the role of platforms such as online courses and virtual reality simulations in making EE more accessible and captivating. The integration of technology in EE has numerous benefits, including enhanced accessibility, heightened awareness, positive behavioral shifts, improved resource efficiency, expanded learning experiences, and enhanced cooperation. In this chapter, the author emphasized various obstacles that hinder the complete realization of technology in EE, including less outside engagement, privacy issues, economic considerations, and more factors. To overcome these problems, it is necessary to engage in strategic planning and allocate resources effectively. The author proposed that crucial measures for addressing problems involve guaranteeing data security, offering top-notch content, and striking a balance between digital technologies and conventional outdoor teaching.

Chapter 8 emphasizes the need for young individuals to acquire and implement ecological literacy in their everyday lives. The aim of this study is to modify a measurement tool in order to assess the levels of ecological literacy among high school students. The authors have conducted the study on high school students using a survey method, employ confirmatory factor analyses. The researchers state that the scale effectively measures ecological literacy by assessing five factors: ecological behavioral literacy, information literacy, emotional literacy, ethical literacy, and mindfulness literacy. The measurement is considered thorough and dependable.

The ninth chapter delves into the degree to which the high school biology curriculum in Turkey incorporates ecological literacy. The study indicates that the Turkish educational system does not sufficiently cover and address ecological literacy themes across different grade levels. The authors emphasize that in order to foster individuals who possess a deep understanding of the environment, it is necessary to reorganize the curriculum to incorporate more extensive and morally-guided ecological material, with a focus on highlighting the significance of integrating ecological knowledge into science instruction.

In Chapter 10, the incorporation of environmental education in Bangladesh was examined by utilizing educational technology and digital media. The authors discuss the advantages and difficulties of integrating these technologies into the curriculum to enhance students' understanding of environmental concerns. This chapter highlights the significance of ensuring that environmental information is easily accessible and captivating, utilizing technology to improve learning and motivate individuals to take action.

Chapter 11 emphasizes the importance of including air pollution and associated subjects in school curricula. The revised academic requirements in Pennsylvania prioritize the integration of environmental education across all disciplines, including science, technology, engineering, and ethical considerations. The author asserted that by instructing students on the subject of air pollution, including its origins and effects on public health, the curriculum strives to cultivate knowledgeable and accountable individuals who can actively promote sustainable behaviors.

These book chapters collectively demonstrate a worldwide acknowledgment of the significance of environmental education in promoting a sustainable future. The approaches they showcase encompass a wide range of disciplines, including science, literature, art, and technology. The ultimate goal of these strategies is to improve environmental literacy. The studies emphasize the necessity of employing multidisciplinary methodologies, pioneering pedagogical techniques, and international partnerships in order to adequately tackle environmental issues. Through the incorporation of environmental education across different academic subjects and the utilization of a wide range of resources, educators may provide students with the necessary information and abilities to effectively contribute to the creation of a sustainable global environment.

Ultimately, this book highlights the crucial importance of continuous research and development in the field of environmental education. It is necessary for educators, policymakers, and stakeholders to stay updated on new technology and teaching methods that help improve environmental literacy. This book promotes an unwavering commitment to advancing innovation in educational methods, emphasizing the significance of adjusting to evolving environmental circumstances and societal demands. The book envisions a future where global education systems deeply integrate environmental education, enabling the next generation to make significant contributions towards a more sustainable world.

Lizoon Nahar
Indiana University of Pennsylvania, USA

Nada Tayem
Indiana University of Pennsylvania, USA

ENDNOTE

1. In 2023, Lizoon Nahar received an Alumni Ties grant from the U.S. Department of State to conduct a high school environmental project in the USA, India, and Bangladesh.

Section 1
Advancing Ecological Literacy Through Independent Projects

This section examines case studies of independent projects from diverse educational settings across the globe to illustrate the implementation of these practices and their impact on student engagement and understanding of environmental issues. Additionally, it provides insights into effective strategies for integrating environmental education across various subjects and grade levels, emphasizing the importance of collaboration among educators and stakeholders in promoting environmental literacy among students.

Chapter 1

Assessing the Influence of a High School Environmental Project on Teacher Readiness Across Borders

Lizoon Nahar
https://orcid.org/0000-0002-7209-8823
Indiana University of Pennsylvania, USA

EXECUTIVE SUMMARY

The educational system should cultivate a generation committed to environmental stewardship and equipped to combat climate change. While individuals of all ages can glean knowledge about climate issues from media sources, educators hold a unique capacity to instill environmental literacy in their students who need adequate preparation and support. This research utilized an unstructured direct observation method to explore the impact of an environmental initiative on empowering educators involving 15 high school teachers and 150 students from the USA, India, and Bangladesh. Data analysis was conducted through the lens of the Environmental Education Strategies Framework. The findings indicated that effective strategies for conveying information and building understanding supported teachers in empowering students with improved skills to conduct sustainable actions. This chapter also highlights the challenges faced by the coordination team for project execution and shares recommendations for all stakeholders for future planning.

DOI: 10.4018/979-8-3693-1710-5.ch001

INTRODUCTION

The current alarming climate change conditions around the world highlight the urgent need for a society that is educated in environmental issues and prepared to take action to mitigate the impact of climate change. Education plays a crucial role in raising awareness about environmental degradation, climate change challenges, and sustainable practices. The education sector may provide opportunities to instill environmental literacy from a young age. Despite this, there is a global scarcity of environmental awareness programs targeted at young children. This gap is significant because developing environmental literacy early in life ensures that future generations can make informed, responsible decisions and take sustainable actions when faced with environmental crises (Ma et al., 2023; Sell et al., 2023). Unfortunately, many children worldwide complete their secondary education without having received adequate environmental education because of the lack of a formal curriculum. Recognizing this, the United Nations and various organizations advocated for systematically integrating environmental education (EE) into school curricula to build a solid foundation of environmental concepts. This integration is essential for students' future understanding and actions as responsible citizens (UNESCO, 2014).

Schools are pivotal in shaping students' attitudes toward the environment, enabling them to become active advocates for climate change mitigation through increased climate literacy and awareness of climate change-related trends (United Nations Educational, Scientific, and Cultural Organization, 2019). Including climate change and other environmental issues in the curriculum is essential. Universities also face the challenge of incorporating diverse perspectives on sustainability, which requires systemic and comprehensive thinking and new teaching methods (Ma et al., 2023). In this situation, preparing our younger generation to handle the climate crisis requires empowering them with strong environmental literacy skills, which are currently not part of most school curricula.

The need for enhanced environmental literacy, teaching climate change, and helping students understand its implications is no longer just the domain of scientists. Teachers must integrate climate change elements into their instructional plans, enabling students to learn how they can contribute to addressing climate change issues within their capacity. Students must understand that climate change has far-reaching consequences for the earth and human lives and that it plays a crucial role in mitigating its adverse effects. Although many teachers are taking the initiative to incorporate climate change awareness into the curriculum, these efforts are insufficient (Corpus et al., 2022). We need more intensified and systematic approaches to nurture responsible, environmentally conscious future citizens.

Given the lack of opportunities for younger students to learn about local environmental issues from curricula, promoting environmental literacy through independent projects is crucial. This paper will discuss a practical example of a project that employed multiple strategies to empower 15 high school teachers and 150 high school students with environmental literacy. Over six months, the project engaged participants in various activities, including workshops, field studies, and community outreach programs. One of the core strategies of the project was to equip high school teachers with the tools and resources needed to integrate environmental literacy into their curricula effectively. This empowerment allowed teachers to become facilitators of change, guiding their students through the learning process and encouraging them to take ownership of local environmental initiatives. Continuous professional development sessions empowered teachers by training them in the latest environmental education methodologies and providing them with practical resources to support their teaching.

For students, the project offered a unique opportunity to engage in real-world environmental issues. They participated in activities such as biodiversity assessments, water quality testing, and waste management projects. These hands-on experiences not only enhanced their understanding of environmental concepts but also enabled them to apply their knowledge in practical settings. By the end of the project, students had developed actionable plans to address specific environmental issues within their communities, demonstrating their ability to translate learning into impactful action. The experiences from this project offer environmental educators, researchers, and administrative heads valuable insights into how to raise awareness about environmental issues among students. To determine the effectiveness of the 'Know Nature: Local to Global' project, the following research questions were formulated:

RQ 1: How did the project "Know Nature: Local to Global" empower high school teachers to take action to enhance their students' environmental literacy?

RQ 2: What challenges did the project face during implementation?

LITERATURE REVIEW

In this section, I synthesized studies that discuss the influence of educational environmental programs on students and the concept of environmental literacy. Then I provided the highlights of the components of a successful environmental program. I concluded by outlining the Environmental Education Strategies framework (Monroe et al., 2007).

Outcomes of Environmental Programs in School Settings

An educational, environmental program focuses not only on understanding the environment but also on actively contributing to its preservation (Glacking & King, 2020; Bercasio, 2022). Several studies have highlighted the increasing importance of educational institutions in driving sustainable development. For instance, Zafar et al. (2023) stated that the education level of its residents influences the environmental quality of an area. Higher education helps people rethink their impact on the planet and addresses important sustainability issues (Hensley, 2020). These findings underscore the importance of integrating environmental education into school lessons to foster a generation capable of effectively addressing environmental challenges. However, because of the multidisciplinary nature of the environmental education field, with its various approaches and metrics and complex interrelationships, obtaining a comprehensive global picture of environmental education evolution and the impact of different approaches remains challenging (Díaz-López et al., 2023).

Environmental Literacy

Environmental literacy is crucial in education for creating a society of informed and empowered citizens capable of tackling serious socio-environmental issues (Díaz-López et al., 2023). It encompasses environmental awareness, problem-solving skills for environmental issues, relevant values, motivation, skills, and a readiness to take action (Roth, 1992). According to Hollweg et al. (2011), environmental literacy includes the knowledge, attitudes, motivation, cognitive abilities, skills, and behaviors necessary for making effective environmental decisions. People who possess environmental literacy are motivated to improve the well-being of people, societies, and the global environment, and they actively engage in civic life. There is growing recognition in the academic community of the need to invest in education to develop environmental literacy. Education plays a role in reducing carbon emissions, and developing a comprehensive policy framework for sustainable development is a challenge faced by many countries worldwide (Zafar, 2020). International experiences show a strong link between education and sustainable development, highlighting the essential role of environmental literacy in achieving sustainability (Wu et al., 2023).

Characteristics of Effective Environmental Educational Programs

Effective climate education comprises four key components: community dynamics, effective teacher engagement, stakeholder involvement, and empowering learners with skills (Athman & Monroe, 2000). Community dynamics and collaborative learning emphasize collective action and diverse perspectives, contributing to locally adapted and globally relevant climate education strategies (Datta, 2024; Tynan, 2021). Embedding content in a local context allows learners to explore and experience their surroundings, fostering a sense of wonder and place (Setianingrum, 2023).

For sustainable development content to be effective, teachers must understand, interpret, and internalize its principles, as they are pivotal agents of social change and the implementers of sustainability education (Darling-Hammond et al., 2020; Perry & Metzger, 2023; Sell et al., 2023). Successful environmental education programs involve stakeholders in all stages, from development to evaluation, ensuring the program meets the community's needs (Verhelst et al., 2023; Monroe, 1999). Ultimately, these programs empower learners with skills and a sense of responsibility, progressing from awareness to action, to address and prevent environmental issues.

Theoretical framework: Environmental Education Strategies by Monroe et al. 2007

Martha Monroe and colleagues propose strategies for effective environmental education which are adapted from the frameworks by Fien et al. (2001) and Scott and Gough (2003) to enhance their applicability in educational and communicative contexts. This modified framework categorizes interventions into four objectives: conveying information, building understanding, improving skills, and enabling sustainable actions within accepted definitions of environmental education.

Convey Information

This category involves one-way transmission of information to provide missing facts and instructions, or raise awareness about a specific topic in environmental education. It relies on mediums like lectures, textbooks, and videos to disseminate factual, conceptual, or procedural knowledge. While essential, this approach can limit participation, as learners may not be involved in content selection or distribution.

Build Understanding

Here, information transmission is two-way, engaging audiences in developing their own mental models to understand concepts, values, or attitudes. Educational strategies include guided nature walks, workshops, and inquiry learning activities. The educator primarily drives the process, but actively seeks audience feedback. This category helps uncover misconceptions about complex issues.

Figure 1. Environmental Education Strategies Framework (Monroe et al., 2007)

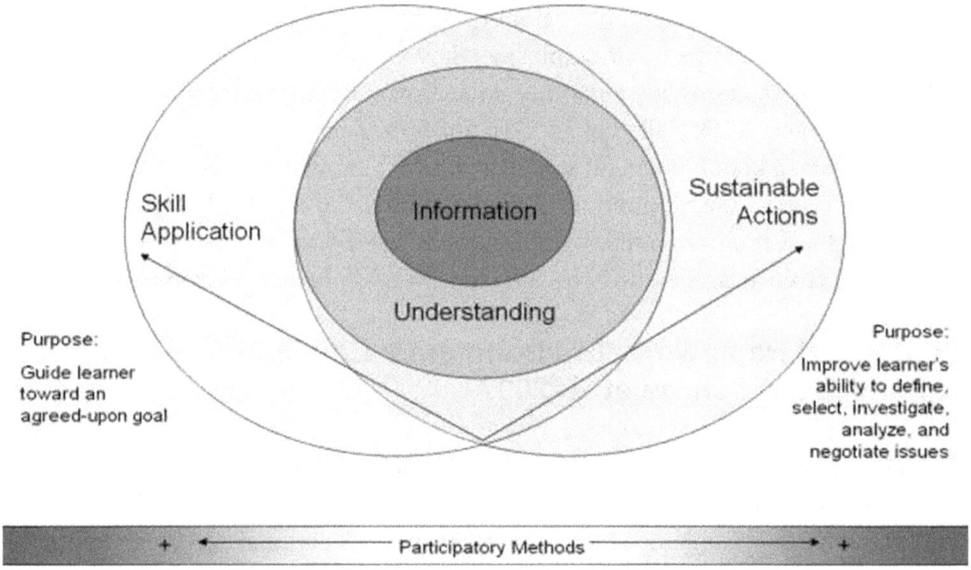

Improve Skills

Programs in this category aim to build skills that enhance or change behavior. Learners apply or implement skills through hands-on training or activities like service learning. The focus is on practical application, often incorporating theories of behavior change. The community generally agrees upon the ultimate outcome, like clean water or resource conservation.

Enable Sustainable Actions

Educational strategies aim to transform learners, issues, educators, and organizations by critically addressing problems. The process involves collaboration between educators and learners in defining goals and methods. Stakeholder involvement is crucial, and outcomes are unique products of those involved. Examples include community-based environmental education and collaborative learning, fostering empowerment, and long-term problem-solving aligned with goals of education for sustainable development.

It delineates three categories of interventions: information provision, communication, and mediation. Information provision involves one-way communication aimed at conveying facts and values, while communication entails two-way engagement with agreement on facts and values. Mediation involves facilitating learning through a two-way exchange that questions and explores disputed facts, values, and assumptions. Additionally, strategies like self-directed learning, coaching, and mentoring contribute to capacity building.

STUDY CONTEXT AND PARTICIPANT SELECTION

One of the emerging needs in the educational landscape is to foster environmental awareness among young students. Providing effective environmental education at the K-12 level is challenging, as it requires teachers to possess multidisciplinary knowledge and schools to offer intensive support. The project "Know Nature: Local to Global" aims to create a project template for three high schools in Indiana, Pennsylvania (USA), West Bengal (India), and Rangpur (Bangladesh). This initiative was to engage students with local environmental problems and develop their awareness of global environmental issues. The Alumni Ties Small Grant Project "Know Nature: Local to Global" aimed to foster awareness and understanding of environmental issues by connecting local communities with global perspectives. This project empowered 15 high school teachers and 150 students to conduct studies on locally significant environmental issues. The main focus of the studies was to promote sustainability, conservation efforts, and biodiversity awareness, with the goal of empowering individuals and communities to take action for a healthier planet.

The project's goal is to offer educational opportunities for high school teachers and students at these schools to collaborate, design, and implement projects addressing locale-specific environmental issues with global significance. The objective is to empower them, along with their peers and other stakeholders, in environmental literacy. Additionally, the project team published student-created project reports as a magazine within the school communities to raise awareness of both local and global environmental issues.

The project was designed to meet the following three objectives:

1. To develop skills among 15 High School teachers from three schools in the USA, Bangladesh, and India in designing and conducting environmental projects following the Monroe et al. 2007 framework.
2. To develop skills among 150 High School students of Class IX-XII from three schools in the USA, Bangladesh, and India in designing and conducting locale-specific environment projects objective (1) of global significance and digital display of the Project report with a critical awareness message.
3. The participating students, under the guidance of the mentor teachers, will be able to generate standard digital awareness material on local environmental issues and maintain the designated webpage of the project two years from the project's completion to sustain the project's impact.

The schools were selected based on convenience sampling. We chose five teachers (Biology/ Environmental science/ others) from each school to train a diverse range of subjects in environmental literacy. The selection process prioritized committed and passionate teachers. The project selected at least 50% of female teachers to equip them to combat gender inequity in environmental issues. Teachers were encouraged to select students based on their willingness, regardless of their economic, cultural, and racial backgrounds. Therefore, the teachers chose students based on their willingness, regardless of their economic, cultural, and racial backgrounds. Among the 150 students, a significant number (at least 50%) of girls were selected to develop environmental awareness since girls and women are the worst sufferers of ecological degradation, climate crisis, and allied problems. Students from grades seven to twelve participated in this project. Teachers from 3 schools in Indiana, Pennsylvania, West Bengal, India, and Dhaka, Bangladesh, have agreed to take part in this project and design and implement environmental activities.

Figure 2. Project Timeline

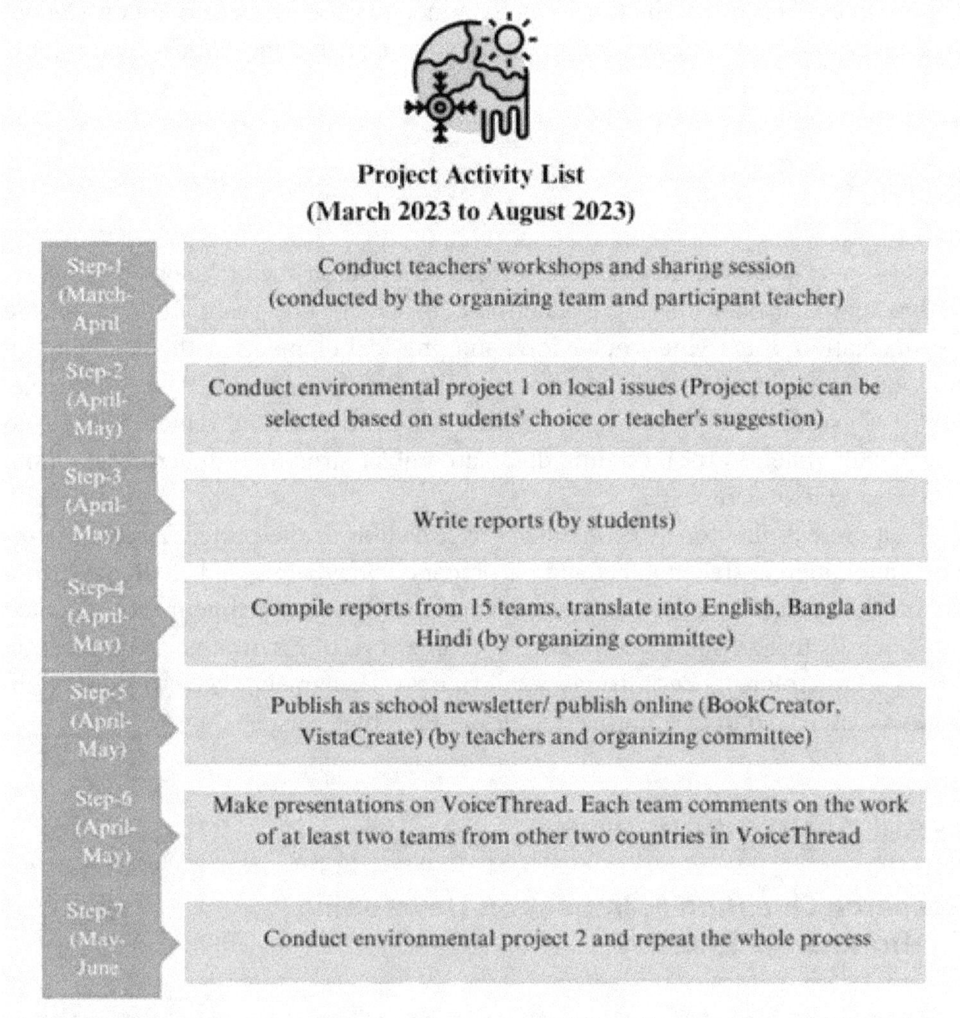

Know Nature: Local to Global seeks to establish a project framework for three high schools located in Indiana, Pennsylvania, West Bengal, India, and Rangpur, Bangladesh. 150 students and 15 teachers actively participated in this educational endeavor. The primary aim was to engage students in these schools with local environmental challenges while fostering an understanding of broader global environmental issues. Additionally, the project aimed to support high school educators by providing them with the necessary tools and resources to identify local environmental issues and develop corresponding projects for their students through teachers' training and sharing sessions. Furthermore, the project team compiled and published project reports created by students as a magazine, which was distributed within the school communities. This dissemination strategy aimed to raise awareness not only about local environmental concerns but also about their interconnectedness with global environmental issues.

METHOD

This study uses the unstructured direct observation method, which is an effective approach for gathering evaluation data. Direct observation allows for the documentation of activities, behaviors, and physical elements without relying on individuals' willingness or ability to respond to questions (Taylor-Powell & Steele, 1996). This study employed unstructured direct observation, which follows no conceptual strategies for recording data and, unlike structured direct observation, yields qualitative data.

As a project director, I recorded my observation of the project activity, communication among stakeholders and participants, drawbacks, and challenges from March to August 2023, over the six-month period of the project timeline. I conducted data analysis using Monroe et al. 2007's Framework of Environmental Education Strategies to explore four constructs of this framework, namely, Convey Information, Building Understanding, Improved Skills and Enabled Sustainable Action.

RESULT AND DISCUSSION

Research Question 1: Impact on Developing Environmental Literacy

To address Research Question 1, which examines how the project "Know Nature Local to Global" empowered high school teachers and students to take action on environmental issues, I analyzed the documented observation data according to the four themes outlined in the Environmental Program Strategies framework by Monroe et al. (2007).

Convey Information

The project supported teachers in two principal ways. First, it provided them with valuable information through training sessions, enabling mutual learning through sharing sessions. Second, it disseminated resources via the project website. The crucial role of the three country coordinators, all with teaching backgrounds, was evident in efficiently communicating with teachers about online training sessions, sharing website links, providing regular updates, and assisting teachers in instructing students on forming groups and engaging them in environmental activities.

The project coordination team, participating teachers, and speakers convened for virtual sessions using the Zoom platform. Initially, participating teachers were briefed on the project's goals, objectives, main outline, and the environmental project curriculum framework. They were also acquainted with environmental education practices and policies guided by the United Nations aimed at fostering sustainability. The training session included instructions on utilizing tech-based applications such as iNaturalist, Google Lens, Google Maps, and others for environmental projects. Moreover, teachers gleaned insights from firsthand experiences shared by other Fulbright environmental educators during the training session.

The project website offered project guidelines, training materials, recorded training videos, sample environmental project examples, environmental activity ideas, and more for both teachers and students. This website served as a platform for participants to showcase their work to others. Additionally, the project coordination team utilized social media, specifically Facebook, to disseminate news and project updates, maintaining enthusiasm among teacher-student teams throughout the project duration.

The above-mentioned features were helpful for conveying information about the goal and objectives of the projects and resources for teachers to participate in the project activities involving their students in lesson activities on local environmental issues.

Building Understanding

Teachers organized students into groups and implemented various strategies to facilitate understanding. They employed their preferred techniques for communicating with students, forming teams, and planning work distribution within each group. At the onset of the project, teachers tasked their students with identifying a local environmental issue. They instructed students to explore local media sources, such as television news, radio, and newspapers, to find recent reports on environmental issues. Different groups focused on distinct areas of environmental education. For instance, a team from School A in Indiana, Pennsylvania, chose to survey bird species and investigate the impact of climate change on local avian populations. Meanwhile, a team from School A in Rangpur, Bangladesh, examined how rooftop gardens in Rangpur could offer green solutions to urban challenges.

I noticed that there has been a strong influence on how teachers from different countries work with their students. For example, American teachers who worked on local bird species asked students to make a PowerPoint slide on the local birds, their scientific classification, and other information, such as distribution and diet. In contrast, in Bangladesh, students first visited a bird sanctuary and took photos of birds. Later, the teacher asked students to identify those species using the photos through technology, namely iNaturalist.

The country coordinators of each country act as a bridge between the teachers and the school authorities. This step helps in building understanding among the different stakeholders of the project. The country coordinators also communicate with local organizations for permission for field trips. For example, a team from School B, West Bengal, India collaborated with a local university for their activity. On the other hand, a team from Bangladesh interviewed local administrative officers to collect data to enhance their understanding of the topic they have been investigating. Various approaches help to build understanding among the project team, the school, and the local community.

Improved Skills

The project emphasizes students' participation levels, ranging from passive reception of information to active engagement. This shift aims to encompass all existing strategies, though it may not cover them completely. Nonetheless, it introduces significant concepts for modern environmental education. To equip students with the skills necessary to identify environmental challenges and increase their knowledge about local species, the following steps were taken:

- Conducting surveys of local species for various activities involving diverse living organisms, thereby enhancing students' understanding of local flora and fauna.
- Educating students about survey techniques for different species, as well as data collection and analysis methods.
- Providing training on iNaturalist, an online platform for species identification, which allowed students to engage with digital tools such as Google Earth and project websites to explore digital content.
- Facilitating collaboration among students through platforms like Google Classroom and the school's learning management software, leading to significant digital platform exposure from March to September 2023.

- Guiding students in writing reports on environmental activities conducted, following scientific report writing conventions.
- Encouraging students to present their studies and share them with non-participant students in the school, thereby enhancing their presentation skills.

We noticed skill acquisition varies from country to country, school to school, and teacher to teacher. Not all groups did the same activities. However, all students gained inquiry skills to explore local environmental issues and learned at least one technological tool to assist them in project activities.

Enabled Sustainable Actions

Once the teachers and students were informed about the goal and procedure and equipped with skills to identify important local environmental issues, they conducted various activities as a part of sustainable action. 3 teams from 2 schools in Bangladesh, 2 teams from two schools in the USA, and 4 teams from 2 schools in India conducted activities on environmental issues. Figure 3 depicts the detailed activities conducted by students during the two phases of the project.

Figure 3. Local Environmental Initiatives Led by Teacher-Student Teams for Sustainability

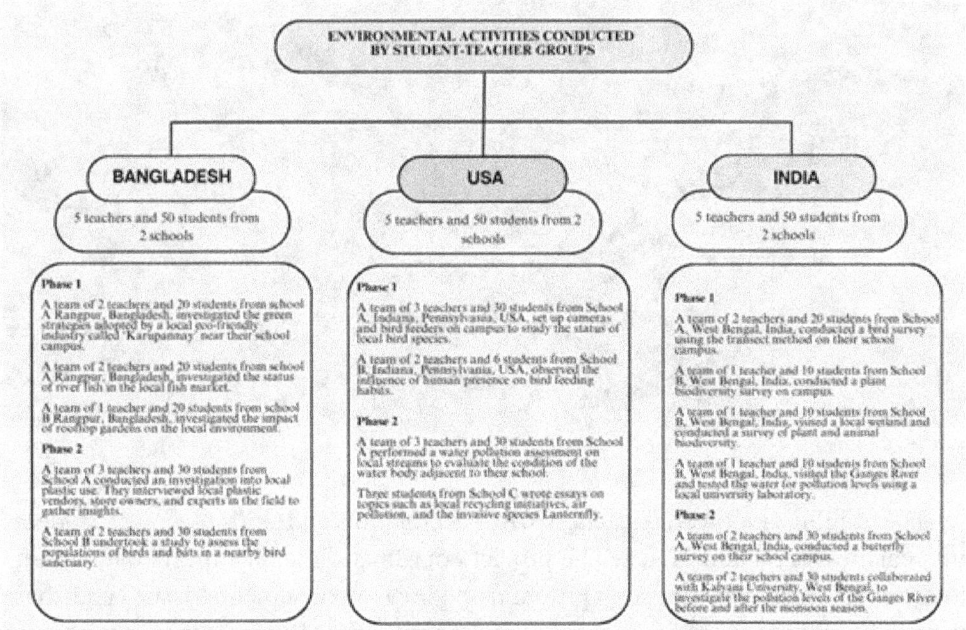

Besides teacher-students' activities, which enhanced students' environmental literacy, the project has employed some strategies to ensure long-term sustainability. The project team started the public dissemination of the work by students. They collected all reports from the student-teacher teams, compiled them, and published them in a magazine. They also collaborated with language experts to translate the documents into Hindi and Bangla so that students in India and Bangladesh could read and learn about their peers' contributions to the project in their own language. The project coordination team also circulated the printed magazines to local libraries so that other students and community members could read about the project's activities and outcomes, aiming to foster greater ecological awareness across all the schools involved. This approach allowed the students to take pride in having their reports published, thereby encouraging them to work on environmental issues more. It also encouraged non-participant students to get involved in any future environment-related programs at school. The magazines inspired students to learn about environmental challenges and how their peers from other countries are taking action against them, providing a global perspective on climate change issues.

Figure 4. Student Engagement Throughout the Project

The training empowered teachers with several tools and skills to engage students in local environmental issues. The project coordination team expects the teachers to use their skills for more such programs on local environmental issues and share them with other educators in their school community. During the Zoom sharing session, the school principals showed interest in supporting teachers' endeavors in future activities to keep students engaged in local environmental projects. All participating teachers expressed their interest in sharing future project work globally, with a commitment to spreading awareness of local environmental issues through the project website (https://knownaturelocal2global.weebly.com/).

The project coordination team plans to maintain the project website for the next three years after the end of the project. The website shares all project templates, training sessions, and students' original works as resources. Furthermore, in order to make the project accessible to environmental educators worldwide, the project coordination team also made the project website available in the public domain.

Research Question 2: Challenges

We examined research question 2, which explored the challenges faced by the participants and other stakeholders while implementing the project during implementation. We employed thematic analysis of the documented observation data. The project coordination team suffered from several challenges while conducting this environmental project.

Multiple factors were taken into consideration while planning the teachers' training and sharing sessions. However, choosing a suitable online session time was a constant challenge. We had to prioritize the teacher's convenience over the speaker's schedule. We had to miss several eminent speakers we planned for this project because of schedule unavailability. However, we used the US State Alumni network to connect with environmental educators and other experts. One major challenge was managing the time differences across different locations. For example, in Bangladesh and India, the sessions were held in the evening, whereas it was an early morning for the teachers in the USA. Saturday was chosen for the sessions since it is the only common weekend day for all three countries. However, Saturday mornings were not ideal for American teachers because it was their resting time after a week of classes. Additionally, some sessions had to be cut short because Bangladeshi teachers were fasting during Ramadan, making it difficult for them to attend sessions in the evening soon after breaking their fast. Furthermore, teachers from different countries missed some sessions because of festivals and national events. For instance, one session clashed with the New Year celebration, preventing some Indian teachers from attending.

Making resources available was important so that teachers could see the resources that can help them learn about environmental projects, current climate crises, and how to involve students in their projects. However, choosing the right platform for making resources available was difficult because teachers in the US are less fond of Facebook pages, which are more popular among Bangladeshi and Indian teachers.

Reaching schools was difficult in the USA, whereas in Bangladesh and India, it was easy. We found it difficult to get US schools permission to survey students. Initially, the principal agreed, but later the teachers felt uncomfortable about conducting a survey on climate change issues. They are concerned that parents might object to climate change issues, whereas they are okay with their children doing environment-related programs. My university professors (Indiana University of Pennsylvania) helped me with this problem.

Teachers faced challenges while conducting environmental activities. For example, the US teachers who planned to do some activities during the summer school time did not get responses from students, which caused a delay in report writing and publishing magazines. With the help of the Indiana International Hospitality Center, a local organization in Indiana, Pennsylvania, that supports international collaboration, we asked individual students to write essays on local environmental issues and thus mitigate the problems. However, teachers' support for sharing the published documents and sharing posters and banners for educating the greater community was wonderful.

In India, the schools missed several opportunities to visit the field and conduct investigations as per their plans, thus postponing the dates for those events. They also could conduct their activities later than the planned time frame. For example, the teachers designed an activity for students to conduct a butterfly survey. However, during monsoon, the survey group became unsafe because of snakes, therefore the teachers waited several months before they let their students conduct the activity.

In Bangladesh, the countrywide teacher strikes from July to August 2023 hampered the smooth operation of the project activities. Teachers conducted the activities at the eleventh hour. To overcome challenges in all three countries, constant communication with all stakeholders, flexibility to change the existing environmental activities, and local collaborators' support were instrumental.

We worked with diverse participants from three countries. Teachers from the USA are well ahead in report writing and designing scientific and environment-related investigations than their counterparts from Bangladesh and India. Indian teachers showed better competence than Bangladeshi teachers. In terms of enthusiasm and motivation to participate in projects, Bangladeshi teachers showed the highest level of interest than others, whereas American teachers were least interested. American teachers expressed concern that parents might dislike the concept of Environmental Education, while no such observation was made in the two other countries. The combination of project participants' diversity brought both challenges and complementarity.

Different community organizations supported multiple challenges faced by the project. Indiana University of Pennsylvania, supported by some purchased software, helped to get access to local schools. CHESPAX environmental education program,

Maryland provided Stream Survey instruments and curriculum documents Kalyani University, West Bengal, India, provided their lab space and instruments to the participating students to support their project. GEIST International Foundation, Bangladesh, helped to spread the news and invited more environmental teachers to join the session for a more enriching sharing experience.

CONCLUSION AND RECOMMENDATIONS

To enhance environmental awareness among young children, individual projects initiated by teachers or external organizations can complement formal education. It can successfully promote environmental literacy if the environmental project is planned considering the four constructs of the Environmental Education Framework or Monroe et al. (2007). However, there is still a possibility that such projects may encounter challenges if not meticulously planned.

This project encountered limitations, including unmet goals such as surveys, webinars, and an award ceremony due to time constraints and scheduling conflicts. To address this, future projects should span a whole year, allowing ample time for planning and execution. Moreover, facilitating collaboration among participating teachers is essential for their professional development. Creating a platform for interaction can enhance learning outcomes and foster a sense of community among educators. Also, looking for collaboration with local organizations for sharing resources and resource persons, and gaining access proved to be phenomenal in this project. Therefore, I recommend that future project planning prioritize this aspect of conduct activities with ease.

Time management proved challenging, especially coordinating activities across different time zones. Ensuring synchronous participation of teachers and students from diverse locations is essential for effective engagement. Future project directors should prioritize synchronization to maximize effectiveness and inclusivity. Securing permission from schools, particularly in the United States, can be difficult due to misconceptions surrounding climate change education. Therefore, proactive communication with schools, preferably a year in advance, is crucial to gaining their support. Additionally, navigating bureaucratic processes, such as obtaining permission from school districts in the United States versus taking it from individual school principals in Bangladesh and India, requires careful consideration.

Ultimately, educators bear the responsibility of involving students in environmental projects to promote environmental literacy and empower them to address climate change challenges. By integrating environmental education into classrooms or collaborating on external projects, students can become informed and proactive citizens, capable of safeguarding the planet and mitigating environmental crises.

Scientific Ethics Declaration

This project has been approved by the Indiana University of Pennsylvania Institutional Review Board for the Protection of Human Subjects (IRB Number: 23-021–SCH DIST).

Acknowledgments or Notes

This research data was collected from the Alumni TIES project, 'Know Nature: Local to Global'. Alumni TIES is sponsored by the US Department of State with funding provided by the US Government and supported in its implementation by World Learning, in partnership with the Office of Alumni Affairs of the Bureau of Educational and Cultural Affairs (ECA).

REFERENCES

Athman, J. A., & Monroe, M. C. (2001). Elements of effective environmental education programs.

Bercasio, R. R. O., & Remolacio, L. K. Q. (2022). Mainstreaming environmental education in the teacher education curriculum in the Philippines. *International Journal of Evaluation and Research in Education*, 11(3), 1552–1564. 10.11591/ijere.v11i3.21748

Corpuz, A. M., San Andres, T. C., & Lagasca, J. M. (2022). Integration of environmental education (EE) in teacher education programs: Toward sustainable curriculum greening. Problems of Education in the 21st Century, 80(1), 119.

Darling-Hammond, L. (2020). Accountability in teacher education. *Action in Teacher Education*, 42(1), 60–71. 10.1080/01626620.2019.1704464

Datta, R. (2024). Relationality in indigenous climate change education research: A learning journey from indigenous communities in Bangladesh. *Australian Journal of Environmental Education*, 1–15. 10.1017/aee.2024.13

Díaz-López, C., Serrano-Jiménez, A., Chacartegui, R., Becerra-Villanueva, J. A., Molina-Huelva, M., & Barrios-Padura, Á. (2023). Sensitivity analysis of trends in environmental education in schools and its implications in the built environment. *Environmental Development*, 45, 100795. 10.1016/j.envdev.2022.100795

Fien, J., Scott, W. A. H., & Tilbury, D. (2001). Education and conservation: Lessons from an evaluation. *Environmental Education Research*, 7(4), 379–395. 10.1080/13504620120081269

Glackin, M., & King, H. (2020). Taking stock of environmental education policy in England – the what, the where, and the why. *Environmental Education Research*, 26(3), 305–323. 10.1080/13504622.2019.1707513

Hensley, N. (2020). Educating for sustainable development: Cultivating creativity through mindfulness. *Journal of Cleaner Production*, 243, 118542. 10.1016/j.jclepro.2019.118542

Hollweg, K. S., Taylor, J. R., Bybee, R. W., Marcinkowski, T. J., McBeth, W. C., & Zoido, P. (2011). Developing a framework for assessing environmental literacy. Washington, DC: North American Association for Environmental Education.

Ma, L., Shahbaz, P., Haq, S. U., & Boz, I. (2023). Exploring the moderating role of environmental education in promoting a clean environment. *Sustainability (Basel)*, 15(10), 8127. 10.3390/su15108127

Monroe, M. C., Andrew, E., & Biedenweg, G. (2007). A framework for environmental education strategies. *Applied Environmental Education and Communication*, 6(3-4), 205–216. 10.1080/15330150801944416

Monroe, M. C., & Richard, R. (2019, June 3). Plate, Annie Oxarart, Alison Bowers & Willandia A. Chaves (2017): Identifying effective climate change education strategies: a systematic review of the research. *Environmental Education Research*, 25(6), 791–812. Advance online publication. 10.1080/13504622.2017.1360842

Parry, S., & Metzger, E. (2023). Barriers to learning for sustainability: A teacher perspective. *Sustainable Earth Reviews*, 6(1), 2. 10.1186/s42055-022-00050-3

Roth, C. E. (1992). Environmental literacy: its roots, evolution and directions in the 1990s.

Scott, W., & Gough, S. (2003). Rethinking relationships between education and capacity—building: Remodelling the learning process. *Applied Environmental Education and Communication*, 2(4), 213–219. 10.1080/15330150390241409

Sell, M., Schiera, R., Siddique, Z., White, S., & Babal, J. (2023). Decision Literacy, Multimodal Storytelling, and the Digital Storygame Project. *English Journal*, 112(3), 28–35. 10.58680/ej202332226

Setianingrum, D. A., Matahari, D. B., Jumadi, J., & Wilujeng, I. (2023). Development of science e-book containing gamelan's local wisdom based on steam-poe to facilitate students' love of local culture. *Jurnal Penelitian Pendidikan IPA*, 9(6), 4791–4800. 10.29303/jppipa.v9i6.3760

Taylor-Powell, E., & Steele, S. (1996). Collecting evaluation data: Direct observation. *Program Development and Evaluation. University of Wisconsin-Extension.*

Tynan, L. (2021). What is relationality? Indigenous knowledge, practices and responsibilities with kin. *Cultural Geographies*, 28(4), 597–610. 10.1177/14744740211029287

Verhelst, D., Vanhoof, J., & Van Petegem, P. (2023). School effectiveness for education for sustainable development (ESD): What characterizes an ESD-effective school organization? *Educational Management Administration & Leadership*, 51(2), 502–525. 10.1177/1741143220985196

Wu, Y., Wan, J., & Yu, W. (2023). Impact of environmental education on environmental quality under the background of low-carbon economy. *Frontiers in Public Health, 11*, 1128791.

Zafar, M. W., Qin, Q., & Zaidi, S. A. H. (2020). Foreign direct investment and education as determinants of environmental quality: The importance of post Paris Agreement (COP21). *Journal of Environment*

APPENDIX

Teaching Note

Questions and Answers

1. Why should educational institutions welcome individual environmental projects to enhance their students' environmental literacy?

Schools should embrace individual environmental projects to boost students' environmental knowledge. These projects often involve experienced environmental facilitators, which regular teachers may not be. They also provide educational resources and funding for teachers and students. Additionally, these projects can complement the curriculum, making learning more relevant and impactful.

2. What strategies does the project 'Know Nature: Local to Global' employ to empower teachers to instill environmental literacy in their students?

The project empowers teachers to instill environmental literacy in their students by providing training sessions, sharing valuable information, and offering resources through a dedicated website. Country coordinators, with teaching backgrounds, effectively communicate updates, share online training sessions, and assist teachers. Virtual sessions on platforms like Zoom introduce teachers to project goals, environmental education practices, and tech-based applications for environmental projects. The website provides guidelines, training materials, and examples of environmental projects, while social media is used to maintain engagement and enthusiasm among participants.

3. What strategies did teachers and students of 'Know Nature: Local to Global' adopt to take long term sustainable actions to make the project learning impactful?

To ensure long-term sustainability, the project broadcasted students' work by collecting reports, compiling them, and publishing them in a magazine. The English reports were translated into Hindi and Bangla to make them accessible to students in India and Bangladesh. The printed magazines were distributed to local libraries, increasing ecological awareness among students and the community. In addition to this, the training empowered teachers with tools and skills to involve students in local environmental issues, with school principals supporting future activities.

4. How does this project 'Know Nature: Local to Global' help students build skills?

The project helps students build skills by actively engaging them in various activities that enhance their understanding and practical abilities. Students conducted surveys of local species and learned about flora and fauna. They receive education on survey techniques, data collection, and analysis. Training on digital tools like iNaturalist, Google Earth, and other platforms allows students to explore and analyze environmental content. Students are guided in writing scientific reports and presenting their findings, improving their report-writing and presentation skills. While skill acquisition varies, all students develop inquiry skills to explore local environmental issues and gain experience with at least one technological tool.

5. How does publishing students' work as a magazine helps students?

This dissemination approach motivated students by instilling pride in their published work and providing both participants and non-participant students with a global perspective on climate change, showcasing how students from different countries are taking actions. It also encouraged students to take part in future environmental projects.

Epilogue and Lessons Learned

The initiatives and outcomes of the 'Know Nature: Local to Global' project can have a big impact. The project established a template for environmental education that is both robust and adaptable, allowing for future expansion. In addition, the project helped teachers learn innovative ways to teach about the environment by providing training and sharing resources. Teachers are ready to involve students in important environmental projects, promoting ongoing awareness and action in their classrooms. The project website will remain a valuable resource for teachers and students. It offers a repository of guidelines, training materials, and examples of successful environmental projects. By bringing together teachers from the USA, Bangladesh, and India, the project fostered a sense of global community and mutual learning. This international collaboration has established a network of educators who can continue to share resources, ideas, and best practices for environmental education. The connections made during the project could cause future collaborations and ongoing cultural exchanges.

The project's activities, including the creation of a magazine featuring student work and the publication of reports in multiple languages, have increased environmental awareness among a broader audience. This approach encourages students and educators to continue exploring and addressing environmental challenges. The experiences gained from the project offer valuable lessons for future environmental education initiatives. The project's emphasis on sustainability and the creation of long-term resources aim to instill a lasting commitment to environmental stewardship. By encouraging teachers to share their experiences and by providing ongoing access to educational materials, the project supports the continued growth of environmental education beyond the project's duration.

I learned several valuable lessons while conducting the high school environmental project 'Know Nature: Local to Global'. Technology played a crucial role in the project's success by enabling collaboration and providing essential teaching tools. Training sessions via Zoom were vital for sharing information on environmental and climate change issues. The project website served as a comprehensive resource hub, offering guidelines, training materials, recorded sessions, and examples of environmental projects. Social media was also instrumental in maintaining participant enthusiasm. While technology provides benefits, I should prioritize fostering teacher-to-teacher interactions for mutual benefit. Enhanced opportunities for teachers to collaborate and share experiences could further strengthen the project's impact. Managing online sessions across different time zones and dealing with scheduling conflicts, especially during cultural and national events, proved challenging. Flexibility and constant communication were essential in overcoming these obstacles. Understanding and respecting cultural differences was crucial. Variations in enthusiasm, report writing, and investigation skills among teachers from different countries highlighted the importance of providing tailored support to meet the needs of each team. The challenges faced, such as coordinating across time zones and securing school permissions, along with the strategies employed to overcome them, provide insights for improving future projects. These lessons will help in designing more effective and inclusive environmental education programs.

Additional Resources for Practitioners

1. https://www.shaverscreek.org/conservation-and-research/citizen-science-projects/

2.https://chespxblogs.com/?fbclid=IwAR2S0Ud6ihMeX8kacBFrM5RrJyXrJz7nRZ-8cIhIBLwco-PNEKvC8FCwjQk

3. https://sites.google.com/view/iscs-curriculum-page/environmental-science/?pli=1&fbclid=IwAR07VJ38JZSfelTfnUQ0VY3foMkoxH0DspjzIjcxcBio-YojjOgd_z0vN08
4. https://www.exploratorium.edu/snacks/subject/human-impacts?fbclid=IwAR1ROpHZHBFeYMhrBqrQJkutK3DmVqs4I9w5VpeRZEdx3ZE24Wi3QxZNzf8

5. https://eepro.naaee.org/resource
6. https://www.baybackpack.com/state-initiatives/pennsylvania
7. https://cdn.naaee.org/about/ee/environmental-literacy-framwork
8. https://oecdedutoday.com/green-at-fifteen-schools-support-climate/
9. https://www.boisestate.edu/sps-environmental/experiential-learning/capstone-projects/
10. https://www.nationalbiodiversityteachin.com/news/human-impact-and-biodiversity-in-the-world?fbclid=IwAR29ok7xhf17YklCUj7DNmN-6WfRSKr8vGq_py5peiOs1GegqIPaiSjeqSU
11. https://www.nationalbiodiversityteachin.com/news/birds-and-biodiversity?fbclid=IwAR2dHBt5nPCC3GmlBiRSWJyNecVVRqt4RfO2VzXG-ZTiBBJMlFZmzMIW5I8
12. https://www.youtube.com/@bbcearth
13. https://www.sciencebuddies.org/blog/biodiversity-lessons?fbclid=IwAR1yCl9M-bVq6MCuYgObJfUfdvox2GyMbzD2euwxsnVTfhSGW6Sa-5VXhkI
14. https://scied.ucar.edu/interactive/simple-climate-model?fbclid=IwAR17C-HrD_9Y1R9PQqmyF_T9vBkwJeUsEy_Na1niACUKbg5dCMXYcqs4XeU
15. https://celfeducation.org/
16. https://myemail.constantcontact.com/Local-and-Global-Connections-for-Students.html?soid=1102885321240&aid=3cDfI2siK9aQ

Chapter 2
Immersive Agricultural Education:
Gamifying Learning With Augmented Reality and Virtual Reality

Maryam Bigonah
Auburn University, USA

Fatemeh Jamshidi
Auburn University, USA

Daniela Marghitu
Auburn University, USA

EXECUTIVE SUMMARY

This chapter introduces the exciting world of cutting-edge technologies in controlled environment agriculture (CEA) and their potential impact on high school education. The objective of this chapter is to establish the groundwork for transforming education through the creation of state-of-the-art augmented reality (AR) and virtual reality (VR) learning platforms, and research within the domain of CEA. These innovative technologies provide immersive learning experiences, allowing high school students and new entrepreneurs with diverse capabilities to explore new agricultural technologies in a virtual environment before implementing them in the world. While AR and VR technologies hold significant promise, there are notable challenges in effectively integrating AR and VR technologies within educational and CEA settings. These challenges highlight the need for thoughtful integration, aiming to enrich learning experiences. Through this chapter, the authors examine various

DOI: 10.4018/979-8-3693-1710-5.ch002

ideas that have successfully harnessed the benefits of AR and VR in education and discuss a proposed framework.

INTRODUCTION

Agriculture, one of the world's oldest industries, is experiencing a profound transformation due to advancements in technology. Among the most promising innovations are Augmented Reality (AR) and Virtual Reality (VR), which offer new ways to educate and train individuals in various agricultural practices. Controlled Environment Agriculture (CEA) is a multifaceted and complex field which involves growing crops in controlled environments such as greenhouses and indoor farms, and it particularly benefits from these advanced technologies.

The primary objective of this research is to establish CEA as a sustainable food production system capable of efficiently generating nutritious foods in a low-carbon economy. By seamlessly integrating agricultural practices with advanced AR and VR technologies, this initiative enhances high school students' and new entrepreneurs' understanding of STEM concepts within an educational setting. Aligned with the Individuals with Disabilities Education Act (IDEA), and Universal Design for Learning (UDL) Guidelines, the platform ensures equitable access to STEM education, fostering an inclusive environment for students of diverse abilities and learning styles. Furthermore, the research seeks to expand market opportunities for CEA farmers and entrepreneurs by providing them with cutting-edge technologies and comprehensive learning resources. To achieve the goals, the research focuses on three primary questions:

1. How effective are AR and VR learning modules and the greenhouse in enhancing understanding and use of CEA concepts aligned with Next Generation Science Standards (NGSS) for secondary school students?
2. How do gamified elements and dynamic stochastic gaming probabilistic concepts impact the engagement and learning experiences of students in CEA education?
3. How well do the developed learning modules cater to the diverse needs of learners, including those with disabilities, in alignment with Section 504 of the Rehabilitation Act of 1973, IDEA, and Universal Design for Learning Guidelines?

In the following chapter, we explore the integration of agricultural education and emerging technologies to foster high school students' understanding of sustainable food production in the literature review section. We begin by discussing the role of agricultural education and the technologies that support sustainable practices.

Following this, the chapter delves into the use of AR, VR, and Extended Reality (XR) technologies to enrich the learning experience. An introduction to CEA and the Green-Reimagining of Agriculture in Controlled Environment (Grace) Research Project provides context for the Grace Educational Platform Strategy. Detailed sections explore the interactive panels on the platform and its accessibility features, ensuring inclusivity for all students. Potential challenges are addressed, followed by a comprehensive evaluation of the research through qualitative and quantitative assessments. The chapter concludes with a discussion on the implications of the findings and outlines directions for future work.

LITERATURE REVIEW

Introduction to Modern Agricultural Learning

The Evolution of Agricultural Education

Agricultural education has continually evolved over the past century due to the advancement of technological and social changes. This review synthesizes key developments in agricultural education, with a focus on recent literature from the past few years.

Early Focus on Production and Rural Development

Formal agricultural education originated in the late 19th century focused predominantly on advancing production practices and rural development (True, 1929). The 1914 Smith-Lever Act established cooperative extension to disseminate best farming methods (Rasmussen, 1989). As such, early curricula and textbooks focused heavily on topics like soil science, machinery, and livestock operations to train future farmers (Welton & Mallaro, 1915).

Vocational Training and the Rise of the FFA

The 1917 Smith-Hughes Act established vocational agricultural education (Hillison, 1996), leading to thousands of high school programs centered on hands-on career preparation through supervised farming projects. This spurred the 1928 founding of Future Farmers of America (now National FFA Organization) to support youth leadership and interest in production careers (Tenney, 1977).

Adapting to Technology and Industry Changes

As agriculture industrialized and mechanized throughout the 20th century, education broadened to serve evolving industry needs. Post-secondary programs expanded across wider disciplines like economics, policy, business, sustainability, and global agriculture (Williams, 1991). By the 1960s, curricula incorporated more science, technology, and addressed contemporary issues like food security and climate change impacts (Bellah & Dyer, 2009).

Recent Developments and Continuing Challenges

Several studies starting in 2020 reaffirm agricultural education's ongoing broadening beyond traditional production (Baker et al., 2021; Vincent & Villa, 2021). For instance, more undergraduates now major in high-demand fields like business, communications, or food science than animal science or agronomy (Tummons et al., 2021). However, other research shows secondary programs still emphasize production topics (Thieman et al., 2020).

Looking ahead, agricultural educators face various challenges. For one, teachers struggle balancing hands-on learning and student engagement with rising online and distance formats (Sorensen et al., 2021). Moreover, teacher preparation programs often lack sufficient training for instructing on emerging technologies like data analytics, Artificial Intelligence (AI), or precision agriculture that drive modern innovation (Baker et al., 2021). There remain calls for expanded curriculum and new classroom innovations addressing contemporary issues like climate change, global food security, and sustainability (Tummons et al., 2021).

In conclusion, formal agricultural education originated focused on production but has greatly expanded over 125+ years in scope and purpose. While production careers and hands-on learning remains vital to many secondary programs, post-secondary and industry needs demand broader knowledge across science, technology, business, and social sciences. As agriculture continues evolving in an increasingly urbanized, technology-driven, and climate-impacted world, further adaptations in agricultural education will be imperative for preparing qualified graduates across diverse career pathways in the global food system.

Emerging Technologies: AR, VR, AI, and Gamification

In today's rapidly evolving world, the incorporation of cutting-edge technologies in educational settings has become increasingly important. Recent research highlights their potential benefits in various industries like manufacturing (Masood et al., 2022), tourism (Samaddar et al., 2022), healthcare (Eckert et al., 2022), social media and communication (Sayadi et al., 2024) etc. However, studies and trials were needed to determine the use and efficacy of AR and VR technologies due to their rapid expansion (Krishnan et al., 2022). Al-Ansi et al. (2023) and Zhao et al. (2023) discuss the increasing interest in using AR and VR in education. They examine how these technologies can create immersive learning experiences and highlight their potential impact on educational practices.

Furthermore, AR and VR are sustainability enhancing, as educators can interact with their students remotely through e-learning in conjunction with AR and VR (Griffiths et al., 2019). These immersive technologies offer unique opportunities for agricultural education, allowing students to engage in hands-on learning experiences that were previously inaccessible. For instance, VR has been applied as an educational method in various industries, but there has been limited research on its use in the agriculture domain (Yang et al., 2021). Recent research published in Frontiers in Psychology highlights the positive influence of AI and VR technologies on student learning outcomes, particularly in art education, emphasizing the benefits of deep learning practices facilitated by these technologies (Rong et al., 2022). Additionally, a recent analysis published on Wiley Online Library focuses on the role of AR, VR, and AI games in teaching science subjects, providing a comprehensive discussion on how these technologies can be leveraged to improve science education (Rossetto et al., 2023).

However, a few notable studies have showcased the potential of AR and VR in agriculture education. For example, Kim et al. developed an educational simulator using VR technology for improved farmer education compared to traditional two-dimensional screens. This VR simulator provided an immersive and interactive experience that enhanced technical training for the maintenance of agricultural machinery. In addition to VR, the use of AR in agricultural education has also been explored. AR and VR technologies have been shown to enhance learning skills, engagement, and cognition in various fields, including engineering education (Faridi et al., 2020).

Research has delved into the use of AR and VR in medical education, highlighting how these technologies can drive adoption through provider-centered design, ultimately benefiting learners (Zweifach & Triola, 2019). Also, investigations have focused on the potential of AR and VR technologies to promote wellbeing in older adults, emphasizing the importance of contextual and evaluation frameworks for critical reviews in this area (Lee et al., 2019).

The literature has examined the implications of AR and VR on teaching-learning processes in engineering courses, revealing that students who engage with these technologies tend to learn more and express higher satisfaction compared to traditional teaching approaches (Anjos et al., 2021). Junaini et al. (2022) have explored the use of AR and VR games for occupational safety and health training, indicating a significant growth in their utilization for safety training and rehabilitation purposes.

Đurđević et al. (2022) have also investigated the development of AR applications, such as augmented reality video applications, showcasing the continuous evolution and innovation in this field. Studies have also explored the integration of AR technologies in educational projects, highlighting the benefits of merging virtual aspects with the real world to create immersive learning experiences (Iatsyshyn et al., 2020).

Caria et al (2020) designed wearable AR equipment to guide farmers in collecting soil samples, demonstrating the potential of AR in agricultural practices. This aligns with the broader trend in education, where emerging technologies like VR and AR are being leveraged to offer more immersive learning experiences to students, thus enhancing their comprehension and retention of educational content. Recent developments, such as the VR simulator for technical training in the maintenance of agricultural machinery, indicate a growing interest and recognition of the potential benefits that VR can offer in agricultural education. Moreover, the combined potential of both VR and AR technologies to contribute to the automation and intelligence of agriculture, as evidenced in the concept of Agriculture, highlights the transformative impact these technologies can have on the future of agricultural education.

Table 1 provides a comparison of existing AR and VR applications in engineering education, highlighting the diverse range of possibilities for incorporating these technologies into agricultural education. Furthermore, the use of AR and VR in agricultural education contributes to the automation and intelligence of agriculture, allowing for more efficient and sustainable farming practices. Moreover, the utilization of VR in agriculture education goes beyond technical training and can also be applied to other aspects of farming (Yang et al., 2021). These technologies not only enhance technical training and knowledge acquisition but also promote engagement, critical thinking, and problem-solving skills among students. In summary, the use of AR and VR in agricultural education has been limited but shows great potential.

Ecological Literacy and Agricultural Education

The concept of ecological literacy in agriculture involves understanding the intricate relationships between agricultural practices and the environment, promoting sustainable farming methods, and fostering environmental stewardship among farmers. Ecological literacy in agriculture emphasizes the importance of integrating ecological principles into agricultural education to enhance farmers' understanding of ecosystem dynamics, biodiversity conservation, and sustainable land management practices (Fretwell, 2017; Hubert et al., 2000; Reilly et al., 2022).

Incorporating ecological literacy into agricultural education is crucial for promoting sustainable farming practices and environmental stewardship. Farmers who possess a strong foundation in ecological principles are more likely to adopt sustainable farming techniques, such as agroecology and organic farming, which contribute to biodiversity conservation and soil health. By equipping individuals with knowledge about ecological concepts, agricultural education programs can empower farmers to make informed decisions that benefit both their operations and the environment (Yu et al., 2022). Additionally, enhancing ecological literacy through agricultural education can lead to improved resource management practices, reduced environmental degradation, and increased resilience to climate change impacts in agricultural systems.

The positive impact of incorporating ecological literacy into agricultural education curricula has been highlighted in various research studies. One such study by David et al. (2008) delves into the concept of agricultural literacy and discusses different approaches to promoting agricultural literacy in K-12 classrooms. It emphasizes the importance of systematically integrating agricultural education to enhance ecological understanding and promote sustainable practices among students. Additionally, Fretwell (2017) explores the significance of ecological literacy in fostering a sustainable relationship with nature. The study underscores the need to revise industrial-driven ideologies and emphasizes the value of ecological literacy in education for establishing a lasting partnership with the environment.

Furthermore, initiatives like the Planet Bee Foundation emphasize the significance of eco-literacy in schools to cultivate a sense of environmental stewardship among students (Planet Bee Foundation, n.d.). By educating the younger generation about ecological principles, schools can play a vital role in shaping future agricultural practices that prioritize sustainability and environmental protection.

The Need for Transformation in Agricultural Learning

In recent years, there has been a growing recognition of the need for transformation in agricultural learning methods. Traditional methods of agricultural education, such as classroom lectures and hands-on training, have limitations in terms of engaging learners and providing them with experiential learning opportunities.

Immersive technologies, such as AR, VR, and gamification, are emerging as potential tools to address these limitations and enhance agricultural learning experiences. Scavarelli et al.' (2023) conducted a literature review on the educational use of AR and VR within social learning spaces. They found that these technologies have the potential to engage learners, provide realistic and immersive experiences, and enhance knowledge retention. Additionally, Akçayır and Akçayır (2017) performed a systematic review highlighting the advantages and challenges of using AR in education. They identified that AR could improve learners' motivation, engagement, and understanding of complex concepts in agricultural education. Furthermore, Garzón and Acevedo (2019) conducted a meta-analysis of 64 quantitative research papers and found that AR has a medium-sized effect on learning gains in agricultural education. In a similar vein, Dunleavy and Dede (2014) conducted a literature review on instructional AR and VR systems across different fields. They found that these technologies show promise in enhancing task learning and providing interactive and immersive experiences for learners in the agricultural field. Moreover, the combination of gamification with these immersive technologies further enhances agricultural learning experiences. For instance, several studies have explored the combination of gamification with social networks, AR, VR, and mobile appliances in agricultural learning. These studies have demonstrated that integrating gamification mechanisms with technology-presented multimedia can reduce learners' anxiety and enhance their engagement in learning agricultural concepts, particularly in mathematics. These technologies provide learners with realistic and interactive experiences, allowing them to explore and engage with agricultural concepts in a virtual environment. This transformative approach to agricultural learning through AR, VR, and gamification technologies has the potential to revolutionize the way agricultural knowledge is taught and acquired. VR technology offers a novel approach to agricultural learning by providing learners with simulated environments that mimic real-life farming scenarios. These virtual environments allow learners to practice and develop their skills in a safe and controlled setting. Furthermore, the integration of AR technology in agricultural learning enables learners to overlay digital content onto the physical world, providing a rich and interactive learning experience. These technologies also allow for the gamification of agricultural learning, incorporating game elements such as points, rewards, and competition to enhance learner motivation and engagement.

AR IN AGRICULTURAL EDUCATION

Fundamentals of AR Technology

AR technology enhances the real-world environment with computer-generated perceptual information, overlaying digital data onto the physical world. Unlike VR, which creates an entirely artificial environment, AR augments the existing world with additional information. Recent developments in sensor technology, computer vision, object recognition, and display systems have greatly advanced the capabilities and applications of AR technology. These advancements allow for a more immersive and interactive experience, opening up new possibilities in various fields including agriculture (Billinghurst, 2021).

AR Applications in Crop Management and Animal Husbandry

The integration of AR in agriculture is revolutionizing crop management and animal husbandry. Recent studies have focused on the development of AR systems to aid in precise and efficient agricultural practices. For instance, a study by Smith et al. (2020) demonstrates the use of AR for real-time crop monitoring and disease detection. This AR system overlays vital information about crop health directly onto the farmer's field of view, allowing for immediate and informed decision-making. Another research by Chatterjee et al. (2021) explores the application of AR in animal husbandry, where AR headsets provide farmers with instant access to animal health metrics and history, enhancing the care and management of livestock.

Case Studies: Successful AR Implementation in Agriculture

Case studies of AR implementation in agriculture highlight its effectiveness and potential. A study by Wen et al. (2020) implemented an AR system to assist with precision planting, showing a significant increase in planting efficiency and accuracy. The AR system provided visual guides to ensure optimal seed placement, leading to improved crop yields. Another case study by Bekiaris et al. (2021) discusses the use of AR in greenhouse management. The AR application allowed users to visualize environmental data and plant growth information superimposed on the actual plants, facilitating better resource management and plant care.

Challenges and Opportunities

While AR technology offers promising opportunities for agriculture, it also presents challenges. One of the main challenges is the need for robust and accurate tracking and alignment of virtual and real-world elements, which is critical for effective AR applications. Additionally, there are concerns regarding the usability and accessibility of AR technology for all farmers, especially those in remote or resource-limited areas. However, the continuous advancements in AR technology, including improved hardware and software solutions, are gradually overcoming these challenges. Researchers are exploring new methods to enhance the accuracy and user-friendliness of AR systems in agriculture.

In conclusion, AR technology holds significant promise for advancing agriculture through enhanced crop management and animal husbandry. Recent studies and case implementations demonstrate the potential of AR to revolutionize this field, making agricultural practices more efficient, informed, and sustainable. However, addressing the challenges related to tracking accuracy, usability, and accessibility is crucial for the widespread adoption and success of AR in agriculture.

VR IN AGRICULTURAL LEARNING

Understanding VR: Immersive Technology for Education

VR is an immersive technology that has gained significant attention in the field of education. It provides a computer-generated environment that simulates physical presence and enables users to interact with the environment in a realistic way. VR technology has the potential to enhance learning experiences by providing a more engaging and interactive environment for students. In the context of agricultural learning, VR can be used to create realistic simulations of farming environments, allowing students to gain practical experience without the need for physical resources. Recent research by Akçayır and Akçayır (2017) highlights the potential of VR in education, emphasizing its ability to create immersive and interactive learning environments. The study suggests that VR can improve student engagement and motivation, leading to better learning outcomes. Similarly, Wu et al. (2019) discuss the benefits of VR in education, emphasizing its ability to provide experiential learning opportunities particularly valuable in fields like agriculture.

VR for Training in Agricultural Techniques

In agricultural learning, VR technology can be utilized for training in various techniques and machinery operation. By creating realistic simulations of farming equipment and processes, VR can provide hands-on training experiences for students. This can be particularly valuable for teaching complex agricultural techniques and machinery operation in a safe and controlled environment. A study by Sopegno et al. (2016) explores the use of VR for training in agricultural machinery operation, highlighting its potential to improve learning outcomes and reduce the risk of accidents associated with traditional training methods. The authors emphasize the importance of realistic simulations in providing effective training experiences for agricultural students.

The Role of VR in Environmental Education

VR technology can also play a significant role in environmental and sustainability education within the agricultural context. By creating immersive simulations of environmental scenarios, VR can help students understand the impact of agricultural practices on the environment and explore sustainable farming techniques. Research by Cho et al. discusses the use of VR applications for visualizing scientific data of an ecological model, allowing users to explore the impacts. The study emphasizes the ability of VR to create realistic environmental simulations that can enhance students' understanding of complex ecological concepts.

VR in Agricultural Learning Contexts

When evaluating the impact of VR in agricultural learning contexts, it is important to consider its effectiveness in improving learning outcomes, student engagement, and practical skills acquisition. Recent studies have shown promising results in terms of the benefits of VR technology in agricultural education. A meta-analysis by Lee et al. (2021) evaluates the impact of VR in educational settings, including its effectiveness in improving learning outcomes and student engagement. The study provides evidence of the positive impact of VR on learning, highlighting its potential to enhance agricultural education. In conclusion, VR technology holds great potential for transforming agricultural learning by providing immersive and interactive experiences, enhancing training in agricultural techniques and machinery operation, promoting environmental and sustainability education, and improving learning outcomes in agricultural contexts.

Applications of Extended Reality (XR) in Agriculture, Livestock Farming, and Aquaculture

In this section of the systematic review, we aim to critically analyze the applications of extended reality (XR) in agriculture, livestock farming, and aquaculture education. The review will focus on understanding the potential benefits, challenges, and future prospects of XR technologies in agriculture education.

Anastasiou et al. (2023) presents a detailed review of the applications of extended reality (XR) in the agricultural sector, specifically focusing on livestock farming and aquaculture. The authors provide insights into the potential uses of XR technologies such as AR and VR in improving various aspects of agricultural practices. The review discusses the utilization of XR for training, simulation, monitoring, and decision support in livestock farming and aquaculture. Additionally, the authors highlight the potential benefits of XR in enhancing productivity, efficiency, and sustainability in these sectors.

The review by. Anastasiou et al. (2023) also addresses the challenges and limitations associated with the adoption of XR in agriculture, livestock farming, and aquaculture. These include technological barriers, cost implications, and the need for specialized skills for effective implementation. Furthermore, the authors discuss the prospects of XR technologies in these domains, emphasizing the potential for continued innovation and advancement to address current limitations and expand the scope of applications.

In the context of applying Extended Reality (XR) technologies across various domains, a comprehensive analysis reveals a predominant inclination towards their utilization in the educational sector, accounting for approximately half of all applications (50%). This significant proportion underscores the emerging role of XR as a transformative tool in education. Following closely, the use of XR for decision-making processes constitutes a substantial portion, representing 39.7% of its applications. This highlights the technology's effectiveness in enhancing critical and strategic decision-making scenarios.

Moreover, the application of XR in system development specific to the primary sector, as well as in action-oriented research, is also notable, though to a lesser extent. These areas account for 6.9% and 3.4% of XR usage, respectively. This distribution suggests a burgeoning interest in leveraging XR for more diverse and specialized fields, beyond its dominant educational and decision-making uses.

The data, graphically depicted in Figure 1, offers a clear visual breakdown of these percentages, providing an intuitive understanding of how XR technology is being integrated across different applications. The graph not only illustrates the current state of XR deployment but also hints at future trends. The predominance of XR in education suggests its potential as a key tool for learning and teaching in the years to come. Simultaneously, the notable usage of XR in commercial settings, primarily for monitoring purposes, indicates a growing recognition of its value in business operations and strategy formulation. This comprehensive overview of XR applications presents a compelling case for the technology's versatility and its increasing importance in both educational and commercial landscapes.

Figure 1. Percentage of XR Technology used Per Application

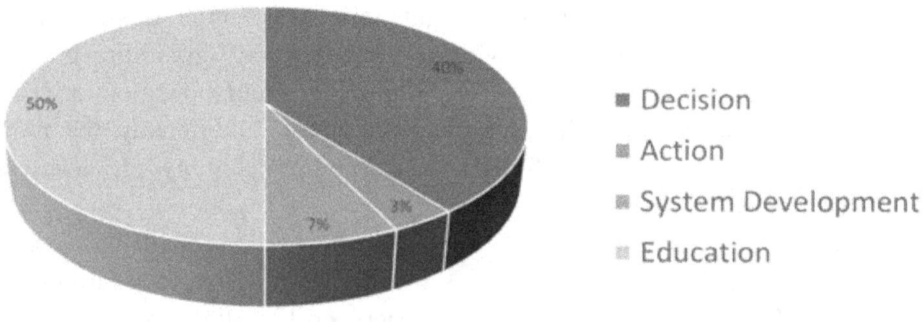

Table 1 in this systematic review organizes the research studies into categories based on their main focus, the type of Extended Reality (XR) technology they use, the devices for viewing XR, and the controllers for interacting with XR. This table makes it easier to see the range of XR applications, the technologies being used, and the preferences for viewing and controlling XR in different research areas. It helps to understand current trends and gaps in XR research and is a useful tool for anyone interested in how XR is being used and studied.

Table 1. Classification of research studies per focus area, application, XR type, XR viewing device and XR controller

Research Study	Focus Area	Application	XR Type	XR Viewing Device	XR Controller
(Kim et al., 2021)	Agriculture	Education	VR	HMD	Handheld with Motion Controller Wearable
(Bekiaris et al., 2021)	Agriculture	Decision Action	AR	Portable Device	Portable Device
(Ponnusamy et al., 2021)	Agriculture	Decision	AR	HMD	Handheld Controller
(Larbaigt & Lemercier, 2021)	Agriculture	Decision	AR	HMD	Voice Controller HMD Controller
(Garzon et al., 2020)	Agriculture	Education	AR	Portable Device	Portable Device Controller
(Garzon, Baldiris, Acevedo, & Pavon, 2020)	Agriculture	Education	AR	Portable Device	Portable Device
(Carruth, Hudson, Fox, & Deb, 2020)	Agriculture	Education	VR	Monitor	PC Controllers
(Shaleendra et al., 2019)	Agriculture	Decision	AR	Portable Device	Portable Device Controller
(Phupattanasilp & Tong, 2019)	Agriculture	Decision	AR	Monitor	PC Controllers
(Soujanya, Roopa, & Babu, 2019)	Agriculture	Monitoring Action	VR	HMD	Portable Device Controller
(Rosmansyah, Achiruzaman, & Hardi, 2019)	Agriculture	Education	VR	Monitor	PC Controllers
(Zheng & Campbell, 2019)	Agriculture	Decision	AR	Portable Device HMD	PC Controllers Handheld with Motion Controller HMD Controller Portable Device
(Cheng, Kang, Cai, Li, & Sun, 2018)	Agriculture	Education	VR	HMD	Handheld with Motion Controller
(Huang, Deng, Fei, & Hu, 2017)	Agriculture	Decision	VR	Monitor	PC Controllers

continued on following page

Table 1. Continued

Research Study	Focus Area	Application	XR Type	XR Viewing Device	XR Controller
(Nigam, Kabra, & Doke, 2011)	Agriculture	Decision	AR	Portable Device	Portable Device
(Yongyuth, Prada, Nakasone, Kawtrakul, & Prendinger, 2010)	Agriculture	Education	VR	Monitor	PC Controller
(Beck, 2009)	Agriculture	Education	VR	Monitor	PC Controllers
(Cheung et al., 2008)	Agriculture	Education	VR	Monitor	PC Controllers

Introduction to Controlled Environment Agriculture

Controlled Environment Agriculture (CEA), which involves cultivating food crops in greenhouses and indoor spaces, is projected to experience a fivefold increase in market size over the next decade (Krishnan et al., 2020). It is often referred to as a sustainable solution and the future of food (Benke & Tomkins, 2017), particularly with the emergence of high-tech vertical farms illuminated by vibrant LED lights. This method of agriculture can be implemented in urban areas, providing a consistent supply of fresh, local produce throughout the entire year. Therefore, it plays a vital role in enhancing food security and reducing the carbon footprint associated with transportation.

Despite the advantages, CEA infrastructure poses a significant challenge as a massive energy consumer and is fundamentally incompatible with a low-carbon economy in its current form. A comprehensive review of life cycle assessments has demonstrated that greenhouse gas (GHG) emissions associated with produce grown in heated greenhouses are, on average, 5.6 times higher than those from field-grown produce (Clune et al., 2017). In addition, the outlook for indoor agriculture is anticipated to be even more problematic. The major contributors to this high energy consumption are climate control, artificial lighting, and the use of chemical fertilizer. Consequently, a fundamental reimagination of CEA is essential to transform it into a sustainable approach to food production. Moreover, we cannot disregard the importance of nurturing the skills and knowledge of the younger generation while stepping into the realm of CEA. By integrating educational initiatives designed for the youth, we have the opportunity to cultivate a sustainable and knowledgeable workforce, empowering them to spearhead innovation in CEA practices.

Introduction to Green-Reimaging of Agriculture in Controlled Environment (Grace) Research Project

Grace is an interdisciplinary collaboration among multiple land-grant institutions: Auburn University, the University of Florida, Tuskegee University, Michigan State University, and the University of Georgia, along with departments in agriculture, computer science, and biosystem engineering. Researchers from each institution contribute specialized knowledge in state-of-the-art technologies and methodologies within their respective fields, encompassing plant biology, horticulture, biosystems engineering, computer science, and economics (shown in Figure 2). This collaboration is funded by the National Institute of Food and Agriculture (NIFA) [Award NO. 2023-68012-38994].

The study focuses on three integral components – extension, education, and research – that collectively contribute to CEA practices. Central to this exploration are the interconnected themes of lower carbon intensity, increased efficiency, reduced demand, and the strategic shift in consumer and producer behavior. In the following sections, each component will be thoroughly examined and discussed in detail.

Extension

Extension services play an essential role in fostering sustainable practices within CEA. Existing producers are trained in low-carbon techniques through workshops and publications, facilitating the exchange of knowledge and experiences. Moreover, extension services extend their reach to train the next generation of CEA practitioners by developing gamified online learning modules for high school students and new entrepreneurs. The reciprocal relationship between lower carbon intensity and increased efficiency is evident as the dissemination of efficient practices contributes to a reduction in the carbon footprint of CEA.

Research

Research initiatives contribute significantly to the synergy between lower carbon intensity and increased efficiency in CEA. By employing bio-energy sources for heat and power, utilizing waste sources of CO2, and irrigating with wastewater, researchers address the dual challenge of lowering carbon intensity and increasing efficiency. Modeling and AI further support better control, while innovations such as improved plant temperature tolerance and the adoption of a seasonal cropping approach contribute to a reduction in demand for resources.

Education

Education initiatives focus on raising awareness and fostering behavioral shifts in both consumers and producers. Informative marketing strategies are employed to educate consumers about the carbon footprint of produce, supplemented by online calculators. Understanding consumer preferences and willingness to pay is explored among diverse populations. Educational efforts extend to training high school students and new CEA entrepreneurs through gamified online learning modules, fostering a new generation of practitioners committed to sustainable practices.

Figure 2. The Schematic of Grace

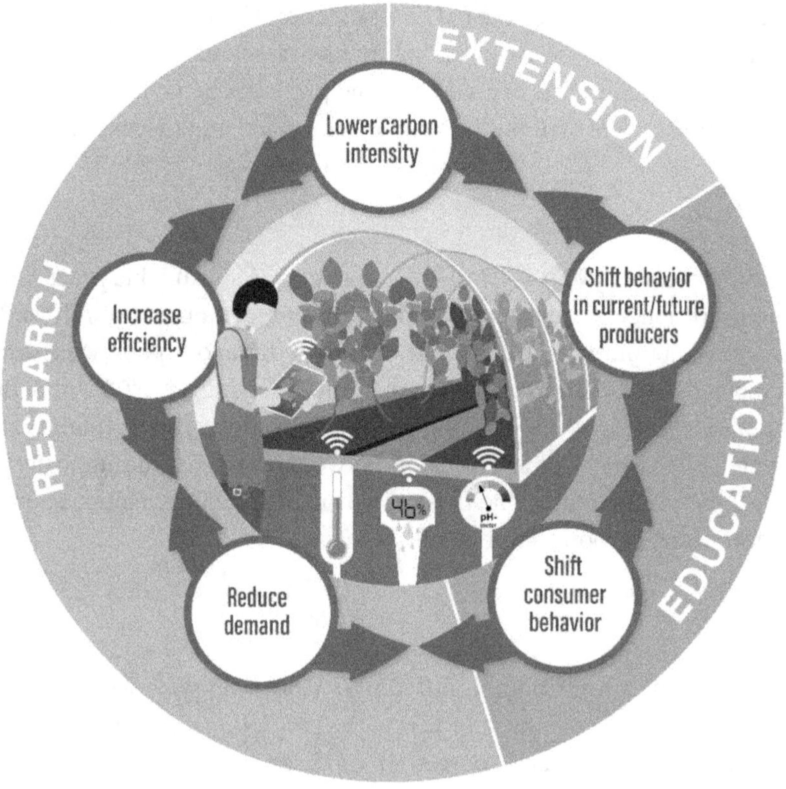

PROBLEM OF TRADITIONAL CLASS METHODS

Many conventional teaching approaches often face difficulty making complex concepts accessible and captivating for students, particularly in demanding subjects like CEA (Olapiriyakul & Scher, 2011). Several studies have compared traditional classroom teaching with online learning, highlighting various aspects of the two approaches. These studies provide valuable insights into the challenges of conventional teaching methods and the comparisons between traditional and online learning.

Paul and Jefferson (2019) published a study in Frontiers in Education that compares student performance in an online environmental science course with that in face-to-face course from 2009 to 2016. The research found that technological advancements have improved the speed and accessibility of distance education, making online instruction a more viable option. The study also emphasized that online and traditional education methods share many similarities, with students still required to attend class, grasp the material, and fulfill their assignments.

Another study, published in the Journal of Higher Education, examined the concepts of traditional classroom and online learning. It emphasized the importance of understanding these concepts from a philosophical point of view and highlighted the role of interpersonal relationships among students and teachers in the traditional classroom setting (Hassan et al., 2014).

Furthermore, a study in Sage Journals from 2022 compared university students' performance in traditional learning to that of online learning during the COVID-19 pandemic. The research analyzed the lessons learned from faculty's perspectives and provided insights into the differences between traditional and online learning during the pandemic. It presents a mixed picture of the effectiveness of online learning compared to traditional learning. However, it is important to consider the diverse factors that can influence the outcomes of such comparisons, including the nature of the content, the characteristics of the learners, and the specific context of the study (Alzahrani, 2022).

GRACE EDUCATIONAL PLATFORM STRATEGY

Grace educational platform stands out as one of the prominent solutions to tackle the challenges of conventional teaching approaches. The platform, as a dynamic part of the extension and education component of Figure 2, leverages innovative techniques such as cutting-edge AR, VR, and AI in CEA to engage learners. It will go beyond traditional educational methods, employing gamified online learning modules that engage and educate high school students and new CEA entrepreneurs.

The platform incorporates insights from ongoing research, offering a dynamic and up-to-date educational experience.

The Grace platform comprises both a website and a mobile application. Students can utilize these platforms to access gamified AR and VR learning and training sections. Figure 3 provides an overview of the platform's architecture, including a shared database across the Web and Mobile applications. This design allows students and teachers to reach their personalized dashboards through the website or the application. The development of the website and application utilizes ASP.NET and Unity3D tools, renowned cross-platform frameworks created by Microsoft for constructing dynamic online applications. This study employs a dynamic virtual environment through which students can explore significant subjects in ecology, microbial interactions, heating, ventilation, air conditioning (HVAC) control, and plant genetics.

Figure 3. The Grace Platform Outline

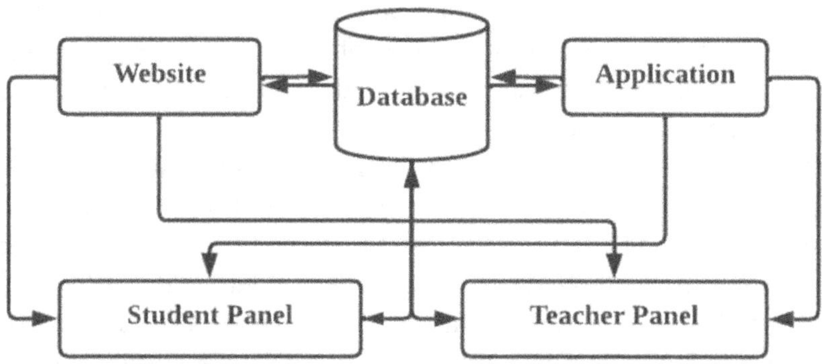

The study strategically visualizes essential modules that form the backbone of comprehensive learning. These modules encompass critical topics such as water quality assessment, insect pest management, pH level evaluation, precipitation-related concerns, and nutrient imbalances. The water quality assessment module immerses learners in understanding the intricate dynamics of water quality within controlled environments, emphasizing the significance of maintaining optimal conditions for plant growth. Simultaneously, the insect pest management module equips students with essential skills in identifying, preventing, and mitigating pest-related challenges. The inclusion of pH level evaluation sheds light on the crucial role of maintaining the right pH balance in nutrient solutions, contributing to plant health and vitality. Addressing precipitation-related concerns ensures learners are adept at managing water supply in CEA. Lastly, the nutrient imbalances module delves into the delicate equilibrium necessary for plant nutrition that emphasizes the role of nutrients in fostering healthy and productive crops. Through a visually immersive approach, these modules provide a holistic understanding of key facets.

To enrich the learning journey, the dynamic stochastic gaming probabilistic (DSGP) concept (Szajowski, 2019) and gamification techniques (Metwally et al., 2021) are being applied. DSGP and gamification techniques involve the application of stochastic games and probabilistic transitions in a dynamic game setting, along with the integration of game elements into non-game contexts to enhance engagement and learning. Stochastic games, introduced by Shapley (1953), are repeated games with probabilistic transitions played by one or more players in a sequence of stages. Players select actions and receive payoffs based on the current state and chosen activity, and the game moves to a new random state based on the previous state and the chosen actions. Stochastic games have applications in various fields such as economics, evolutionary biology, and computer networks (Szajowski, 2019). The gamification techniques aim to enhance the learning experience by incorporating game elements into the modules, potentially increasing motivation and engagement (Metwally et al., 2021; Szajowski, 2019). Therefore, utilizing DSGP and gamification techniques in this study can provide a comprehensive and engaging greenhouse learning journey by making decisions based on random chance events and leveraging probabilistic concepts.

In this study, we will also employ avatar-guided learning experiences. The virtual tour guide avatar can navigate through virtual environments, stop at specific points of interest, and provide information to users. The concept involves using a guide avatar to lead users through virtual venues. It can be programmed to stop at designated spots, provide audio information, and answer user questions. This technology has applications in different virtual industries such as tourism, education, training, and etc. Avatar-guided learning experience in Grace educational platform will involve the use of a virtual avatar as a tour guide within a virtual greenhouse environment. It can navigate the greenhouse, provide information, and potentially answer student questions. Within this immersive setting, an interactive tour guide avatar will enhance the learning experience and engage students in decision-making processes related to a greenhouse system.

Goals of the Grace Educational Platform

The research aims to establish CEA as a sustainable food production system capable of efficiently generating nutritious foods in a low-carbon economy. To achieve this objective, the platform seamlessly integrates agricultural practices with technology to enhance high school students and new entrepreneurs understanding of

STEM concepts in an educational setting. This initiative aligns with the principles outlined in Section 504 of the Rehabilitation Act of 1973 (McMenamin & Zirkel, 1998), the Individuals with Disabilities Education Act (IDEA) (Weiss & Mettrick, 2015), Universal Design for Learning (UDL) Guidelines (Rose et al., 2018). By upholding these principles, the platform guarantees fair access to STEM concepts, creating an educational environment that is encouraging and available to students with diverse capabilities and learning styles.

The research also endeavors to broaden market opportunities for CEA farmers and entrepreneurs by equipping them with cutting-edge technologies through the platform and facilitating learning in CEA. Our aim is to enhance their overall sustainability and efficiency in agricultural production by preparing them with these technological tools and knowledge. Therefore, farmers can make informed decisions, optimize resource utilization, and implement sustainable practices. In this manner, the immediate success of CEA farmers and the long-term resilience of the entire CEA industry will be guaranteed.

Grace Educational Platform and AR, VR, and AI Technologies

In this section, we aim to comprehensively explore the integration of Grace Platform with AR, VR, and AI technologies and the reasons behind our selection of them, highlighting their specific roles within the context of our research.

AR Technology: Enhancing Real-world Understanding

AR technology overlays digital content onto the real-world environment and creates an augmented learning experience that goes beyond textbooks and traditional methods. Grace Educational Platform leverages AR to enhance students' understanding of complex CEA concepts by providing interactive 3D models, simulations, and real-world applications. Imagine students exploring the intricacies of planting or plant irrigation in real-time through AR-enhanced lessons. Through the incorporation of AR, this study fosters a dynamic learning environment where students can interact with virtual objects.

VR Technology: Immersive Learning Journeys

VR technology transports students to immersive virtual environments, allowing them to engage in realistic simulations and scenarios. Grace recognizes the potential of VR in bridging the gap between theory and application, preparing students for the challenges of real-world problem-solving in the field of CEA. It embraces VR as a tool to create immersive learning journeys in the realm of CEA.

AI Technology: Personalized and Adaptive Learning

In this study, AI is a strategic method to create adaptive learning paths, allowing students to progress at their own speed. The study understands that each student has a unique learning journey and facilitates the customization of educational content to meet individual needs. The method involves an initial assessment of each student's learning style, preferences, and proficiency levels, which is then integrated into the AI algorithms. This comprehensive understanding of the distinctive learning journey of each student enables the system to tailor educational content, ensuring that it aligns with individual needs and maximizes relevance.

EXPLORING PANELS ON GRACE EDUCATIONAL PLATFORM

Grace's platform incorporates four main panels: an admin panel, an educator panel, a student panel, and general learner panel. Each panel serves distinct roles, contributing to the platform's comprehensive functionality and adaptability across various user roles within an educational setting.

Admin Panel

Administrative access plays a pivotal role in efficient system management, ensuring a smooth and organized learning environment. This section explores the admin panel, focusing on key functionalities, including educator management, course management, content management, user management, and settings. Figure 4 displays the admin dashboard on the website.

Figure 4. Admin Dashboard on the Website

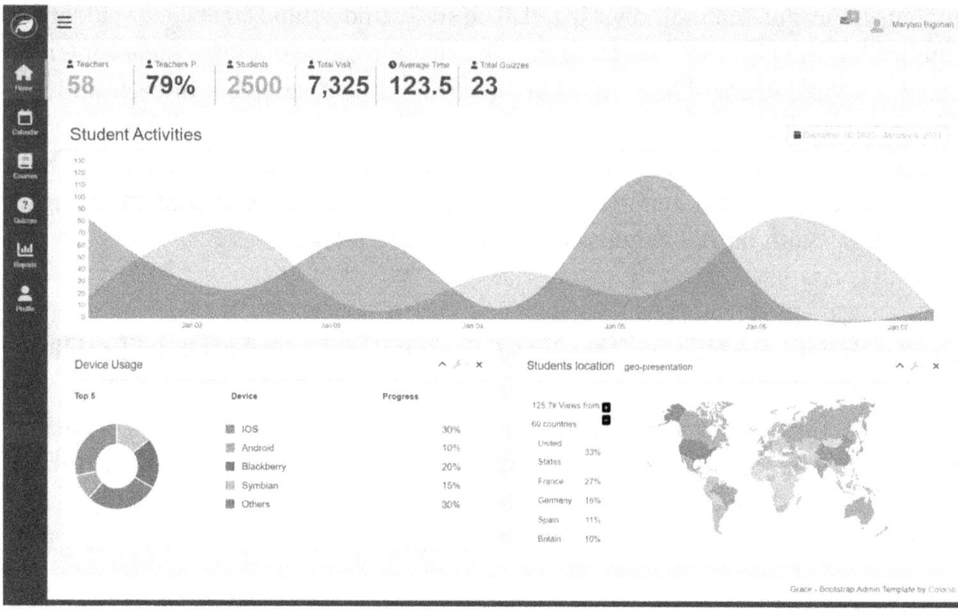

Course Management

Course creation and management are central components of educational administration. Admins wield the power to create, edit, and delete courses. The platform facilitates the assignment of educators to specific courses and allows them to have an effective teaching approach. Furthermore, administrators can set detailed course schedules and durations, ensuring the educational programs align with predefined timelines. This level of control over courses empowers administrators to align educational programs with organizational objectives and student needs.

Educator Management

Administrators with higher access levels can manage educators within the platform. It involves the ability to add, edit, and delete educator accounts. The system allows administrators to efficiently assign courses to educators and manage teaching schedules, providing flexibility in course distribution and educator allocation. This level of control ensures optimal utilization of human resources and a well-organized educational structure.

Additionally, all registered educators and their added courses need to be approved in the platform by administrators. The approval of educators plays a pivotal role in the sustainability of any educational initiative, particularly in the context of the educational platform. Through the approval of educators, administrators can establish a robust framework for maintaining and elevating the quality of educational content.

Content Management

Efficient content management lies at the heart of delivering a rich and engaging learning experience. Administrators possess the authority to upload, organize, and manage various types of educational content, including text, images, videos, and documents. They have the capability to enhance the interactivity of content through features such as clickable images, flying-over images, and drag-and-drop elements, different type of charts, thereby creating a more dynamic and engaging learning environment for students. This multifaceted approach not only broadens the range of interactive elements within the content but also ensures that educational materials are accessible to all students, promoting an inclusive and enriching learning atmosphere.

Student Management

Administrators hold a leading role in student management, encompassing tasks such as adding new students, editing student profiles, and even deleting student accounts when necessary. Additionally, administrators oversee the maintenance of user accounts and promptly address any technical issues that may arise. This proactive involvement guarantees seamless access to educational content and establishes a supportive environment conducive to effective learning in the realm of CEA.

User Management

User management functionalities are crucial for maintaining a secure and organized educational platform. The administration panel incorporates robust user management functionalities, allowing for the addition, modification, and deletion of user accounts across various roles, including administrators, educators, and students. This level of control extends to user roles and permissions management, ensuring that each user has appropriate access rights. The ability to finely tune user roles enhances the security and integrity of the educational platform.

The Platform Settings

The settings section provides administrators with the tools to customize the appearance and functionality of the educational platform. It includes general and personal settings such as the site name, logo, changing password, profile information, etc. The settings empower administrators to create a cohesive online learning environment.

Real-time and non-Real-time Educational Quizzes

One of the fundamental functionalities of the administrator panel is its capability for real-time and non-real-time educational quiz management. This feature enables them to create, deploy, and manage quizzes, fostering interactive learning experiences. In addition, it can help educators gauge student understanding instantly and adjust instructional approaches based on real-time assessment data. The integration of real-time and non-real-time quiz management not only enhances the immediacy of feedback but also promotes active student participation and comprehension.

Educator Panel

The educator panel emerges as a dynamic and powerful tool for educators, empowering them to elevate the educational experience through an array of advanced functionalities. This section delves into the key features of the educator panel, elucidating its role as a powerful facilitator for educators seeking to enhance the quality and engagement of their educational content.

Monitoring Students

The educator panel goes beyond traditional teaching methods by incorporating comprehensive student monitoring tools. Educators can actively monitor student progress, track performance trends, and identify areas that may require additional attention (shown in Figure 5). Real-time monitoring facilitates a proactive teaching approach, allowing educators to intervene promptly if students encounter challenges or require additional support. The ability to monitor student engagement and understanding empowers educators to tailor their instructions.

Figure 5. Monitoring Students on the Website

Assigning a Student to Courses

An educator can add registered students, who have enrolled in their courses to the class. By leveraging this function, educators can efficiently manage their class rosters, ensuring that all registered students have access to the designated courses. The educator's authority to add registered students facilitates a smooth and organized educational experience.

Student Panel

Student panel aims to provide an engaging, personalized, and effective learning experience for students and adult learners interested in expanding their knowledge of plants and agriculture. It comprises the following essential elements.

Informal Learning Modules

The informal learning modules within the educational platform are meticulously designed to offer students a transformative and engaging approach to comprehend intricate concepts. Far from traditional instructional methods, these modules are dynamic, fostering an environment where students actively participate in their learning journey. Through interactive multimedia elements, such as visually compelling graphics and real-world case studies, students are not merely recipients of information but active participants in their educational experience. The dynamic nature of these modules ensures that complex concepts are presented in relatable contexts and makes them more accessible to students and adult learners with diverse capabilities. The incorporation of gamification elements will add an extra layer of excitement, turning the learning process into an enjoyable and motivating adventure. This approach encourages curiosity, critical thinking, and a deeper understanding of the subject matter, ultimately contributing to a holistic educational experience that extends beyond traditional classroom boundaries. Figure 6 displays a sample of a course slide.

Figure 6. Sample Course Slide on the Website

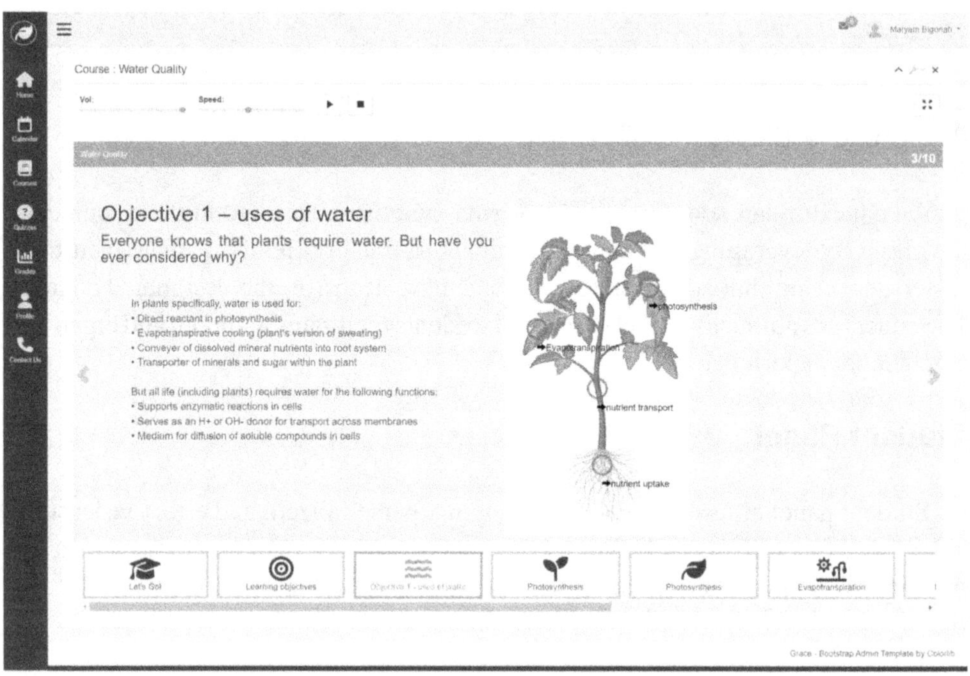

Real-time and non-Real-time Educational Quizzes

The inclusion of both real-time and non-real-time educational quizzes in the Grace platform enhances learning outcomes through varied assessment strategies (Figure 7). Real-time quizzes offer immediate feedback, allowing students to gauge their understanding promptly, while non-real-time quizzes provide flexibility for self-paced review. These quizzes contribute to a comprehensive learning experience, aligning with diverse learning preferences and promoting a deeper engagement with the educational material.

Figure 7. Quiz List on the Website

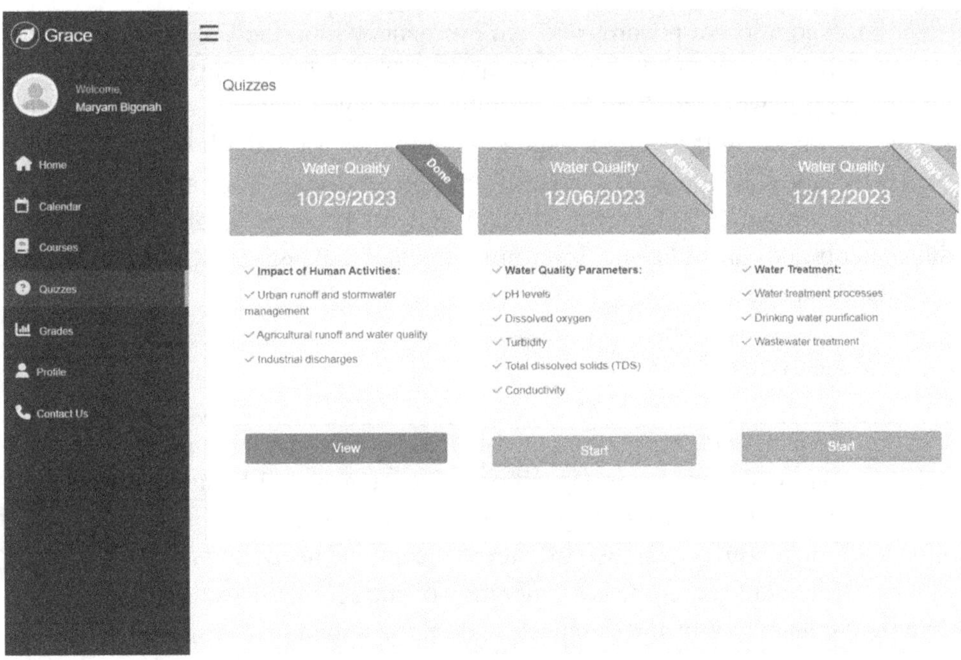

A Gamified (Metwally et al., 2021) Avatar Tour Guide

The gamified avatar tour guide transforms the educational journey into an interactive and captivating experience. By integrating gamification principles, students will be guided through a virtual tour by an engaging avatar, making the learning process enjoyable and memorable. This innovative approach not only enhances student engagement but also leverages the motivational aspects of gamification to create a dynamic and immersive learning environment within the Grace platform.

Tracking Progress

Students can review their achievements, identify areas for improvement, and track overall progress. This feature promotes self-awareness and goal setting, increasing a sense of accountability and encouraging continuous improvement. By providing students with insights into their learning path, progress tracking enhances the personalized and adaptive nature of the educational experience.

Rewards and Achievements

The incorporation of rewards and achievements within the Grace platform will create a motivating and positive learning environment by recognizing and celebrating students' accomplishments. This feature will be displayed during gameplay and after each quiz, comparing the student's score within the class scale. It encourages sustained effort and contributes to a sense of achievement, supporting and encouraging the learning community. The recognition of students' progress through rewards and achievements adds an additional layer of motivation and engagement to the overall educational experience.

General Learner Panel

In this research endeavor, one of the objectives is to cultivate and empower the upcoming cohort of practitioners, specifically focusing on individuals venturing into entrepreneurship. The methodology revolves around the development of gamified online learning modules designed to cater to the unique needs and challenges faced by them. The targeted outcome is a generation of entrepreneurs well-versed not only in the fundamentals of business but also equipped with a profound understanding of cutting-edge technologies relevant to CEA. To facilitate this transformative educational experience, we design a dedicated panel for them. This panel, which is identical to the student panel, serves for individuals to enroll in courses specifically. However, approval from administrators is not required for access. It can contribute to the broader objective of cultivating a skilled and knowledgeable workforce in the realm of entrepreneurship and CEA.

ACCESSIBILITY

Accessibility is a critical aspect of designing websites and applications. It guarantees that all students can access and utilize the platform, regardless of their physical or cognitive abilities. Legal compliance is another compelling reason to mandate accessibility in various countries through laws like the Americans with Disabilities Act (ADA). Furthermore, accessible design enhances the user experience for everyone, offering features like text alternatives and adaptable navigation that benefit students in different contexts.

Accessibility in AR and VR applications for agriculture is not merely a technical requirement but a fundamental aspect that ensures equitable access to educational content. However, to realize the full potential of these technologies, it is essential to consider the diverse needs of students, including those with varying physical and cognitive abilities.

In this research, we incorporate several pivotal features to enhance the accessibility of websites and mobile applications to ensure a more inclusive digital experience for individuals with diverse abilities. Among these essential features is voice-guided navigation, allowing users to traverse content seamlessly through spoken instructions. Moreover, a commitment to inclusivity is demonstrated by providing alternative text descriptions for visual elements, aiming to empower users with visual impairments or those relying on screen readers to fully comprehend images and visual content.

The functionalities embedded in the educational platform play a pivotal role in elevating the learning experience for both students and educators. With a robust content management system, the platform facilitates seamless organization and deployment of educational materials, ensuring that relevant and interactive resources are readily available. The platform's multifaceted impact extends beyond accessibility and engagement to encompass cost-efficiency, streamlining educational processes and optimizing resource utilization.

ESTIMATED PROBLEMS

This endeavor is not without its challenges. One significant hurdle is the technical limitations inherent in AR and VR development, including hardware compatibility issues, software bugs, and constraints in device capabilities. Overcoming these obstacles requires testing and optimization to ensure smooth functionality across different platforms and devices. Additionally, creating engaging and educational content suitable for the target audience while navigating the complexities of AR and VR technology poses a challenge. Content development must strike a balance between accuracy, educational standards, and immersive interactivity to maximize learning outcomes.

Moreover, ensuring equitable access to AR and VR technology presents another notable challenge. Not all students may have access to the necessary devices or reliable internet connections, potentially limiting the reach and effectiveness of the application. Addressing issues of accessibility and inclusivity is essential to ensure that the benefits of AR and VR education are available to all students, regardless of their socioeconomic status or geographical location. Furthermore, educator training and support are crucial for successful implementation in classrooms. Educators may require guidance on integrating AR and VR technology into their teaching practices effectively. Providing resources, tutorials, and ongoing assistance can empower educators to leverage the application to enhance student learning experiences. By proactively addressing these challenges, the study can contribute to the advancement of AR and VR education in agriculture and foster a more engaging and inclusive learning environment for high school students.

RESEARCH EVALUATION

Several studies, such as those by Benedick, Robert, & Le Traon (2021) and Baratimehr, Moosavi, & Tahayori (2023), have examined the effectiveness and suitability of different systems. AR and VR educational systems are no exception to this scrutiny, emphasizing the essential nature of conducting both qualitative and quantitative evaluations to assess their performance. In this research, a comprehensive evaluation framework will be utilized to rigorously assess the impact and effectiveness of AR and VR learning modules and the virtual gamified greenhouse, developed for secondary school students in the context of CEA. The evaluation process involves a multifaceted approach, addressing the following research questions:

- Question 1: How effective are AR and VR learning modules and the greenhouse in enhancing understanding and use of CEA concepts aligned with Next Generation Science Standards (NGSS) for secondary school students?
- Question 2: How do gamified elements and dynamic stochastic gaming probabilistic concepts impact the engagement and learning experiences of students in CEA education?
- Question 3: How well do the developed learning modules cater to the diverse needs of learners, including those with disabilities, in alignment with Section 504 of the Rehabilitation Act of 1973, IDEA, and Universal Design for Learning Guidelines?

The evaluation will be conducted using formal surveys and informal feedback from regional high school students and teachers. These instruments aim to gauge the effectiveness of the AR and VR learning modules and greenhouse in enhancing students' understanding and application of CEA concepts aligned with NGSS.

The evaluation procedure comprises two pivotal phases. Initially, the Module Testing and Fine-Tuning phase involves rigorous testing and refinement of the developed modules, drawing insights from formal surveys and informal feedback from regional high school students and teachers. This iterative approach ensures that the learning resources undergo continuous improvement, aligning seamlessly with educational objectives and effectively resonating with the target audience. Subsequently, the National Dissemination and Ongoing Monitoring phase, entails the nationwide release of the online modules through a server hosted at Auburn University. During this period, meticulous tracking of usage metrics will occur, and users will be actively encouraged to contribute feedback.

To assess the impact, replicate baseline data collection will be performed, including the adoption status of CEA Best Management Practices (BMPs). Data will be analyzed using pre-post comparison and paired t-test analysis to identify any statistically significant differences.

The variables under evaluation are of paramount significance as they directly influence the enhancement of students' comprehension, engagement, and application of CEA concepts. They encompass the effectiveness of learning modules and the impact of gamified elements. The primary variable of interest of this research is the effectiveness of the AR and VR learning modules, along with the greenhouse, in facilitating students' understanding and utilization of CEA concepts aligned with NGSS. This variable is pivotal to the overall success and relevance of the developed educational resources.

Recurring evaluation practices form an integral part of the research methodology, encompassing two key components. Firstly, Annual Process Evaluation involves a comprehensive assessment of programmatic efforts, outputs, and impacts in alignment with the project proposal. This evaluation considers various aspects such as producer participation, practice application, outreach and training activities, and summarized outputs realized across the project team. The second component employs a Simple and Consistent Framework designed to elicit feedback and measure impacts across various dimensions, including knowledge gain, behavior change, and economic outcomes. This adaptable framework extends its application to all outreach activities, encompassing online platforms such as the website, videos, courses, discussion forums, and newsletters. Moreover, it is integral to face-to-face demonstration projects conducted at stakeholder collaborative sites and university centers. This unified approach ensures a thorough and consistent evaluation across diverse project components, fostering continuous improvement and informed decision-making throughout the project's duration.

Qualitative Assessment

In this section, we will discuss the qualitative assessments, which can be listed as follows:

1. Effectiveness of AR and VR Learning Modules
   - Evaluate the enhancement of understanding and usage of CEA concepts through AR and VR modules.
   - Assess alignment with NGSS for secondary school students.
   - Qualitatively analyze content clarity, engagement, and relevance.
2. Impact of Gamified Elements
   - Examine the influence of gamified elements and dynamic stochastic gaming on student engagement.
   - Explore how these elements enhance the overall learning experience in CEA education.
   - Qualitatively assess the integration of gamification in achieving educational objectives.
3. Catering to Diverse Learner Needs
   - Investigate how well the developed learning modules address the needs of diverse learners.
   - Consider inclusivity for students with disabilities, aligning with Section 504 of the Rehabilitation Act of 1973, IDEA, and Universal Design for Learning Guidelines.
   - Qualitatively assess the adaptability and accessibility of the modules.
4. Iterative Module Testing and Fine-Tuning
   - Evaluate the effectiveness of the iterative testing and refinement process.
   - Understand how feedback from high school students and teachers contributes to continuous improvement.
   - Qualitatively assess the responsiveness of the modules to educational objectives.
5. National Dissemination and Ongoing Monitoring
   - Assess the effectiveness of nationwide release and server-hosted online modules.
   - Qualitatively analyze user feedback obtained during the monitoring phase.
   - Understand the adaptability of the modules to a diverse national audience.

6. Recurring Evaluation Practices
   - Evaluate the Annual Process Evaluation for programmatic efforts, outputs, and impacts.
   - Assess the Simple and Consistent Framework for feedback across various dimensions.
   - Qualitatively analyze the effectiveness of evaluation practices in guiding project decisions.

Quantitative Assessment

Here, we aim to discuss quantitative assessments, and we can present them as follows:

1. Surveys and Feedback Analysis
   - Analyze formal surveys to quantitatively measure the effectiveness of AR and VR modules.
   - Quantify feedback from high school students and teachers to identify trends and areas for improvement.
2. Usage Metrics and Adoption Status
   - Track usage metrics during the National Dissemination phase.
   - Quantitatively assess the adoption status of CEA BMPs.
   - Use quantitative data for pre-post comparison and paired t-test analysis.
3. Baseline Data Collection
   - Quantitatively measure baseline data.
   - Use statistical analysis to identify statistically significant differences.
4. Variables Under Evaluation
   - Assess variables influencing students' comprehension, engagement, and application of CEA concepts.
   - Use statistical analysis to understand the significance of learning modules and gamified elements.

In conclusion, the evaluation section is designed to provide a robust and systematic assessment of the developed learning modules, ensuring their alignment with educational objectives, responsiveness to diverse learner needs, and contribution to the advancement of CEA education.

DISCUSSION AND FUTURE WORK

This book chapter discusses CEA as a revolutionary approach to food cultivation, anticipating a remarkable fivefold increase in market size over the next decade. For this purpose, the integration of ecological literacy into agricultural education is paramount to foster sustainable farming practices and environmental stewardship. Empowering farmers with a solid understanding of ecological principles enhances their ability to make informed decisions that benefit both their operations and the environment. In addition, combining ecological literacy with agricultural education facilitates a deeper comprehension of the intricate relationships between agriculture, the environment, and society. The approach promotes sustainable practices and enhances resilience to climate change in agricultural systems. Therefore, this way not only promises a consistent supply of fresh, local produce but also stands as a potential solution for urban areas seeking sustainable agriculture throughout the year. However, the current energy-intensive nature of CEA infrastructure poses a significant challenge, especially in the context of a low-carbon economy.

The interdisciplinary collaboration known as Grace, involving multiple land-grant institutions, addresses this challenge through a comprehensive study focusing on extension, education, and research in CEA practices. It seeks to solve the challenges of CEA infrastructure by envisioning a sustainable food production system that aligns with the principles of a low-carbon economy. The research aims to train the next generation of practitioners through gamified online learning modules tailored for high school students and new entrepreneurs. It strives to enhance market opportunities for CEA farmers by equipping them with cutting-edge technologies and fostering a deep understanding of CEA practices.

At the heart of Grace's initiatives is the innovative Grace Education Platform, designed to revolutionize learning in CEA. This dynamic platform integrates cutting-edge technologies such as AR, VR, and AI to engage learners and establish CEA as a sustainable food production system. The unique features of the Grace Education Platform align with its overarching goals. It seeks to make complex concepts accessible and captivating for learners in the demanding field of CEA, thereby emerging as a solution to challenges posed by conventional teaching approaches. The incorporation of AR, VR, and AI technologies facilitates enhancing real-world understanding, immersive learning journeys, and personalized and adaptive learning. The platform will employ the DSGP concept, gamification techniques, and a virtual tour guide avatar to enrich the learning journey, providing students with a unique and engaging educational experience.

Grace platform is organized into four main panels: admin, educator, student panels, and general learner panel. Each panel serves distinct roles, contributing to the platform's comprehensive functionality and adaptability across various user roles. The admin panel oversees educator management, course management, content management, user management, real-time and non-real-time educational quizzes, and platform settings. The educator panel focuses on student monitoring and assign students to courses, while the student panel engages learners through informal learning modules, quizzes, gamified avatar guides, and progress tracking.

A crucial aspect of Grace's educational endeavor is its commitment to accessibility. Recognizing the significance of inclusivity, the research incorporates pivotal features into the platform to enhance accessibility for users with diverse abilities. Voice-guided navigation and alternative text descriptions for visual elements contribute to fostering inclusivity. This emphasis on accessibility aligns with the broader goal of making the educational platform accessible to everyone.

While this study employed formal surveys with predefined sets of words to collect feedback from students and teachers, an important area for future research is the assessment and analysis of such qualitative feedback. In this domain, novel methodologies like computing with words and perceptual computing could provide powerful tools that have not yet been widely applied (Mendel & Wu, 2010; Mendel, 2018; Mehr, Tahayori, & Moosavi, 2024). These methodologies involve computing directly with linguistic expressions from natural language rather than just numerical data. Leveraging these approaches could enable more nuanced and contextual interpretations of the rich qualitative feedback gathered through surveys. Such analyses could yield deeper insights into the experiences and perspectives of students and teachers, complementing the findings from the formal survey responses. The synergies between traditional survey methods and these emerging computational techniques present an exciting avenue for enhancing our understanding of educational processes and outcomes.

In addition, previous research has demonstrated integrating chatbots in education can significantly enhance the learning experience for students by offering immediate assistance, quick access to information, and personalized learning materials tailored to individual needs (Okonkwo, C. W., & Ade-Ibijola, A., 2021). Utilizing chatbots has been proven to boost student engagement, improve learning outcomes, and streamline administrative processes within educational institutions (Labadze, L., Grigolia, M., & Machaidze, L., 2023). Transformative potentials of chatbots in educational settings shows that incorporating this technology can be helpful for the student.

Future work should also focus on developing design guidelines, accessibility features, and risk mitigation strategies tailored to the unique needs and vulnerabilities of individuals with neurodevelopmental disorders, ensuring their privacy, safety, and security while leveraging the immersive potential of XR, AR or VR technologies for effective learning (Jones et al. 2023).In Anticipation, the incorporation of these cutting-edge technologies promises to revolutionize the educational landscape.

While AR and VR evolve, their potential to bridge theoretical knowledge with practical application in agriculture learning becomes increasingly apparent. This chapter lays the groundwork for future exploration, anticipating insights that will shape the way students engage with agriculture education, fostering a dynamic and immersive learning environment.

Integrating AR and VR technologies into high school education, particularly within CEA, can create immersive and interactive learning experiences that transcend traditional teaching methods. Therefore, these technologies provide a unique opportunity to bridge theoretical knowledge with real-world applications in agriculture.

REFERENCES

Akçayır, M., & Akçayır, G. (2017). Advantages and challenges associated with augmented reality for education: A systematic review of the literature. *Educational Research Review*, 20, 1–11. 10.1016/j.edurev.2016.11.002

Al-Ansi, A. M., Jaboob, M., Garad, A., & Al-Ansi, A. (2023). Analyzing augmented reality (AR) and virtual reality (VR) recent development in education. *Social Sciences & Humanities Open*, 8(1), 100532. 10.1016/j.ssaho.2023.100532

Alzahrani, M. (2022). *Traditional Learning Compared to Online Learning During the COVID-19 Pandemic: Lessons Learned From Faculty's Perspectives*. Sage Journals. 10.1177/21582440221091720

Anastasiou, E., Balafoutis, A. T., & Fountas, S. (2023). Applications of extended reality (XR) in agriculture, livestock farming, and aquaculture: A review. *Smart Agricultural Technology*, 3, 100105. 10.1016/j.atech.2022.100105

Anjos, F., Rocha, L., Silva, D., Pacheco, R., & Pinheiro, D. (2021). Impacts of the application of virtual and augmented reality on teaching-learning processes in engineering courses. *International Journal of Virtual and Personal Learning Environments*, 12(1), 1–19. 10.4018/IJVPLE.291541

Baker, A., Bramer, W. F., & Bobbitt, A. (2021). Agricultural literacy outcomes: A systematic literature review. *Journal of Agricultural Education*, 62(2), 214–229.

Baratimehr, K., Moosavi, M. R., & Tahayori, H. (2023). Measures for evaluating IT2FSs constructed from data intervals. *Applied Soft Computing*, 136, 110084. 10.1016/j.asoc.2023.110084

Beck, H. W. (2009). Lyra Virtual World Environment (Lyra VWE): Educational Applications in Agriculture and Natural Resources. In *7th World Congress on Computers in Agriculture Conference Proceedings*, 22-24 June 2009, Reno, Nevada (p. 1). American Society of Agricultural and Biological Engineers.

Bekiaris, I., Leonidis, A., Korozi, M., Stratakis, C., Zidianakis, E., Doxastaki, M., & Stephanidis, C. (2021). GRETA: Pervasive and ar interfaces for controlling intelligent greenhouses. In Proceedings of the *17th International Conference on Intelligent Environments (IE)* (pp. 1–8). IEEE. 10.1109/IE51775.2021.9486584

Bellah, K. A., & Dyer, J. E. (2009). Attitudes and Stages of Concern of Elementary Teachers Toward Agriculture as a Context for Teaching across Grade Level Content Area Standards. *Journal of Agricultural Education*, 50(2), 12–25. Advance online publication. 10.5032/jae.2009.02012

Benedick, P. L., Robert, J., & Le Traon, Y. (2021). A systematic approach for evaluating artificial intelligence models in industrial settings. *Sensors (Basel)*, 21(18), 6195. Advance online publication. 10.3390/s2118619534577398

Benke, K., & Tomkins, B. (2017). Future food-production systems: vertical farming and controlled-environment agriculture. Sustainability: Science. *Practice and Policy*, 13(1), 13–26.

Billinghurst, M. (2021). Grand challenges for augmented reality. *Frontiers in Virtual Reality*, 2, 12. 10.3389/frvir.2021.578080

Caria, M., Todde, G., Sara, G., Piras, M., & Pazzona, A. (2020). Performance and usability of smartglasses for augmented reality in precision livestock farming operations. *Applied Sciences (Basel, Switzerland)*, 10(7), 2318. 10.3390/app10072318

Carruth, D. W., Hudson, C., Fox, A. A. A., & Deb, S. (2020). User interface for an immersive virtual reality greenhouse for training precision agriculture. In Chen, J. Y. C., & Fragomeni, G. (Eds.), *Virtual, Augmented and Mixed Reality. Industrial and Everyday Life Applications* (pp. 35–46). Springer International Publishing. 10.1007/978-3-030-49698-2_3

Chatterjee, S., Rana, N. P., Dwivedi, Y. K., & Baabdullah, A. M. (2021). Understanding AI adoption in manufacturing and production firms using an integrated TAM-TOE model. *Technological Forecasting and Social Change*, 170, 120880. 10.1016/j.techfore.2021.120880

Cheng, C., Kang, L., Cai, S., Li, J., & Sun, R. (2018). Virtual display and interactive experience platform of farming culture based on unity3D. *IFAC-PapersOnLine*, 51(11), 637–642. 10.1016/j.ifacol.2018.08.125

Cheung, K. K., Jong, M. S., Lee, F. L., Lee, J. H., Luk, E. T., Shang, J., & Wong, M. K. (2008). FARMTASIA: An online game-based learning environment based on the VISOLE pedagogy. *Virtual Reality (Waltham Cross)*, 12(1), 17–25. 10.1007/s10055-008-0084-z

Cho, Y., & Park, K. S. (2023). Designing Immersive Virtual Reality Simulation for Environmental Science Education. *Electronics (Basel)*, 12(2), 315. 10.3390/electronics12020315

Clune, S., Crossin, E., & Verghese, K. 2017. Systematic review of greenhouse gas emissions for different fresh food categories. *Journal of Cleaner Production, 140*, 766-783.

Đurđević, S., Golubović, G., Maričić, K., & Vladić, G. (2022). Development of augmented reality video application. 10.24867/GRID-2022-p32

Eckert, M., Volmerg, J. S., & Friedrich, C. M. (2019). Augmented Reality in medicine: Systematic and bibliographic review. *JMIR mHealth and uHealth*, 7(4), e10967. 10.2196/1096731025950

Faridi, H., Tuli, N., Mantri, A., Singh, G., & Gargrish, S. (2020, October 4). A framework utilizing augmented reality to improve critical thinking ability and learning gain of the students in Physics. *Computer Applications in Engineering Education*. https://scite.ai/reports/10.1002/cae.22342

Fretwell, S. (2017). Cultivating ecological literacy and rethinking our connections to nature (Master's thesis). *Georgia Southern University, Jack N. Averitt College of Graduate Studies*. Retrieved from Digital Commons@Georgia Southern.

Garzón, J., & Acevedo, J. (2019). Meta-analysis of the impact of Augmented Reality on students' learning gains. *Educational Research Review*, 27, 244–260. 10.1016/j.edurev.2019.04.001

Garzon, J., Acevedo, J., Pavon, J., & Baldiris, S. (2020). Promoting eco-agritourism using an augmented reality-based educational resource: A case study of aquaponics. *Interactive Learning Environments*, 1–15. 10.1080/10494820.2020.1712429

Garzon, J., Baldiris, S., Acevedo, J., & Pavon, J. (2020). Augmented reality-based application to foster sustainable agriculture in the context of aquaponics. In *Proceedings of the 2020 IEEE 20th International Conference on Advanced Learning Technologies (ICALT)* (pp. 316–318). IEEE. 10.1109/ICALT49669.2020.00101

Griffiths, S., Wong, M. S., Kwok, C. Y. T., Kam, R., Lam, S. C., Yip, T. L., Heo, J. Y., Chan, B. S. B., Xiong, G., & Lu, K. (2019, July 24). Exploring Bluetooth Beacon Use Cases in Teaching and Learning: Increasing the Sustainability of Physical Learning Spaces. https://scite.ai/reports/10.3390/su11154005

Hassan, A., Abiddin, N. Z., & Yew, S. K. (2014). The Philosophy of Learning and Listening in Traditional Classroom and Online Learning Approaches. *Higher Education Studies*, 4(2), 19. 10.5539/hes.v4n2p19

Hillison, J. (1996). The Origins of Agriscience: Or Where Did All That Scientific Agriculture Come From? *Journal of Agricultural Education*, 37(4), 8–13. 10.5032/jae.1996.04008

Huang, R. L., & Nejdl, W. (Eds.). (2011). *Advances in Web-Based Learning – ICWL* (pp. 34–43). Springer. 10.1007/978-3-642-22763-9_4

Hubert, D., Frank, A., & Igo, C. (2000). Environmental and agricultural literacy education. *Water, Air, and Soil Pollution*, 123(1/4), 525–532. 10.1023/A:1005260816483

Iatsyshyn, A., Kovach, V., Lyubchak, V., Zuban, Y., Piven, A., Sokolyuk, O., & Shyshkina, M. (2020). Application of augmented reality technologies for education projects preparation. *Cte Workshop Proceedings, 7*, 134-160. 10.31812/123456789/3856

Jones, D., Ghasemi, S., Gračanin, D., & Azab, M. (2023). Privacy, Safety, and Security in Extended Reality: User Experience Challenges for Neurodiverse Users. In Moallem, A. (Ed.), Lecture Notes in Computer Science: Vol. 14045. *HCI for Cybersecurity, Privacy and Trust. HCII 2023*. Springer., 10.1007/978-3-031-35822-7_33

Junaini, S., Kamal, A., Hashim, A., Shaipullah, N., & Truna, L. (2022). Augmented and virtual reality games for occupational safety and health training: A systematic review and prospects for the post-pandemic era. *International Journal of Online and Biomedical Engineering (Ijoe)*, 18(10), 43–63. 10.3991/ijoe.v18i10.30879

Kim, R., Kim, J., Lee, I., Yeo, U., Lee, S., & Decano-Valentin, C. (2021). Development of three-dimensional visualization technology of the aerodynamic environment in a greenhouse using CFD and VR technology, Part 2: Development of an educational VR simulator. *Biosystems Engineering*, 207, 12–32. 10.1016/j.biosystemseng.2021.02.018

Krishnan, L., Poorani, S., & Sawmya, S. (2022, October 25). Emerging Trends in Training Knowledge Workers; *New Economy*. https://scite.ai/reports/10.5296/ijld.v12i4.20413

Krishnan, S., Robb, W., Manweiler, K., & Wise, M. (2020). Growing Beyond the Hype: Controlled Environment. *Agricultures (Montrouge)*, 2G.

Labadze, L., Grigolia, M., & Machaidze, L. (2023). Role of AI chatbots in education: Systematic literature review. *International Journal of Educational Technology in Higher Education*, 20(1), 56. 10.1186/s41239-023-00426-1

Larbaigt, J., & Lemercier, C. (2021). An evaluation of the acceptability of smart eyewear for plot diagnosis activity in agriculture. *Ergonomics in Design*. Advance online publication. 10.1177/10648046211018541

Lee, L., Kim, M., & Hwang, W. (2019). Potential of augmented reality and virtual reality technologies to promote wellbeing in older adults. *Applied Sciences (Basel, Switzerland)*, 9(17), 3556. 10.3390/app9173556

Masood, T., & Egger, J. (2020). Augmented reality in support of Industry 4.0—Implementation challenges and success factors. *Robotics and Computer-integrated Manufacturing*, 58, 181–195. 10.1016/j.rcim.2019.02.003

McMenamin, M. M., & Zirkel, P. A. (1998). Section 504, the Americans with Disabilities Act, and Public School Employment: A Legal Primer. *Journal of Personnel Evaluation in Education*, 11(3), 243–253. 10.1023/A:1007929221417

Mehr, K. B., Tahayori, H., & Moosavi, M. R. (2024). Information preserving approach for encoding words into interval type-2 fuzzy sets. *Granular Computing*, 9(3), 55. 10.1007/s41066-024-00479-1

Mendel, J., & Wu, D. (2010). *Perceptual computing: Aiding people in making subjective judgments*. John Wiley & Sons. 10.1002/9780470599655

Mendel, J. M. (2018). The perceptual computer: The past, up to the present, and into the future. *Informatik-Spektrum*, 41(1), 15–26. 10.1007/s00287-018-1088-z

Metwally, A. H. S., Nacke, L. E., Chang, M., Wang, Y., & Yousef, A. M. F. (2021). Revealing the hotspots of educational gamification: An umbrella review. *International Journal of Educational Research*, 109, 101832. 10.1016/j.ijer.2021.101832

Nigam, A., Kabra, P., & Doke, P. (2011, October). Augmented Reality in agriculture. In 2011 IEEE 7th International Conference on Wireless and Mobile Computing, Networking and Communications (WiMob) (pp. 445-448). IEEE.

Olapiriyakul, K., & Scher, J. M. (2011). A Methodology for Integrating Traditional Classroom Learning with Contemporary Online Learning.

Paul, J., & Jefferson, F. (2019). A Comparative Analysis of Student Performance in an Online vs. Face-to-Face Environmental Science Course From 2009 to 2016. *Frontiers of Computer Science*, 1, 7. Advance online publication. 10.3389/fcomp.2019.00007

Phupattanasilp, P., & Tong, S.-R. (2019). Augmented reality in the integrative internet of things (AR-IoT): Application for precision farming. *Sustainability (Basel)*, 11(9), 2658. 10.3390/su11092658

Planet Bee Foundation. (n.d.). The importance of eco-literacy in schools. Retrieved from https://www.planetbee.org/planet-bee-blog/the-importance-of-eco-literacy-in-schools

Ponnusamy, V., Natarajan, S., Ramasamy, N., Clement, C., Rajalingam, P., & Mitsunori, M. (2021). An IoT-enabled augmented reality framework for plant disease detection. *RIA*, 35(3), 185–192. 10.18280/ria.350301

Powell, D., Agnew, D., & Trexler, C. (2008). Agricultural literacy: Clarifying a vision for practical application. *Journal of Agricultural Education*, 49(1), 85–98. 10.5032/jae.2008.01085

Rasmussen, W. D. (1989). *Taking the University to the People: Seventy-five Years of Cooperative Extension.* Iowa State University Press.

Reilly, C., Stevenson, K., Warner, W., Park, T., Knollenberg, W., Lawson, D., Brune, S., & Barbieri, C. (2022). Agricultural and environmental education: A call for meaningful collaboration in a U.S. context. *Environmental Education Research*, 28(9), 1410–1422. 10.1080/13504622.2022.2040431

Rong, Q., Lian, Q., & Tang, T. (2022). Research on the influence of AI and VR technology for students' concentration and creativity. *Frontiers in Psychology*, 13, 767689. Advance online publication. 10.3389/fpsyg.2022.76768935401322

Rose, D. H.. (2018). Accurate and Informative for All: Universal Design for Learning (UDL) and the Future of Assessment. In Elliott, S., Kettler, R., Beddow, P., & Kurz, A. (Eds.), *Handbook of Accessible Instruction and Testing Practices.* Springer., 10.1007/978-3-319-71126-3_11

Rosmansyah, Y., Achiruzaman, M., & Hardi, A. B. (2019). A 3D multiuser virtual learning environment for online training of agriculture surveyors. *Journal of Information Technology Education*, 18, 481–507. 10.28945/4455

Rossetto, A. G. de M., Martins, T. C., Silva, L. A., Leithardt, D. R. F., Bermejo-Gil, B. M., & Leithardt, V. R. Q. (2023). An analysis of the use of augmented reality and virtual reality as educational resources. *Computer Applications in Engineering Education*, 31(6), 1761–1775. Advance online publication. 10.1002/cae.22671

Samaddar, K., & Mondal, S. (2024). AR and VR-based travel: A responsible practice towards sustainable tourism. *International Journal of Tourism Cities*, 10(1), 105–128. 10.1108/IJTC-05-2022-0135

Sayadi, N., & Gómez-Zará, D. (2024). Fostering Inclusive Virtual Reality Environments: Discussing Strategies for Promoting Group Dynamics and Mitigating Harassment. *University of Notre Dame, USA.* https://doi.org//arXiv.2405.0591710.48550

Scavarelli, A., Teather, R. J., & Arya, A. (2023, May). Exploring Selection and Search Usability Across Desktop, Tablet, and Head-Mounted Display WebXR Platforms. In *2023 9th International Conference on Virtual Reality (ICVR)* (pp. 505-514). IEEE.

Shaleendra, T., Wishvamali, B. T., Gunarathne, N., Hareendran, S., & Abeygunawardhana, P. (2019). Simulation of the Influence of environmental factors related to Greenhouses using Augmented Reality. In *Proceedings of the 2019 International Seminar on Research of Information Technology and Intelligent Systems (ISRITI)* (pp. 462–467). IEEE. 10.1109/ISRITI48646.2019.9034572

Shapley, L. S. (1953). Stochastic Games. *Proceedings of the National Academy of Sciences of the United States of America*, 39(10), 1095–1100. 10.1073/pnas.39.10.109516589380

Sopegno, A., Calvo, A., Berruto, R., Busato, P., & Bocthis, D. (2016). A web mobile application for agricultural machinery cost analysis. *Computers and Electronics in Agriculture*, 130, 158–168. 10.1016/j.compag.2016.08.017

Sorensen, T. J., Lambert, M. D., & McKim, A. J. (2021). A statewide examination of secondary agricultural educators' ability and willingness to teach climate change topics. *Journal of Agricultural Education*, 62(1), 213–229.

Soujanya, N., Roopa, Y. M., & Babu, D. K. (2019). Virtual reality meets IoT Through telepresence. In *Proceedings of the 2019 3rd International Conference on Trends in Electronics and Informatics (ICOEI)* (pp. 972–976). IEEE. 10.1109/ICOEI.2019.8862746

Szajowski, K. (2019). Stochastic Games and Learning. In Baillieul, J., & Samad, T. (Eds.), *Encyclopedia of Systems and Control* (pp. 1–7). Springer London. 10.1007/978-1-4471-5102-9_33-3

Tenney, A. W. (1977). The FFA at 50: A Golden Past—A Bright Future. *Phi Delta Kappan*, 58(9), 658–662.

Thieman, E. B., Henry, A. L., & Kitchel, T. (2020). Resilient agricultural educators: Teacher motivations and adaptations under Covid-19. *Journal of Agricultural Education*, 61(4), 238–254.

True, A. C. (1929). *A history of agricultural education in the United States, 1785-1925* (No. 36). US Government Printing Office.

Tummons, J. D., Langley, G. C., Reed, N. A., & Paul, M. P. (2021). Contemporary practices in secondary agricultural education: A national study. *Journal of Agricultural Education*, 62(4), 88–103.

Vincent, S. K., & Villa, J. (2021). Agricultural communications undergraduate enrollment growth, curriculum development, and faculty perceptions. *Journal of Applied Communications*, 105(2).

Weiss, C. L. A., & Mettrick, J. E. (2015). Individuals with Disabilities Education Act (IDEA). In Reynolds, C. R., & Fletcher-Janzen, E. (Eds.), *Encyclopedia of Cross-Cultural School Psychology* (pp. 542–545). Springer., 10.1007/978-0-387-71799-9_212

Welton, D. A., & Mallaro, A. A. (1915). *The Teaching of Agriculture in the High School*. Vocational Education Association of the Middle West.

Wen, F., Sun, Z., He, T., Shi, Q., Zhu, M., Zhang, Z., Li, L., Zhang, T., & Lee, C. (2020). Machine learning glove using self-powered conductive superhydrophobic triboelectric textile for gesture recognition in VR/AR applications. *Advancement of Science*, 7(14), 2000261. 10.1002/advs.20200026132714750

Williams, D. L. (1991). Focusing Agricultural Education Research: Strategies for the Discipline. *Journal of Agricultural Education*, 32(3), 7–12. 10.5032/jae.1991.01007

Wu, B., Yu, X., & Gu, X. (2020). Effectiveness of immersive virtual reality using head-mounted displays on learning performance: A meta-analysis. *British Journal of Educational Technology*, 51(6), 1991–2005. 10.1111/bjet.13023

Yang, X., Shu, L., Chen, J., Ferrag, M. A., Wu, J., Nurellari, E., … & Huang, K. (2021). A survey on smart agriculture: development modes, technologies, and security and privacy challenges. *IEEE/CAA Journal of Automatica Sinica, 8*(2), 273-302. 10.1109/JAS.2020.1003536

Yongyuth, P., Prada, R., Nakasone, A., Kawtrakul, A., & Prendinger, H. (2010, October). AgriVillage: 3D multi-language internet game for fostering agriculture environmental awareness. In *Proceedings of the International Conference on Management of Emergent Digital EcoSystems* (pp. 145-152). 10.1145/1936254.1936280

Yu, L., Liu, W., Yang, S., Kong, R., & He, X. (2022). Impact of environmental literacy on farmers' agricultural green production behavior: Evidence from rural China. *Frontiers in Environmental Science*, 10, 990981. Advance online publication. 10.3389/fenvs.2022.990981

Zhao, X., Ren, Y., & Cheah, K. S. L. (2023). Leading Virtual Reality (VR) and Augmented Reality (AR) in Education: Bibliometric and Content Analysis From the Web of Science (2018–2022). *SAGE Open*, 13(3), 21582440231190821. Advance online publication. 10.1177/21582440231190821

Zheng, M., & Campbell, A. G. (2019). Location-based augmented reality in-situ visualization applied for agricultural fieldwork navigation. In *Proceedings of the 2019 IEEE International Symposium on Mixed and Augmented Reality Adjunct (ISMAR-Adjunct)* (pp. 93–97). IEEE. 10.1109/ISMAR-Adjunct.2019.00039

Zweifach, S., & Triola, M. (2019). Extended reality in medical education: Driving adoption through provider-centered design. *Digital Biomarkers*, 3(1), 14–21. 10.1159/00049892332095765

KEY TERMS AND DEFINITIONS

Agricultural Education: The process of imparting knowledge and skills related to farming, crop management, livestock care, and sustainable agricultural practices, fostering the development of individuals within the agriculture sector.

Americans With Disabilities Act (ADA): It is a U.S. federal law enacted in 1990 that prohibits discrimination against individuals with disabilities in various areas of public life. This includes employment, public services, transportation, and telecommunications. The ADA aims to ensure equal opportunities and access for people with disabilities by requiring businesses, employers, and government entities to provide reasonable accommodations and eliminate barriers that may prevent individuals with disabilities from participating fully in society. The ADA has significant implications for the design and accessibility of physical spaces, digital platforms, and services to ensure inclusion for everyone, regardless of their abilities.

Augmented Reality (AR): A technology that overlays digital information, such as images, videos, or text, onto the real-world environment, enhancing the user's perception and interaction with the physical surroundings.

Augmented Reality Intervention for Safety Education (ARISE): ARISE is an educational program that utilizes augmented reality to enhance safety education and awareness.

Best Management Practices (BMPs): Best Management Practices are a set of guidelines, strategies, and techniques designed to optimize the efficiency and sustainability of specific processes or activities. In the context of agriculture, particularly Controlled Environment Agriculture (CEA), BMPs are practices that enhance productivity while minimizing environmental impact. These practices may include optimized resource use, waste reduction, and sustainable farming techniques to achieve optimal outcomes in agricultural production.

Controlled Environment Agriculture (CEA): Controlled Environment Agriculture involves the use of technology to create an optimized environment for plant growth, including precise control over temperature, humidity, light, and nutrients.

Dynamic Stochastic Gaming Probabilistic (DSGP): It refers to a computational framework used in gaming that incorporates dynamic and stochastic elements, allowing for unpredictability and randomness within the gaming environment. This approach introduces probabilistic elements to simulate real-world uncertainties, providing a more realistic and dynamic gaming experience.

Immersive Learning: A pedagogical approach that leverages technologies like AR and VR to create engaging and interactive educational experiences, enabling learners to explore and comprehend complex concepts.

Individuals With Disabilities Education Act (IDEA): IDEA is a federal law in the United States that ensures students with disabilities receive a free and appropriate public education tailored to their individual needs.

Next Generation Science Standards (NGSS): It refers to a set of K-12 science content standards that aim to provide a consistent and high-quality science education for all students. Developed to reflect advancements in science and education, NGSS emphasizes an integrated approach, focusing on core ideas in science and engineering practices to foster a deeper understanding of these disciplines.

Non-Real-Time Quiz: A non-real-time quiz is a form of assessment where participants respond to questions at their own pace, without immediate feedback. This type of quiz is typically not time-bound, allowing participants to take their time in answering questions. Non-real-time quizzes are often used for self-paced learning or as a means of evaluating knowledge without the pressure of real-time constraints.

PLS (Partial Least Squares): a statistical method commonly used in structural equation modeling, to analyze the relationships between variables.

Precision Agriculture: Precision Agriculture involves the use of technology, including AR and VR, to optimize farming practices by precisely managing resources such as water, fertilizers, and pesticides.

Precision Farming: A farming approach that utilizes advanced technologies, such as GPS, sensors, and data analytics, to optimize the management of agricultural resources. Precision farming aims to enhance the efficiency, productivity, and sustainability of farming practices by precisely tailoring inputs, such as water, fertilizers, and pesticides, based on real-time data and specific conditions within individual fields.

Real-Time Quiz: A real-time quiz is an interactive assessment tool where participants, often students, answer questions presented to them in real-time. Unlike traditional quizzes, real-time quizzes provide immediate feedback, allowing participants to see their results and correct answers instantly. This interactive and time-sensitive format enhances engagement and facilitates quick assessment of participants' understanding.

SBAE (School-Based Agricultural Education): SBAE refers to educational programs that integrate agricultural concepts and practices into school curricula to promote agricultural literacy and skills among students.

User Experience (UX): User Experience, often abbreviated as UX, refers to the overall experience and satisfaction a person has when interacting with a product, service, or system. It encompasses a broad range of factors, including usability, accessibility, design aesthetics, and the emotional response of the user. A positive UX is achieved when the user's needs, expectations, and goals align with the functionality and design of the product, resulting in a seamless and enjoyable interaction.

Virtual Reality (VR): An immersive technology that creates a simulated environment, often through the use of headsets or devices, allowing users to interact with and experience a computer-generated three-dimensional world.

APPENDIX

1. Questions and Answers

-What is the overall problem presented in this case?

The overall problem presented in this case is the absence of effective educational tools for Controlled Environment Agriculture (CEA). There's a need for immersive, engaging tools using AR and VR to teach students about CEA's importance, as the current gap in educational resources limits their understanding of its role in sustainable agriculture.

-What are the factors affecting the problem(s) related to this case?

The problem in this case are multifaceted. Firstly, training next generation in CEA with engaging, comprehensive, and immersive tools. Secondly, the issue of accessibility for users with diverse abilities is critical. Ensuring the educational platform is accessible to all users, regardless of their physical or cognitive abilities, is essential.

-What are some of the emerging technologies that should be considered in solving the problem(s) related to the case?

This research utilizes cutting-edge technologies such as AR, VR, Artificial Intelligence (AI), and the Dynamic Stochastic Gaming Probabilistic (DSGP) concepts to solve challenges related to Controlled Environment Agriculture (CEA).

-Discuss managerial, organizational, and technological issues and resources related to this case.

Implementing the CEA educational tool involves challenges in content provision and software development. Collaborating with educational experts, monitoring outcomes, creating an intuitive interface, ensuring system reliability, and maintaining data security are crucial. Additionally, incorporating accessibility features is essential for inclusivity. Addressing these issues will enhance the tool's teaching effectiveness.

-What are the evaluation methods?

The evaluation method includes formal and informal surveys, encompassing structured, semi-structured, and unstructured formats. This assesses the effectiveness of gamified and informal modules among students, new entrepreneurs, and stakeholders during Module Testing and Fine-Tuning. Engagement impact is analyzed via detailed usage metric tracking in the National Dissemination and Ongoing Monitoring phase.

2. Epilogue and Lessons Learned

Please provide an epilogue paragraph in which you discuss the long-range effects of your case. Do not include an epilogue in your actual case study. Next, provide a list of 3-5 lessons, along with a concise explanation of each that, in your opinion, can be learned from your study.

The Grace Education Platform represents a significant leap forward in the realm of CEA education by incorporating advanced technologies such as AR, VR, and AI. This integration not only transforms the learning experience within CEA but also establishes a model for inclusive and accessible education practices. The study will play a pivotal role in laying the groundwork for a sustainable food production system in alignment with the principles of a low-carbon economy because it will consistently train the upcoming generation of practitioners through engaging gamified modules. The immersive learning experiences facilitated by AR and VR technologies will be expected to bridge the gap between theoretical knowledge and practical application in agriculture. In addition, the enduring impact of Grace's initiatives foresees the creation of a dynamic learning environment, shaping the future of student's engagement in agriculture education. This forward-looking approach not only fosters innovation and sustainability but also holds promise for the advancement of food cultivation and related fields.

Some of the designed courses in the platform are as following:

Water Quality Assessment: The module emphasizes the intricate dynamics of water quality within controlled environments, highlighting the critical role it plays in ensuring optimal conditions for plant growth. Students gain a deep understanding of how water quality directly impacts plant health. This lesson underscores the significance of maintaining suitable water conditions to support the thriving growth of crops in controlled environments.

Insect Pest Management: It equips students with essential skills in identifying, preventing, and mitigating pest-related challenges in CEA. By focusing on integrated pest management, students develop a comprehensive approach to dealing with pests. This lesson emphasizes the importance of proactive measures in pest control, fostering sustainable and effective strategies within CEA.

Nutrient Imbalances: This lesson delves into the delicate equilibrium necessary for plant nutrition, emphasizing the role of nutrients in fostering healthy and productive crops. By exploring nutrient imbalances, learners grasp the intricate relationships among essential elements for plant growth. This lesson reinforces the need for a balanced approach to nutrient management, promoting crop health and productivity within the context of CEA.

PH Balance in Nutrient: The inclusion of pH level evaluation sheds light on the crucial role of maintaining the right pH balance in nutrient solutions, contributing to plant health and vitality in CEA. Students understand the direct correlation between pH balance and nutrient absorption by plants. This lesson underscores the significance of proper pH management for nutrient availability, ultimately influencing the overall health and productivity of crops.

Insect Pest Management: This course equips students with essential skills in identifying, preventing, and mitigating pest-related challenges in CEA. By focusing on integrated pest management, students develop a comprehensive approach to dealing with pests. This lesson emphasizes the importance of proactive measures in pest control, fostering sustainable and effective strategies within CEA.

3. List of Additional Sources

Please prepare a list of other resources (e.g., journal articles, books, Web sites) that might be relevant for obtaining additional information either directly related to or similar to your case. List any industries where the described case may be pertinent.

Journal Articles

George, K., Anastasios, M. T., Dimitrios, M., & Christos, C. (2023). The mobile augmented reality acceptance model for teachers and future teachers. Education and Information Technologies, 1-39.

Xu, Z., Adeyemi, A. E., Landaverde, R., Kogut, A., & Baker, M. (2023). A Scoping Review on the Impact of Educational Technology in Agricultural Education. Education Sciences, 13(9), 910.

Leggette, H. R., Rice, A. H., Carraway, C., Baker, M. A., & Conner, N. (2018). Applying Eye-Tracking Research in Education and Communication to Agricultural Education and Communication: A Review of Literature. Journal of Agricultural Education, 59(2), 79-108.

Books

Kamińska, D., Zwoliński, G., Laska-Leśniewicz, A., Raposo, R., Vairinhos, M., Pereira, E., ... & Anbarjafari, G. (2023). Augmented reality: Current and new trends in education. Electronics, 12(16), 3531.

Tilhou, R., Taylor, V., & Crompton, H. (2020). 3D virtual reality in K-12 education: A thematic systematic review. Emerging technologies and pedagogies in the curriculum, 169-184.

Websites

AgriTech Tomorrow. (2024). The Future of Agriculture: AR and VR. Retrieved January 15, 2024, from [https://www.agritechtomorrow.com/].

EdTechReview. (2023). AR and VR in Education: Case Studies and Examples. Retrieved May 23, 2023, from [https://www.edtechreview.in/trends-insights/insights/ar-and-vr-in-education-enhancing-the-learning-experience/].

Chapter 3
Examining the Alignment of Student Explanations With the System Model in Climate Education Using Systems Thinking

Meltem Ceylan Alibeyoglu
Darussafaka Educational Institutions, Turkey

Zeynep Isilay Baskahya
Eötvös Loránd University, Hungary

Gaye Defne Ceyhan
https://orcid.org/0000-0003-1312-3547
Bogazici University, Turkey

Sena Yildiz Degirmenci
Technical University of Munich, Germany

Nihal Gunes Demir
Hisar Foundation Educational Institutions, Turkey

Saadet Etli
Enka Middle School, Turkey

Gorkem Girgin
https://orcid.org/0000-0001-6367-4481
Darussafaka Educational Institutions, Turkey

Emre Goktepe
Systems Thinking Association, Turkey

Busra Karga
https://orcid.org/0000-0001-5356-8482
Eindhoven University of Technology, The Netherlands

Ece Unsal
Darussafaka Educational Institutions, Turkey

Ulkem Yararbas
Ege University, Turkey

DOI: 10.4018/979-8-3693-1710-5.ch003

EXECUTIVE SUMMARY

The climate crisis poses a major threat to sustainability, highlighting the need for climate education that develops students' systems thinking skills regarding dynamic environmental issues. For this reason, this project, funded by the United Nations Development Programme (UNDP), was developed by a research university, a non-governmental organization, and a middle school partnership. As part of the project, this study investigated a ten-week after-school climate education program for 8th-grade students that aimed to improve their understanding of the climate system through simulation and policymaking. Semi-structured interviews were analyzed to assess the program's model alignment with students' explanations. The most frequently mentioned category was the economy, followed by the carbon cycle; population and the greenhouse effect were the least common. The study offers valuable insights into using a systems approach in climate education to comprehend climate complexity and dynamics by considering the importance of economy and population in understanding human-caused climate change.

INTRODUCTION

Most current socio-environmental problems have a complex scientific background and a controversial social structure (Lombardi, 2022). To interpret complex systems, it is essential to consider problems holistically (Riess & Mischo, 2010). However, individuals often struggle to understand and evaluate the complexity of systems (Sweeney & Sterman, 2000). Using a systems thinking approach in teaching is crucial for developing students' scientific thinking and problem-solving skills in dynamic environmental issues (Porter, 2009). Furthermore, it is important to generate 'system citizens' who can effectively deal with complex problems (Assaraf & Orion, 2005; Sweeney & Sterman, 2007).

Systems thinking is crucial for understanding climate science (Shepardson et al., 2014; Sterman, 2001). It requires an understanding of complex and dynamic systems (Liu, 2023). To comprehend how climate systems operate and their complexity, it is necessary to analyze the stock and flow structures of the system, various feedback processes, linear and non-linear relationships between components, and the resulting delays. Therefore, integrating the systems thinking approach into climate change education is crucial (Blatti et al., 2019). The climate crisis is widely regarded as the most significant threat to ecological balance today (Pörtner et al., 2022). It is essential to raise students' awareness of global issues, including the climate crisis, and empower them to propose solutions.

For this reason, we created a climate education program as part of a project funded by the UNDP Acceleration Lab. The program was developed in collaboration with a research university; an NGO focused on systems thinking, and a middle school. The education program aims to provide students with a learning experience to understand the structure and functioning of the climate system. The program also includes a simulation environment where students can participate in policy development decision-making positions. The education program was carried out as an after-school activity with nineteen volunteer 8th-grade students in the fall semester of 2021. The ten-week program comprises three phases. The first stage involved analyzing the scale, changes, and relationships of population, migration, economy, and greenhouse gas emissions worldwide and country-specifically using systems thinking tools. During the second stage, the World Climate Game developed by Climate Interactive was played, which aims to identify the speed and level of action that countries should take on global climate change due to various negotiations, considering their situation. In the third stage, students were introduced to systems thinking tools to aid their understanding of the carbon cycle and the greenhouse effect.

As part of the larger project, this research aims to assess the alignment between the program's objectives and students' explanations of the climate crisis after being taught using the systems thinking approach and the model applied in the education program. The following question will be addressed:

- To what extent do the students' explanations align with the model used in the climate crisis education program?

BACKGROUND

Systems Thinking

The systems approach, which considers the entire environment in which it exists, is a concept that has been introduced previously. This can be seen from the following quotations:

- Aristotle, ca. 400 BC: "For things that have many parts and are not a mass, the whole is something separate from its parts..." (Aristotle, n.d.)
- Heraclitus, ca. 500 BC: "Different waters flow over those who enter the same rivers" (Heraclitus, n.d.).
- Laozi, ca. 400 BC: "Everything moves together and returns to its roots." (Feng, 1983), "This cycle is the universal law." (Alquiros, 2018).

The systems approach is a well-established concept gaining increasing importance and applicability compared to previous periods (Hossain et al., 2020). In today's world, it is widely recognized that the elements of the world system are highly interconnected, making the systems approach even more crucial. This interconnectedness is exemplified by the exponential growth of population and production, commonly referred to as globalization. Around 2000 years ago, the estimated world population was 200 million, with a gross product of US$ 41 billion (Maddison, 2020). In 2022, the population has increased 40-fold to 8 billion, and world production has increased 2900-fold to 120 trillion US dollars. We live in a more crowded and interconnected world than ever before.

Systems thinking is an approach that contributes to a more holistic understanding of complex systems (Senge, 1990). It involves identifying and making sense of the causal relationships between variables within a system (Assaraf & Orion, 2005; Meadows, 2008). System Dynamics complements systems thinking by quantifying these interactions and developing a view of how the system behaves over time (Forrester, 1993; Nuhoglu, 2008). System Dynamics uses computer models and simulation software to analyze complex problems where behavior changes over time. These models reveal less obvious relationships, dynamic complexity, delays, and the consequences of interactions.

Several simulation software programs have been developed to understand how the components that comprise a system's structure and their interrelationships determine the system's behavior (Cox et al., 2019; Naugle et al., 2023). The study used STELLA software (Structural Thinking Experimental Learning Laboratory with Animation) to create models. The software demonstrated that behaviors change over time as the underlying structures and effects change (Assaraf & Orion, 2005). Systems thinking is a complex field that deals with systems containing a vast number and variety of elements and relationships. Achieving mastery in this field requires significant time and effort. Therefore, application trials and adaptation studies of systems thinking have been conducted in classrooms, starting at the preschool level, long before doctoral research in universities.

Systems Thinking in Education

Systems thinking in education is a holistic approach that emphasizes the interconnectedness and dynamics of educational systems (Senge, 2006). It encourages a deeper understanding of how various components within these systems interact and influence each other (Assaraf & Orion, 2010; Evagorou et al., 2009). This approach has been explored through various lenses, including its theoretical foundations, implications for educational leadership and management, applications in curriculum design and teaching, and its potential to address educational inequities (Budak &

Ceyhan, 2023; Bielik et al., 2023; Liu et al., 2023). Research has underscored the importance of systems thinking in fostering adaptive, responsive educational environments and leading effective school reforms (Senge, 2006).

Systems thinking has been applied to design and improve educational programs and systems and to develop students' critical thinking and problem-solving skills (Assaraf & Orion, 2005; 2010; Nguyen & Santagata, 2021; Phillips, 1997). Experts in the field recommend introducing students to systems thinking at an early age (Assaraf & Orion, 2005; Senge, 2006; Sweeney & Sterman, 2000). Research conducted with K-12 students has shown that they exhibit an improvement in their systems thinking skills. They are better able to identify and understand the components and interactions within systems (Assaraf & Orion, 2010; Evagorou, 2009; Guthrie & Fisher, 1999; Hokayem & Gotwalz, 2016; Sommer & Lücken, 2010).

Implementing systems thinking in educational settings poses challenges, including the need for educators to develop professionally and the complexity of shifting traditional paradigms (Ateskan & Lane, 2018). Future research directions may focus on developing strategies for integrating systems thinking more effectively into educational practice, evaluating the impacts of these interventions, and exploring the role of technology in facilitating systems thinking in education (Bielik et al., 2023). Systems thinking offers innovative pathways for achieving more holistic, sustainable, and effective educational reforms as the field evolves. It emphasizes the importance of viewing education through a systemic lens to navigate better and improve the complex learning and teaching landscape.

Environment and Climate Change Education Through Systems Thinking

In recent years, there has been a growing global interest in sustainable development goals and environmental education (Rossini, 2022). Educators have recognized the need to understand and address complex global issues such as climate change through environmental education, including its applications and research (Irwin, 2020). To address global warming and climate change, educating future decision-makers on environmentalism and sustainability issues is crucial (Rossini, 2022).

Researchers argue that the use of system dynamics as a learning approach in education is crucial for addressing critical, complex, and interdisciplinary problems, such as environmental issues (Strapasson et al., 2022; Mambrey et al., 2020; Ballew et al., 2019; Doganca Kucuk & Saysel, 2018). Furthermore, research indicates that students often have a limited understanding of climate as a system (Roychoudhury et al., 2017), hold misconceptions about global warming and climate change (Shepardson et al., 2009), and lack practical applications of environmental education with system dynamics in the classroom (Strapasson et al., 2022).

Studies applying system dynamics in environmental education cover a range of topics, including waste management systems, energy strategies, climate change, sustainability, ecosystems, and global warming (Ballew et al., 2019; Berenjkar et al., 2021; Mambrey et al., 2020; Roychoudhury et al., 2017; Rossini, 2022; Strapasson et al., 2022). Simulations and system dynamics models can significantly contribute to students' environmental awareness and political participation in science education (Evagorou et al., 2009; Nuhoglu, 2020; Strapasson et al., 2022). The literature on system dynamics implementations suggests that this approach is an effective tool for students in terms of problem-solving, collaboration, generating alternative solutions, and analyzing components and processes within the system (Asik & Doganca Kucuk, 2021; Nuhoglu, 2020; Roychoudhury et al., 2017).

The world has faced various challenges in recent years, including climate crisis, water and food scarcity, clean energy, and health issues such as COVID-19. However, addressing these complex systems has proven difficult for many researchers (Doganca Kucuk & Saysel, 2018; Sweeney & Sterman, 2000). Systems thinking is recognized as a vital tool for comprehending climate science and managing climate change. It requires an understanding of complex and dynamic systems (Ballew et al., 2019). Furthermore, systems and system models are emphasized as cross-cutting concepts that aid students in better understanding fundamental ideas in science and engineering (National Research Council, 2013).

Using Games in Environment and Climate Change Education Through Systems Thinking

Various forms of gaming, such as digital, instructional, simulation, serious games, virtual reality, and augmented reality, have the potential to significantly transform educational settings (Oliveria et al., 2023). They facilitate experiential learning and enable learners to acquire knowledge through trial and error. Simulation games can enhance motivation, cognitive development, and critical thinking (Almaki et al., 2023). To improve simulation learning, it is recommended to utilize specific methods such as integrating game use within an instructional program, encouraging a high level of learner activity, and adopting the simulation game as a supplement to other methods (Vlachopoulos & Makri, 2017).

Since the 1950s, games have increased the effectiveness of systems thinking tools developed to understand complex systems (Sterman, 2002). The 'Beer Game' (Martinez-Moyano, 2024) was the first game developed for the system dynamics model, simulating the supply chain fluctuations of enterprises. Over time, games on sustainability have also been developed, which can be played online or face-to-face (Meadows et al., 2016; Sajjadi et al., 2022). Examples such as Fishbanks, CleanStart, and World Climate are explored in the context of renewable resource management, clean tech startup firms, and interactive role-play of global climate negotiations, respectively (Sterman, 2014).

The World Climate Game is similar to the Model United Nations (MUN) activity, commonly organized in high schools and universities, as both are United Nations role-playing games that require prior preparation (Krein, 2023). However, the World Climate Game differs because it primarily focuses on climate change, is supported by a system dynamics model, and conducts negotiations in the participants' native language.

Rooney-Varga et al. (2021) studied 41 World Climate simulation sessions in the US between 2015 and 2020. The study involved 2,080 participants from diverse educational backgrounds and age groups, representing various settings, including formal and informal education for high school, undergraduate, and graduate students, and sessions tailored for educators and other professionals with varying levels of prior knowledge and interest in climate change. According to the study, participants who held individualistic-hierarchical values showed significant improvement in their understanding of climate change, emotional response, and willingness to take action. They also demonstrated a stronger sense of urgency and intention to address climate change than those with communitarian-egalitarian values.

The simulations caused a shift in sociopolitical values, as found by Sterman et al. (2015). Participants who initially held individualistic-hierarchical values showed a significant shift towards communitarian-egalitarianism. These results suggest that World Climate promotes support for evidence-based climate action, even among individuals who are skeptical due to their sociopolitical beliefs. The World Climate simulation has engaged over 100,000 participants in 117 countries, including middle and graduate schools, civic organizations, non-governmental organizations, businesses, the military, and governments (Climate Interactive, 2024).

The Article's Focus

Our study aligns with previous research demonstrating the effectiveness of applying system dynamics in environmental education. For example, Asik and Doganca Kucuk (2021) found that a system dynamics approach improved students' problem-solving, collaboration, and analytical skills in the context of environmental issues. Similarly, Nuhoglu (2020) showed that using system dynamics to model current environmental problems enhanced students' understanding and engagement. However, our study differs from these works by focusing specifically on climate change education and employing a mixed-methods approach to assess the alignment between students' explanations and the system dynamics model used in the education program. Furthermore, our study incorporates simulation games, such as the World

Climate Game, which effectively promotes systems thinking and pro-environmental attitudes (Rooney-Varga et al., 2021; Sterman et al., 2015).

The system models provide a comprehensive approach to recognizing patterns, understanding system behavior, identifying cause-and-effect relationships, comprehending stability and change, identifying malfunctions, and generating solutions (Shepardson et al., 2014). This may enable students to evaluate climate change-related issues with a deeper understanding and solve emerging problems. To understand the complexity of climate systems, students could examine the stock and flow structures, feedback processes, linear and non-linear relationships between components, and time delays (Shepardson et al., 2017; Sterman & Sweeney, 2002). Therefore, it is important to integrate the systems thinking approach as a teaching method in climate change education. This study investigates how the students' explanations developed with the systems thinking approach overlap with the model applied in the education program.

METHOD

Research Design

This study employed the case study research design (Creswell & Poth, 2016), which comprehensively examines a closed system (Merriam & Tisdell, 2016). The objective of this study was to provide a detailed description of the climate education program using a systems thinking approach and to analyze the experiences of the participating students.

Participants

This project involves implementing a ten-week education program as an after-school activity. The program was implemented for 13 female and six male students from a private school who volunteered from the 8th grade (13-14 years old). The research was conducted at the school where most of the researchers worked as teachers, taking into account the implementation and data collection stages. Before the project began, a presentation was given to all 8th-grade students at the school to introduce the project. Interested students were then identified and selected to participate. The reason for selecting 8th-grade students is the science curriculum's focus on climate change outcomes.

Content of the Climate Education Program

The climate education program lasted ten weeks and totaled 13 hours. The entire climate system model is presented in Figure 1 below, and the relevant parts are studied in weekly lessons.

Figure 1. Model for the Climate Education Program

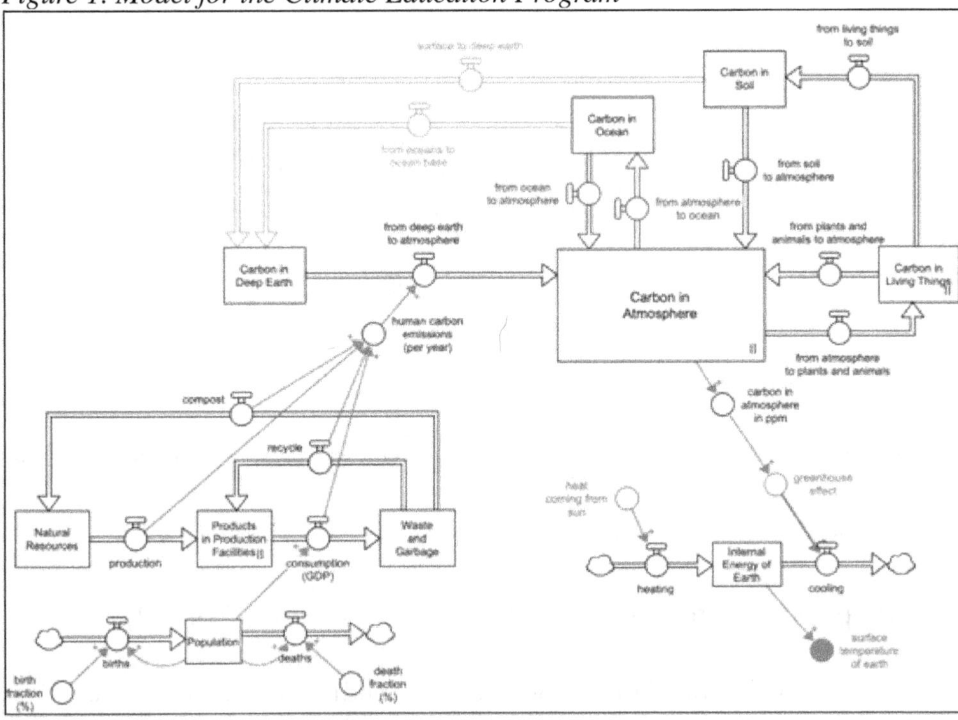

Note: Rectangles in the figure represent the accumulations, stocks, or state of the system, while arrows with valves represent the flows or actions in the system. Circles indicate the factors that can influence the flows. Stocks can only change with flows.

Population and Climate (Week 1)

To facilitate students' comprehension of the interrelationships among the components of a system, the program was initiated by utilizing a model developed in Stella Online software to analyze the population. The model tackled fundamental inquiries such as the factors that influence the population and how to simulate the population's evolution in Turkey over time (Figure 2). Furthermore, the program examined the influence of migration on population dynamics, as well as births and deaths. The discussion covered the relationship between climate change, population, and migration. It also touched on Syria's migration to foreign countries as a result of the civil war.

Figure 2. Migrations by Country Model

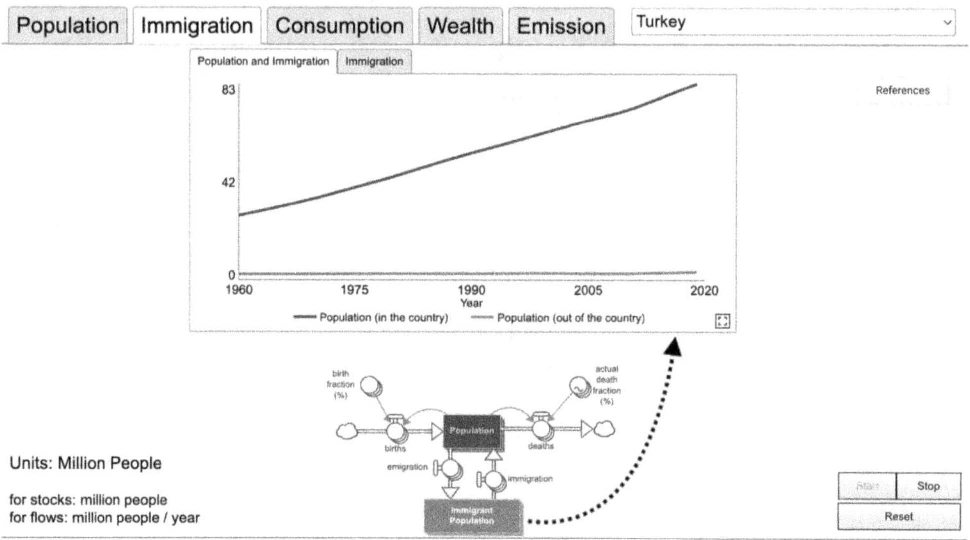

Note: A systems thinking model illustrating the relationship between population, births, deaths, and migration in Turkey.

Consumption, Economy, and Wealth Relationship (Week 2)

The consumption tab of the model was accessed, and the definition of consumption was established through discussion with the students. The concepts of recycling and composting were also deliberated, along with the inquiry of whether an increase in production or consumption each year leads to the utilization of more natural resources (Figure 3). Subsequently, the correlation between economic size and consumption was examined, and student groups were allocated to one of the following countries. The groups compared the consumption data of the USA, Finland, Japan, India, China, Syria, and Turkey to conclude the economic size of the countries.

Figure 3. Consumption by Country Model

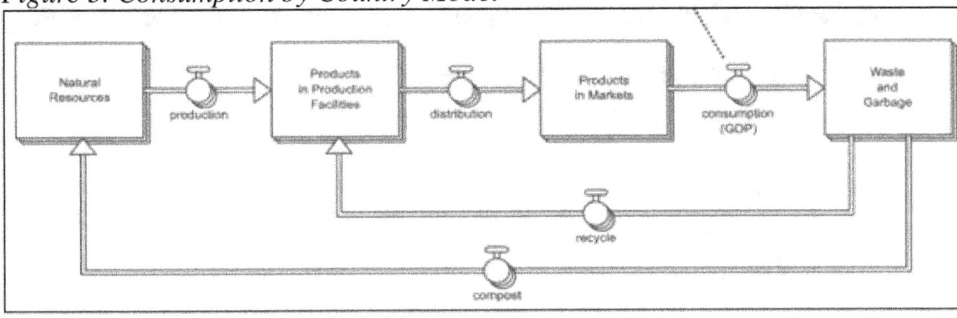

Note: A systems thinking model comparing consumption data across different countries to understand the relationship between economic size and consumption.

Economy, Emissions, and Climate (Week 3)

Figure 4 displays an emissions analysis for the 'World.' The emission factor is the sum of emissions caused by production, distribution, and consumption flows. A correlation between increased emissions and economic growth was discussed, and a graph was plotted on Stella Online to illustrate the relationship between wealth and emissions. Figure 4 shows that emissions change with consumption (GDP) flows. The discussion focused on the potential for increasing wealth without emitting carbon dioxide. The students analyzed examples in the model to explore the relationship between wealth and emissions. In some countries, although wealth increased, emissions did not increase at the same rate. The reasons for this change were discussed with students.

Figure 4. Wealth and Emissions

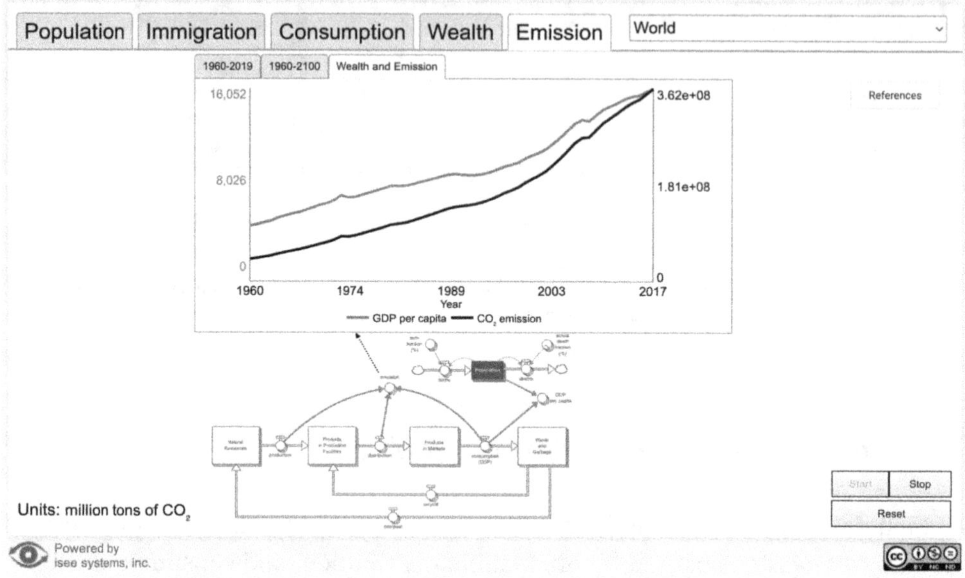

Note: A systems thinking model demonstrates the relationship between wealth (GDP) and carbon dioxide emissions, changing emissions based on consumption flows.

The three lessons mentioned above cover significant concepts, including population, consumption, wealth, migration, and emissions. These concepts serve as a preliminary preparation for the 'World Climate Game' to be played in the next stage, which will utilize the concepts covered in the lessons.

World Climate Game (Week 4)

During the second phase, students participated in a climate negotiation game examining international agreements and the geopolitical aspects of climate change. The objective was to gain insights into the causes of climate change and find solutions (Sterman et al., 2012). In the game, the students assumed the role of a delegate representing six major continents/regions/countries. The objective was to limit global warming to no more than 2.5°C using a simulation tool within a game. Through a role-playing simulation game, students are tasked with making decisions to reduce annual CO2 emissions and ultimately prevent climate change.

Carbon Cycle (Week 5)

The students were tasked with interpreting a graph depicting the behavior of atmospheric carbon levels between 1000 and 2050. They were asked to draw attention to the increase in carbon levels after 1850 and to identify possible reasons for this increase. Before the Industrial Revolution, the students constructed a carbon cycle model using the Stella Online program. The model discussed the various stocks where carbon is stored and the flows between them (Figure 5). The graphs of the sustainable model in the equilibrium state before the Industrial Revolution were analyzed.

Figure 5. Carbon Cycle before the Industrial Revolution

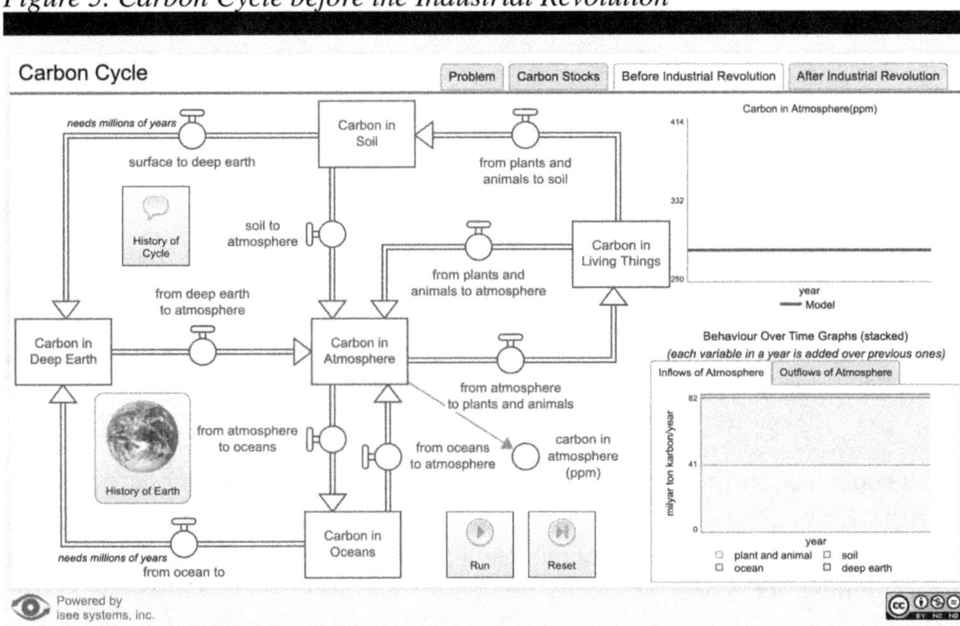

Note: A systems thinking model of the carbon cycle before the Industrial Revolution, depicting the various stocks where carbon is stored and the flows between them.

Carbon Cycle and En-Roads (Week 6)

After the Industrial Revolution, students discussed the point at which the carbon balance was disrupted in the model. They examined how the stock of 'anthropogenic emissions' resulting from the use of fossil fuels would be added to the model and how this would affect the amount of carbon in the atmosphere. To understand the amount of human-caused emissions, we need to examine the relationship between the level of CO2 in the atmosphere and human activities. This connection is particularly relevant to energy production methods that rely heavily on fossil fuels (Figure 6).

Figure 6. Carbon Cycle after the Industrial Revolution

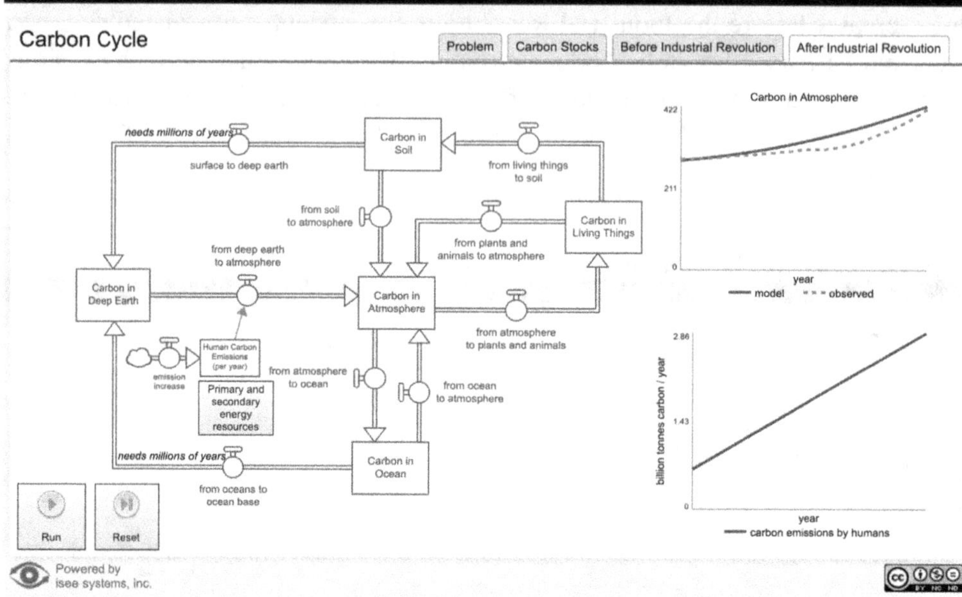

Note: A systems thinking model of the carbon cycle after the Industrial Revolution, illustrating the addition of the 'anthropogenic emissions' stock resulting from fossil fuel use and its impact on atmospheric carbon levels.

The open-access En-Roads simulation (Siegel et al., 2018) was used to observe the behavior of CO2 emissions by energy supply source. The simulation's graph displays the CO2 emissions by source that will occur if we continue to use different energy sources at the current rate.

Greenhouse Effect and Global Warming (Week 7)

At the beginning of the lesson, an experiment was conducted to aid students in comprehending the concept of the greenhouse effect. Two identical cups filled with equal amounts of water at temperatures above 90°C were used. One cup was left open while the other was closed. The temperature changes in the cups were monitored graphically, and the reasons for this occurrence were discussed. The stock-flow-variable relationship was studied to facilitate the modeling of the greenhouse effect using Stella. To comprehend the impact of greenhouse gases on the greenhouse effect, we studied the graphs by revisiting the experiment introduced at the beginning of the lesson. By analyzing the relationship between stocks and flows in the experiment, the students could explain the greenhouse effect and understand that global warming is caused by the earth's inability to cool down due to increased greenhouse gas emissions. Establishing a causal connection between stocks and flows in the model is crucial to achieving this understanding.

En-Roads (Week 8)

Using the EN-ROADS simulation, students developed a proposed solution to mitigate the increase in Earth's surface temperature (Siegel et al., 2018). In groups, they modified EN-ROADS under the headings of energy supply, transportation, structures and industry, growth, land and industry emissions, and carbon removal. They provided a rationale for these changes (Figure 7).

Figure 7. An Example of Student Work in the Solution-Generation Process

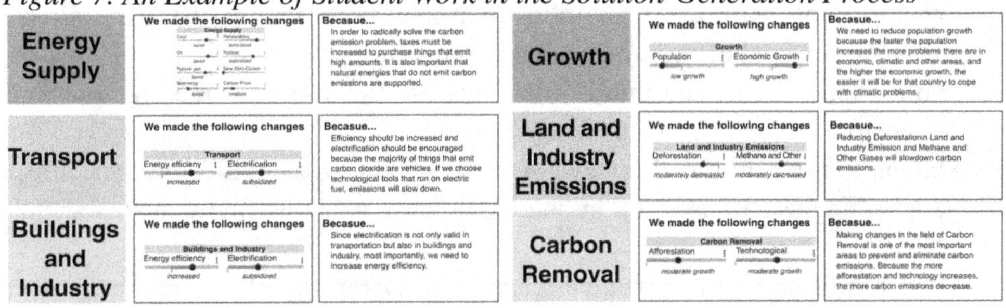

Note: An example of student-generated solutions to mitigate the increase in Earth's surface temperature using the EN-ROADS simulation.

Student Presentations (Week 9)

The students evaluated their solution proposals based on feedback and made necessary changes, both individually and on behalf of the school. After each group presented their proposals and actions, they received feedback from peers and teachers within the allotted time.

Climate Communication Meeting With an Expert (Week 10)

An interactive climate communication meeting was held with an environmental engineer who is an expert on climate change. During the interactive meeting, the measures taken by individuals and governments and the activities carried out were shared, and students were informed about the climate actions in which they were involved.

Data Collection Tools

As a data collection tool, individual interviews were conducted at the end of the education program implementations. The interviews, which lasted approximately 30 minutes, were recorded and transcribed. The interview questions consisted of eleven open-ended questions. The interviews aimed to reveal what the students learned during the program that taught the climate crisis with the systems thinking approach. Interview questions were implemented in the native language of the participants, and the English translations are given below:

1. Why did you want to participate in the climate change education program?
2. What were your expectations from this program?
3. What five concepts come to your mind when considering the systems thinking approach?
4. What did you learn about the climate system, international issues, policies, sustainability, and economy due to this program?
5. Did learning about the climate crisis through the systems thinking approach affect your views on climate change? How?
6. Did learning about the climate crisis from a systems thinking approach affect your willingness to take action on climate change? How (individual, societal, global...)?
7. Did learning about the climate crisis using a systems thinking approach affect your attitudes and behaviors on climate change? How?
8. Did the program lead to a change in your state of hope regarding climate change? Can you explain?
9. Did the program benefit/contribute to you? Can you explain?
    a. Has the program been helpful/contributory in supporting your courses? Can you explain?
    b. Has the program been useful/contributed to the profession you want to acquire? Can you explain?
    c. Did it help/contribute to your personal development? Can you explain?
10. Do you think the program was well planned? What are your suggestions for improving the program?
11. Finally, is there anything you would like to add?

After participating in the education program, the interview aimed to determine if there was a change in the students' opinions, willingness to act, attitudes, behaviors, and hopes regarding the climate crisis. The interview questions investigated the extent to which students' explanations and the model used in the climate crisis education program with a systems thinking approach overlap. The study was conducted with the necessary ethical permissions from the university.

Data Analysis

Thematic content analysis was utilized to analyze individual student interviews conducted at the end of the program. The researchers, the three authors of this study, evaluated project elements as categories and codes before beginning data analysis. Participant responses were analyzed in the context of these codes, and each category was evaluated independently. The program implemented is based on a model (Figure 1) consisting of 10 stocks, 17 flows between these stocks, and seven factors that cause these flows.

The researchers transcribed the oral responses of the students from the individual interviews. They then independently analyzed which elements of the model (Figure 1) corresponded to the students' responses and how frequently they were mentioned. The researchers compared the codes, categories, and frequency of the differing elements and formed a consensus through discussion.

FINDINGS

During the interviews, students emphasized the importance of learning climate change through a systems thinking approach. One student stated:

If you taught the topic directly, it would be complicated. It would be difficult. Because it is a difficult subject, something important, but there are so many factors. Explaining them in a certain way with a systems thinking diagram helped me understand. It made me learn more easily. We talk about the subject, but first, we talk about its effect, why it is like this, and environmental factors. We establish a connection with the system. This helped me understand it more easily. Since I understood more easily, I expressed clearer opinions. I could easily deduce that this was a problem or a factor.

During the interviews, the students utilized 13 out of the 34 model elements at least once. These 13 elements were identified as codes and categorized into four groups: 'economy,' 'carbon cycle,' 'population,' and 'greenhouse effect.' They are listed in Table 1 in descending order of frequency of mention (Table 1).

Table 1. Coding Results of Interviews

Category	*Code*	*frequency*	*%*
Economy	Consumption (GDP)	39	33%
	Recycling	13	11%
	Production	6	5%
	Natural resources	3	3%
	Waste and garbage	1	1%
	Subtotal	62	53%
Carbon cycle	From atmosphere to living things	14	11%
	Human carbon emission	11	9%
	From deep earth to atmosphere	10	8%
	Carbon in atmosphere	1	1%
	Subtotal	36	30%
Population	Population/immigration	11	9%
	Subtotal	11	9%
Greenhouse effect	The surface temperature of the earth	5	4%
	Greenhouse effect	3	3%
	The heat coming from the sun	1	1%
	Subtotal	9	8%
	Total	118	100%

Table 1 indicates that the economy category was the most commonly mentioned in student responses, accounting for 53% (62 out of 118 codes). The economy category consists of five codes, including two stocks (natural resources and waste and garbage) and three flows (consumption (GDP), recycling, and production). During the interviews, natural resource stocks and waste management were mentioned four times, while consumption (GDP) and recycling were mentioned 52 times. These two factors account for 44% of the codes mentioned in all interviews. It is noteworthy that these two factors, consumption and recycling, are the most frequently mentioned ones in the climate education program. One student stated:

> *Actually, at the end of the program, I learned that the climate system is related to international problems because climate change is an international problem, and actually, sustainable solutions need to be found. But if countries can't find something together, the problem can't be solved. And it can be a political problem because not all countries are doing well with each other. They may not get along well with each other. But we need to work together for this crisis, and we need to support it economically.*

The economy category is followed by the carbon cycle based on coding frequency. Of 118 codes, 30% (36 codes) are related to the carbon cycle, including one stock, two flows, and one variable. These codes represent the carbon stock in the atmosphere, flows from the atmosphere to living organisms, flows from the depth to the atmosphere, and the anthropogenic emission variable. During the interviews, the carbon stock in the atmosphere was only mentioned once. In contrast, the flows from the atmosphere to living organisms and deep atmosphere, as well as the anthropogenic emission variable, which causes deep atmosphere flows, were mentioned 35 times. It is important to note that all of these codes are directly related to the carbon stock in the atmosphere. 22% of the 118 codes in the model's carbon cycle section frequently mention 'human carbon emissions' and 'from deep earth to the atmosphere,' emphasizing the impact of anthropogenic actions on the climate crisis. This finding shows the importance of adding the carbon cycle to address its contribution to the climate crisis. One student stated:

I have learned that climate change is important worldwide. I have learned that there are things that each individual can change. We may not be able to change things only by ourselves. With the information I have learned, I have learned that even planting a tree can help. As a problem, the greenhouse effect can be an example. Because CO2 emission is usually high in climate change. When CO2 emissions are high, the greenhouse effect increases. That's why global warming is increasing. Likewise, economically, for example, if we go from carbon dioxide, there may be a cost to be spent on tree plantations or, in terms of utilizing more renewable energy sources, solar panels can be built.

During the interviews, the term 'population' was mentioned 15 times, which accounts for 9% of the discussion. The human population contributes to anthropogenic emissions by driving consumption and production, increasing atmospheric carbon, and the greenhouse effect. It is worth noting that the program on the environment also mentions the human population as an indirect factor, in addition to anthropogenic emissions. One student stated:

For example, when we first entered, I couldn't associate it with population. For example, I learned this when I entered. I had very little knowledge before this project. I thought of melting glaciers and polar bears when something was said about climate change. I did not think of population; I did not know the impact of population on the climate crisis.

The interviews had minimal discussion of the greenhouse effect, with only 9 out of 118 codes (8%) related to it. One of the codes in this category, heat stock and flows worldwide, was not mentioned. However, the causes of these flows, namely the greenhouse effect and heat from the sun and the earth's surface temperature resulting from the heat stock, were discussed. One student stated:

The greenhouse effect can be an example of the problem. Because CO2 emissions are generally high in today's world. When CO2 emissions are high, the greenhouse effect increases. This is why global warming is increasing.

DISCUSSION, CONCLUSION, AND RECOMMENDATIONS

With this project, a climate education program was developed to teach students about the climate crisis by providing a learning experience where they can understand the structure and function of the climate system and a simulation environment where they can take a decision-making position in policy development. The study emphasizes the importance of incorporating systems thinking into climate education, which aligns with the literature regarding the significance of integrating complex systems perspectives into environmental science curricula (Karaarslan Semiz & Teksöz, 2019; Mambrey et al., 2020; Shepardson et al., 2014). Various national and international studies emphasize the importance of holistic teaching in climate education (Barak & Gönençgil, 2020; Rossini, 2022).

This study supports the significance of teaching complex and dynamic environmental issues through the use of a systems thinking approach (Ballew et al., 2019; Doganca Kucuk & Saysel, 2018; Mambrey et al., 2020; Strapasson et al., 2022). Moreover, this study extends beyond existing studies by qualitatively measuring the impact of systems thinking on students' ability to address complex climate-related challenges. This highlights the urgency of revising current educational practices to incorporate systems-based methodologies, thereby equipping the next generation with the tools needed to navigate the complexities of climate change (Bielik et al., 2023; Budak & Ceyhan, 2023). Additionally, our study suggests that integrating systems thinking frameworks into curriculum standards could enhance climate literacy and foster critical thinking skills for addressing global environmental issues. This recommendation could have important policy implications for educational policymakers.

Simulations and system dynamics models contribute to students' environmental and climate awareness (Evagorou et al., 2009; Nuhoglu, 2020; Strapasson et al., 2022). In this study, after the simulation game, with the interactive modeling of the greenhouse effect, the emergence of the dilemma between the consequences of the disruption of the carbon cycle and the economic events that cause it, and the evaluation of the solutions, this dilemma helped students develop scientific reasoning and decision-making skills using controversies in environmental issues (Ballew et al., 2019). The fact that the model used showed the environment (carbon cycle), society (population and migrations), and economy (production, distribution, and consumption flows) as a whole enabled students to establish a relationship between the use of natural resources and production and consumption activities and gain knowledge about the effects of environmental problems and global climate change on the environment, society and economy (Blatti et al., 2019).

Our findings align with and extend previous research on applying systems thinking in climate change education. Consistent with prior studies (Ballew et al., 2019; Doganca Kucuk & Saysel, 2018), a systems thinking approach enhances students' understanding of the complex dynamics and interconnections within the climate system. However, this study expands on previous work by offering a comprehensive qualitative analysis of how students' explanations corresponded with the specific elements of the system dynamics model utilized in the educational program. Our thorough examination of student responses revealed that the model's economic factors and carbon cycle components were the most frequently referenced, indicating that these aspects of the program had a significant impact on student learning.

Additionally, our research adds to the increasing amount of literature that shows the efficacy of simulation games, such as the World Climate Game, in encouraging systems thinking and involvement with climate change concerns (Rooney-Varga et al., 2021; Sterman et al., 2015). By integrating this game into our education program and analyzing student responses, we can provide additional evidence for the effectiveness of this approach in promoting a better understanding of the dynamics of climate change and potential solutions. Overall, the findings emphasize the significance of integrating systems thinking and interactive simulation games into climate change education. Additionally, the study provides new insights into the specific elements of a system dynamics model that may be most relevant for student learning. These contributions can inform the design and implementation of future educational interventions promoting climate literacy and action.

We acknowledge the limitations of our study, including its scope and the diversity of educational contexts. Future research could explore the applicability of systems thinking across different age groups and educational settings to validate and expand upon our findings. The program implemented during the project can be developed as a sample program for the environmental education and climate change courses. Development studies can be carried out to adapt the curriculum according to the grade level of the sixth, seventh, and eighth grades, which are within the scope of the curriculum, by associating it with the achievements of other curricula. More solution-oriented dimensions, such as environmentally friendly technologies, more local dimensions, and more principle-based dimensions, such as environmental ethics, can be added to the education program in line with the objectives of the climate curriculum. In future research, developing effective curriculum and system dynamics pedagogy, determining assessment and evaluation methods, and studying teacher development in climate change and environmental education in science education will contribute significantly to educational research and practice (Roychoudhury et al., 2017).

In conclusion, this study contributes to a deeper understanding of the importance of systems thinking in climate education. We aim to inspire a shift towards more integrated and effective educational strategies by highlighting the benefits and providing a foundation for further research. Such a shift is imperative for preparing students to confront the challenges of climate change, ensuring they possess the knowledge and skills to contribute to a sustainable future.

REFERENCES

Almaki, S. H., Gunda, M. A., Idris, K., Hashim, A. T. M., & Ali, S. R. (2023). A systematic review of the use of simulation games in K-12 education. *Interactive Learning Environments*, 1–25. DOI: 10.1080/10494820.2023.2205894

Aristoteles, M. (2020, June). Book 8, section 1045a. (t.y.). Perseus Digital Library. *Access*, 13. https://www.perseus.tufts.edu/hopper/text?doc=urn:cts:greekLit:tlg0086.tlg025.perseus-eng1:8.1045a

Assaraf, B. O., & Orion, N. (2005). Development of systems thinking skills in the context of Earth system education. *Journal of Research in Science Teaching*, 42(5), 518–560. DOI: 10.1002/tea.20061

Assaraf, O. B. Z., & Orion, N. (2010). System thinking skills at the elementary school level. *Journal of Research in Science Teaching*, 47(5), 540–563. DOI: 10.1002/tea.20351

Ateskan, A., & Lane, J. F. (2018). Assessing teachers' systems thinking skills during a professional development program in Turkey. *Journal of Cleaner Production*, 172, 4348–4356. DOI: 10.1016/j.jclepro.2017.05.094

Ballew, M. T., Goldberg, M. H., Rosenthal, S. A., Gustafson, A., & Leiserowitz, A. (2019). Systems thinking as a pathway to global warming beliefs and attitudes through an ecological worldview. *Proceedings of the National Academy of Sciences of the United States of America*, 116(17), 8214–8219. DOI: 10.1073/pnas.1819310116 PMID: 30962375

Barak, B., & Gönençgil, B. (2020). Dunyada ve Turkiye'de ortaokul ogretim programlarinin iklim degisikligi egitimi yaklasimina gore karsılastirilmasi. *Cografya Dergisi*, (40), 187–201. DOI: 10.26650/JGEOG2019-0039

Berenjkar, P., Li, Y. Y., & Yuan, Q. (2021). The application of system dynamics in different practices of a waste management system. *Environment, Development and Sustainability*, 23(11), 15695–15724. DOI: 10.1007/s10668-021-01362-8

Bielik, T., Jagemann, J., Krell, M., Krüger, D., & Ben-Zvi Assaraf, O. (2023). Using concept maps to evaluate preservice biology teachers' conceptualization of COVID-19 as a complex phenomenon. *Journal for STEM Education Research*, 6, 199–231. DOI: 10.1007/s41979-023-00087-9

Blatti, J. L., Garcia, J., Cave, D., Monge, F., Cuccinello, A., Portillo, J., Juarez, B., Chan, E., & Schwebel, F. (2019). Systems thinking in science education and outreach toward a sustainable future. *Journal of Chemical Education*, 96(12), 2852–2862. DOI: 10.1021/acs.jchemed.9b00318

Budak, U. S., & Ceyhan, G. D. (2023). Research trends on systems thinking approach in science education. *International Journal of Science Education*, 1–18. DOI: 10.1080/09500693.2023.2245106

Cox, M., Elen, J., & Steegen, A. (2019). The use of causal diagrams to foster systems thinking in geography education: Results of an intervention study. *The Journal of Geography*, 118(6), 238–251. DOI: 10.1080/00221341.2019.1608460

Creswell, J. W., & Poth, C. N. (2016). *Qualitative inquiry and research design: Choosing among five approaches*. Sage publications.

Doganca Kucuk, Z., & Saysel, A. K. (2018). Developing seventh-grade students' understanding of complex environmental problems with systems tools and representations: A quasi-experimental study. *Research in Science Education*, 48(2), 491–514. DOI: 10.1007/s11165-017-9620-8

Evagorou, M., Korfiatis, K., Nicolaou, C., & Constantinou, C. (2009). An investigation of the potential of interactive simulations for developing systems thinking skills in elementary school: A case study with fifth-graders and sixth-graders. *International Journal of Science Education*, 31(5), 655–674. DOI: 10.1080/09500690701749313

Forrester, J. W. (1993). System dynamics and the lessons of 35 years. In *A systems-based approach to policymaking* (pp. 199–240). Springer. DOI: 10.1007/978-1-4615-3226-2_7

Guthrie, S., & Fisher, D. (1999). Systems thinking and system dynamics in the CC-STADUS high school project (How high school students become system thinkers). Retrieved 07/05/06, 2006, from http://www. clexchange. org/ftp/documents/Implementation/IM1997-07SDLeveragePoints. Pdf.

Heraclitus, . (2020, June). (Stanford encyclopedia of philosophy). (t.y.). Stanford Encyclopedia of Philosophy. *Access*, 13. https://plato.stanford.edu/entries/heraclitus/#Flu

Hokayem, H., & Gotwals, A. W. (2016). Early elementary students' understanding of complex ecosystems: A learning progression approach. *Journal of Research in Science Teaching*, 53(10), 1524–1545. DOI: 10.1002/tea.21336

Hossain, N. U. I., Dayarathna, V. L., Nagahi, M., & Jaradat, R. (2020). Systems thinking: A review and bibliometric analysis. *Systems*, 8(3), 23. DOI: 10.3390/systems8030023

Irwin, R. (2020). Climate change and education. *Educational Philosophy and Theory*, 52(5), 492–507. DOI: 10.1080/00131857.2019.1642196

Karaarslan Semiz, G., & Teksöz, G. (2019). Sistemsel dusunme becerilerinin tanimlanmasi, olculmesi ve degerlendirilmesi uzerine bir calisma: Kavram haritalari. *Baskent University Journal of Education*, 6(1), 111-126. https://buje.baskent.edu.tr/index.php/buje/article/view/177

Krein, A. T. (2023). *Model United Nations: A Practical Guide*. Springer Nature. DOI: 10.1007/978-3-031-13524-8

Liu, S. C. (2023). Examining undergraduate students' systems thinking competency through a problem scenario in the context of climate change education. *Environmental Education Research*, 29(12), 1–16. DOI: 10.1080/13504622.2022.2120187

Lombardi, D. (2022). Climate crisis mitigation and adaptation: Educational and developmental psychology's responsibility in helping face this threat. *Educational and Developmental Psychologist*, 39(1), 1–4. DOI: 10.1080/20590776.2021.2012834

Maddison Project Database. (2020). Bolt, Jutta, and Jan Luiten van Zanden. Maddison style estimates of the evolution of the world economy. A new 2020 update.

Mambrey, S., Schreiber, N., & Schmiemann, P. (2020). Young students' reasoning about ecosystems: The role of systems thinking, knowledge, conceptions, and representation. *Research in Science Education*, 1–20. DOI: 10.1007/s11165-020-09917-x

Martinez-Moyano, I. J. (2024). History of the Beer Game. *System Dynamics Review*, 40(2), 1–32. DOI: 10.1002/sdr.1767

Meadows, D., Sweeney, L. B., & Mehers, G. M. (2016). *The climate change playbook: 22 systems thinking games for more effective communication about climate change*. Chelsea Green Publishing.

Meadows, D. H. (2008). *Thinking in systems: A primer*. Chelsea Green Publishing.

Merriam, S. B., & Tisdell, E. J. (2016). Designing your study and selecting a sample. *Qualitative research: A guide to design and implementation, 67*(1), 73-104.

MoNE. (2022). Çevre Egitimi ve Iklim Degisikligi Dersi (6, 7 veya 8. Sınıflar) Ogretim Programi. Ankara: Millî Egitim Bakanlığı. http://mufredat.meb.gov.tr/ProgramDetay.aspx?PID=1143. Erişim Tarihi: 21. 01.2023.

National Research Council. (2013). *Next generation science standards: For states, by states.*

Naugle, A., Verzi, S., Lakkaraju, K., Swiler, L., Warrender, C., Bernard, M., & Romero, V. (2023). Feedback density and causal complexity of simulation model structure. *Journal of Simulation*, 17(3), 229–239. DOI: 10.1080/17477778.2021.1982653

Nguyen, H., & Santagata, R. (2021). Impact of computer modeling on learning and teaching systems thinking. *Journal of Research in Science Teaching*, 58(5), 661–688. DOI: 10.1002/tea.21674

Nuhoglu, H. (2008). İlkogretim fen ve teknoloji dersinde sistem dinamigi yaklasiminin tutuma, basarıya ve farkli becerilere etkisinin arastirilmasi. [Doctoral dissertation, Gazi Üniversitesi].

Nuhoglu, H. (2020). The effect of deduction and induction methods used in modeling current environmental issues with system dynamics approach in science education. *Participatory Educational Research*, 7(1), 111–126. DOI: 10.17275/per.20.7.7.1

Oliveira, W., Hamari, J., Shi, L., Toda, A. M., Rodrigues, L., Palomino, P. T., & Isotani, S. (2023). Tailored gamification in education: A literature review and future agenda. *Education and Information Technologies*, 28(1), 373–406. DOI: 10.1007/s10639-022-11122-4

Phillips, R. J. (1997). Can juniors read graphs? A review and analysis of some computer-based activities. *Journal of Information Technology for Teacher Education*, 6(1), 49–58. DOI: 10.1080/14759399700200005

Porter, T., & Córdoba, J. (2009). Three views of systems theories and their implications for sustainability education. *Journal of Management Education*, 33(3), 323–347. DOI: 10.1177/1052562908323192

Pörtner, H. O., Roberts, D. C., Adams, H., Adler, C., Aldunce, P., Ali, E., ... Birkmann, J. (2022). Climate change 2022: Impacts, adaptation and vulnerability. *IPCC Sixth Assessment Report*, 37-118. DOI: 10.1111/padr.12497

Riess, W., & Mischo, C. (2010). Promoting systems thinking through biology lessons. *International Journal of Science Education*, 32(6), 705–725. DOI: 10.1080/09500690902769946

Rooney-Varga, J. N., Hensel, M., McCarthy, C., McNeal, K., Norfles, N., Rath, K., Schnell, A. H., & Sterman, J. D. (2021). Building consensus for ambitious climate action through the World Climate simulation. *Earth's Future*, 9(12). DOI: 10.1029/2021EF002283

Rossini, M. (2022). Teaching children about climate change and sustainability. *The School Science Review*, 103(384), 76–79.

Roychoudhury, A., Shepardson, D. P., Hirsch, A., Niyogi, D., Mehta, J., & Top, S. (2017). The need to introduce system thinking in teaching climate change. *Science Educator*, 25(2), 73–81.

Sajjadi, P., Bagher, M. M., Myrick, J. G., Guerriero, J. G., White, T. S., Klippel, A., & Swim, J. K. (2022). Promoting systems thinking and pro-environmental policy support through serious games. *Frontiers in Environmental Science*, 10, 957204. DOI: 10.3389/fenvs.2022.957204

Senge, P. (2006). Systems citizenship. In *The Leader of The Future 2* (pp. 31–46). Visions, Strategies, and Practices for The New Era.

Senge, P. M. (1990). *The fifth discipline: The art and practice of the learning organization*. Crown Publishing.

Shepardson, D. P., Niyogi, D., Choi, S., & Charusombat, U. (2009). Seventh-grade students' conceptions of global warming and climate change. *Environmental Education Research*, 15(5), 549–570. DOI: 10.1080/13504620903114592

Shepardson, D. P., Roychoudhury, A., Hirsch, A., Niyogi, D., & Top, S. M. (2014). When the atmosphere warms it rains and ice melts: Seventh-grade students' conceptions of a climate system. *Environmental Education Research*, 20(3), 333–353. DOI: 10.1080/13504622.2013.803037

Shepardson, D. P., Roychoudhury, A., & Hirsch, A. S. (2017). *Teaching and learning about climate change*. Routledge Publishing. DOI: 10.4324/9781315629841

Siegel, L. S., Homer, J., Fiddaman, T., McCauley, S., Franck, T., Sawin, E., ... Interactive, C. (2018). *En-ROADS simulator reference guide. Technical Report.*

Sommer, C., & Lücken, M. (2010). System competence–Are elementary students able to deal with a biological system? *Nordic Studies in Science Education*, 6(2), 125–143. DOI: 10.5617/nordina.255

Sterman, J., Franck, T., Fiddaman, T., Jones, A., McCauley, S., Rice, P., Sawin, E., Siegel, L., & Rooney-Varga, J. N. (2015). World Climate: A role-play simulation of climate negotiations. *Simulation & Gaming*, 46(3-4), 348–382. DOI: 10.1177/1046878113514935

Sterman, J. D. (2001). System dynamics modeling: Tools for learning in a complex world. *California Management Review*, 43(4), 8–25. DOI: 10.2307/41166098

Sterman, J. D. (2014). Interactive web-based simulations for strategy and sustainability: The MIT Sloan LearningEdge management flight simulators, Part I. *System Dynamics Review*, 30(1-2), 89–121. DOI: 10.1002/sdr.1513

Sterman, J. D., & Sweeney, L. B. (2002). Cloudy skies: Assessing public understanding of global warming. *System Dynamics Review*, 18(2), 207–240. DOI: 10.1002/sdr.242

Strapasson, A., Ferreira, M., Cruz-Cano, D., Woods, J., do Nascimento Maia Soares, M. P., & da Silva Filho, O. L. (2022). The use of system dynamics for energy and environmental education. *International Journal of Educational Technology in Higher Education*, 19(1), 1–31. DOI: 10.1186/s41239-021-00309-3 PMID: 35106364

Sweeney, L. B., & Sterman, J. D. (2000). Bathtub dynamics: Initial results of a systems thinking inventory. *System Dynamics Review*, 16(4), 249–286. DOI: 10.1002/sdr.198

Sweeney, L. B., & Sterman, J. D. (2007). Thinking about systems: Student and teacher conceptions of natural and social systems. *System Dynamics Review*, 23(2–3), 285–312. DOI: 10.1002/sdr.366

Vlachopoulos, D., & Makri, A. (2017). The effect of games and simulations on higher education: A systematic literature review. *International Journal of Educational Technology in Higher Education*, 14(1), 1–33. DOI: 10.1186/s41239-017-0062-1

ADDITIONAL READING

Daly, H. E. (1990). Toward some operational principles of sustainable development. *Ecological Economics*, 2(1), 1–6. DOI: 10.1016/0921-8009(90)90010-R

Daly, H. E. (1991). *Steady-state economics: with new essays*. Island Press.

Daly, H. E. (1996). *Beyond growth: the economics of sustainable development*. Beacon Press.

Ford, F. A. (1999). *Modeling the environment: an introduction to system dynamics models of environmental systems*. Island Press.

Interactive, C. (2023). *The C-ROADS Climate Change Policy Simulator.* https://www.climateinteractive.org/c-roads/

Meadows, D., & Randers, J. (2012). *The limits to growth: the 30-year update*. Routledge. DOI: 10.4324/9781849775861

Meadows, D., Sweeney, L. B., & Mehers, G. M. (2016). *The climate change playbook: 22 systems thinking games for more effective communication about climate change*. Chelsea Green Publishing.

Meadows, D. H. (2008). *Thinking in systems: A primer*. Chelsea Green Publishing.

Ticotsky, A. (2022). *Now what? A call to action: Environmental systems lessons for upcoming generations*. Creative Learning Exchange.

KEY TERMS AND DEFINITIONS

Climate Change Advocacy: Actions taken by individuals, organizations, or communities to promote awareness, understanding, and action on climate change issues.

Climate Education: Climate education aims to increase students' awareness and understanding of the challenges and solutions related to climate change and equip them with the knowledge and skills necessary to address climate-related issues.

Climate Literacy: A climate-literate individual understands the fundamental principles of the Earth's climate system, including the factors that influence climate and its variability and change over time, and can make informed decisions about climate-related issues.

Holistic Learning Approach: An educational philosophy that recognizes the interconnectedness of different aspects of learning and emphasizes the development of a comprehensive understanding.

Integrated Curriculum: An educational approach integrating knowledge and methods from multiple academic disciplines to create a holistic understanding of a topic or problem.

Sustainability: Sustainability involves meeting present needs without jeopardizing those of future generations. Education for sustainability aims to develop knowledge, skills, and attitudes that enable learners to contribute to a more sustainable society.

Systems Thinking: Systems thinking is an approach that contributes to a more holistic understanding of complex systems by focusing on how various components interact and influence one another as a whole rather than examining parts in isolation.

World Climate Game: The World Climate Game, developed by Climate Interactive, the MIT Sloan Sustainability Initiative, and the UMass Lowell Climate Change Initiative, is an educational simulation where students role-play international climate negotiators analyzing scientific data to make policy decisions aimed at limiting global temperature rise through carbon emissions reductions.

APPENDIX

Questions And Answers

1- What is the overall problem presented in this case?

The overall problem presented in the case is the need to effectively educate students about the complexities of the climate crisis using a systems thinking approach. The goal is to raise awareness, enhance understanding, and equip students with the skills to propose solutions to global climate issues through a comprehensive climate education program. The challenge lies in helping students grasp the interconnectedness of environmental, societal, and economic factors within the climate system and empowering them to analyze these relationships critically to develop effective solutions.

2- What are the factors affecting the problem(s) related to this case?

Educator Development: Educators need to develop professionally to effectively integrate systems thinking into educational practice

Educational Challenges: Students often struggle to grasp the complexity of climate systems and the interplay of various components such as greenhouse gas emissions, population dynamics, and economic activities. Students may hold misconceptions about environmental issues, highlighting the importance of addressing and correcting these misconceptions

Integration of Systems Thinking: Integrating the systems thinking approach into climate change education is necessary to comprehend the complexity of climate systems and their operations. There is a need to improve students' understanding of stock and flow structures, feedback loops, and the dynamic behavior of climate systems.

Educational Program Design: The design and implementation of climate education programs, including simulation environments and specific learning phases, impact students' understanding and alignment with program objectives.

3- What are the managerial, organizational, and technological issues and resources related to this case.

Managerial Issues

Program Coordination: Managing the logistics of a ten-week after-school program involving multiple stakeholders (students, teachers, NGOs, and university researchers).

Evaluation: Assessing the effectiveness of the program in meeting its educational objectives.

Organizational Issues

Collaboration: Ensuring effective collaboration among the research university, NGO, and middle school.

Curriculum Integration: Aligning the climate education program with the existing school curriculum.

Technological Issues

Simulation Games: Utilizing interactive simulation games, such as the World Climate Game, can enhance students' engagement and understanding of climate change dynamics.

Data Visualization Tools: Technology can be used to visualize complex climate data and systems, helping students grasp the interconnectedness of environmental factors.

Online Platforms: Online platforms and digital resources can support remote learning and collaboration, enabling students to engage with climate education content beyond the classroom.

4- What role do different players (decision-makers) play in the overall planning, implementation, and management of the information technology applications?

Educators: Develop and deliver the curriculum, facilitate student engagement, and assess student understanding.

Students: Participate actively in the program, engage with simulations, and apply systems thinking to understand climate issues.

NGO: Provide expertise in systems thinking and support the development and implementation of the program.

University Researchers: Contribute to the design of the program, offer technical support, and conduct research on its effectiveness.

School Administrators: Administrators are responsible for allocating resources for acquiring and maintaining IT applications for climate education programs. They develop policies and guidelines for the use of technology in education, ensuring alignment with educational goals and standards. Administrators provide support and training to educators on effectively utilizing IT applications for teaching climate change concepts.

Technology Specialists: Technology specialists contribute to the technical planning of IT applications, ensuring compatibility, security, and functionality. They assist in the implementation of technology solutions, troubleshoot issues, and provide technical support to educators and students. Technology specialists manage data generated by IT applications, ensuring data security, integrity, and compliance with privacy regulations.

Curriculum Developers: Curriculum developers collaborate with educators to integrate technology applications seamlessly into climate education curricula. They ensure that IT applications align with curriculum objectives, learning outcomes, and standards for effective teaching and learning. Curriculum developers explore innovative ways to leverage technology for enhancing students' understanding of climate systems and environmental issues.

Policy Makers: Potentially use the outcomes of the program to inform educational policies and practices.

5- What are the possible alternatives and pros and cons of each alternative facing the organization in dealing with the problem(s) related to the case?

Intensive Teacher Training on Systems Thinking:

Pros:

Prospective Improvement: Educators equipped with advanced knowledge of systems thinking can effectively integrate this approach into climate change education.

Sustainable Impact: Building educators' capacity can lead to long-term benefits in enhancing students' understanding of climate systems.

Cons:

Resource Intensive: Requires significant investment in training programs, materials, and ongoing support for educators.

Time Consuming: Implementation may take time to yield tangible results in student learning outcomes.

Integration into Regular Curriculum:

Pros:

Reaches more students, sustainable integration.

Cons:

Curriculum adjustments required, potential resistance from educators.

6- What are some of the emerging technologies that should be considered in solving the problem(s) related to the case?

Artificial Intelligence (AI) and Machine Learning:

Application: AI algorithms can analyze large datasets related to climate change, predict future trends, and personalize learning experiences based on student interactions.

Benefits: Supports data-driven decision-making, offers personalized feedback to students, and enables adaptive learning pathways.

Gamification and Simulation Platforms:

Application: Gamified learning platforms and simulation games, such as the World Climate Game, can engage students in decision-making processes related to climate policies and solutions.

Benefits: Encourages active participation, fosters critical thinking skills, and provides a safe space for experimentation and exploration of complex systems.

7- What is the final solution that can be recommended to the management of the organization described in the case? Provide your arguments in support of the recommended solution.

Recommended Solution:

Integrated Climate Education Program with Technology-enhanced Systems Thinking

Arguments in Support of the Recommended Solution:

Holistic Approach: By integrating systems thinking principles into the curriculum and leveraging technology-enhanced learning experiences, the recommended solution offers a comprehensive approach to enhancing students' understanding of climate change and complex environmental systems.

Professional Development: Investing in teacher training and support ensures that educators are equipped with the necessary knowledge and skills to effectively deliver climate education programs, leading to sustained impact on student learning outcomes.

Preparation for the Future: By equipping students with systems thinking skills and climate literacy, the recommended solution prepares them to address the complex challenges of climate change, empowering them to become informed global citizens and advocates for sustainability.

Epilogue And Lessons Learned

Epilogue

In the long term, integrating systems thinking into the regular curriculum can lead to a generation of students who are better equipped to understand and tackle complex global issues such as climate change. This approach can foster critical thinking, problem-solving skills, and a proactive attitude towards environmental stewardship.

Lessons Learned

Collaboration is Key: Successful implementation requires collaboration among various stakeholders, including educators, NGOs, and researchers.

Importance of Systems Thinking: Integrating systems thinking into education helps students understand complex issues holistically and develop problem-solving skills.

Technology as an Enabler: Emerging technologies can enhance learning experiences and make complex concepts more accessible.

Sustainability of Programs: Integrating programs into the regular curriculum ensures sustainability and broader impact.

Continuous Evaluation: Regular assessment and adaptation of the program are essential to meet educational objectives and improve effectiveness.

Chapter 4
Harnessing Art as a Catalyst to Environmental Awareness

Farheen Mahmood
Indiana University of Pennsylvania, USA

Saroash Ali
Independent Researcher, Pakistan

EXECUTIVE SUMMARY

Environmental awareness is crucial for ensuring a sustainable future. This chapter presents how fostering environmental awareness in children is important for securing a safe Mother Earth for future generations. This chapter showcases an initiative implemented through an art-integrated project for environmental awareness across three public schools in Lahore, Pakistan. Beyond strengthening academic and creative skills besides social-emotional learning (SEL), the program also cultivated a deep sense of environmental responsibility in students, teachers, and administrators. Olomopolo Media (Pakistan) launched the global "River of Hope" British Council program to raise awareness about rivers and their deep connection to the cities they flow through. This program used traditional printmaking to raise awareness about river Ravi's problems. It focused on inspiring action for its restoration by presenting the power of art for environmental causes resulting in the culmination of a student-created printmaking exhibition displayed during the Queen's Platinum Jubilee in England.

DOI: 10.4018/979-8-3693-1710-5.ch004

INTRODUCTION

Environmental education fosters an understanding of the intricate balance of our planet's ecosystems and human activities impact on them (Kumari & Thakur, 2023). This awareness is crucial for making informed decisions and achieving a sustainable future. The Earth's climate has undergone constant shifts throughout history. However, the current rate of global warming is unprecedented in the last 10,000 years (IPCC, 2023; *Past Climate*, 2021). Scientific evidence from ancient natural records like ice cores and modern tools like satellites shows that our planet is warming (IPCC, 2023; *Past Climate*, 2021). This warming isn't happening in isolation. The world's ecosystems, the intricate web of life on Earth, are experiencing a rapid transformation unlike anything witnessed before. Human activity is the primary driver behind this change ("Ecosystems and Human Well-Being: Synthesis," 2005). Land conversion for agriculture is a major factor, with vast areas being transformed into cultivated fields at an alarming rate. Fresh water resources are also under strain ("Ecosystems and Human Well-Being: Synthesis," 2005). The construction of dams has significantly altered natural water flow patterns ("Ecosystems and Human Well-Being: Synthesis," 2005). Even coastal ecosystems like mangroves and coral reefs have seen substantial losses in recent decades ("Ecosystems and Human Well-Being: Synthesis," 2005). These changes highlight the interconnectedness of our planet's systems. A warming climate and human actions fundamentally alter the delicate balance of Earth's ecosystems ("Ecosystems and Human Well-Being: Synthesis," 2005). Given the pressing nature of these environmental challenges, equipping school-going children with knowledge is crucial. Educating them about these issues empowers them to become responsible guardians of the Earth. This knowledge will enable them to advocate for change and develop solutions for a sustainable future (Kumari & Thakur, 2023).

Out of the numerous impactful methods of teaching environmental awareness, project-based learning has left profound outcomes among young learners. John Dewey is believed to have developed the idea of project-based learning. Dewey(1976) argued, "Give the pupils something to do, not something to learn; and the doing is of such a nature as to demand thinking; learning naturally results" (1916, p. 191). Project-based learning effectively allows students to explore a topic more deeply through hands-on activities. It fosters active engagement and a deeper understanding of the subject (Cervantes et al., 2015). It also encourages students to apply their knowledge and skills to solve real-world problems. This can include environmental projects like designing sustainable communities or cleaning up local parks. Moreover, project work often involves collaboration and communication, preparing students with essential skills for the future (Cervantes et al., 2015).

When designing project-based learning experiences, art can be a powerful tool. Art allows students to forge into environmental issues on a creative and interactive level. Creative mediums can aid students in expressing their concerns, hopes, and ideas (Papavasileiou et al., 2021). Similarly, through artistic expression, students can raise awareness about environmental issues and inspire others to take action (Flowers et al., 2015). Therefore, incorporating art into learning projects encourages creative thinking and problem-solving skills, which are valuable assets in tackling environmental challenges (Papavasileiou et al., 2021).

An environmental awareness project was led by engaging in art activities at three public schools in Lahore near River Ravi. Ultimately, this synergy between project-based learning and art cultivated a group of critical thinkers with the knowledge, passion, and skills to work cautiously to develop a sustainable future for our planet. Author 2, who served as a program assistant with Olomopolo Media, meticulously documented the entire experience, capturing the essence of the program and its execution. Author 2 provided supporting evidence through pictures to ensure the project's validity. Author 1 took the lead in synthesizing the project information. They gathered all the notes and carefully organized them to write this paper.

BACKGROUND OF RIVER RAVI: LIFELINE OF PAKISTAN'S PUNJAB

The Ravi River is the linchpin of Punjab's agricultural sector, its lifeblood (Gull & Gul, 2017). The extensive network of canals fed by the Ravi irrigates vast swathes of land, providing the vital sustenance that underpins countless communities throughout the Punjab province in Pakistan (Ashraf et al., 2022; Gull & Gul, 2017). Beyond agriculture, the Ravi serves as a critical freshwater source, quenching the thirst of millions in urban centers like Lahore (Gull & Gul, 2017; Shahid et al., 2019). Furthermore, the river fosters a vibrant ecosystem, providing a diverse plant and animal life habitat. This intricate network of life supported by the Ravi creates a truly remarkable ecological tapestry (Gull & Gul, 2017). However, it suffers from severe pollution, particularly in urban areas of Lahore (Asia Water Wire, 2006; Gull & Gul, 2017; Shahid et al., 2019). Untreated industrial waste, agricultural runoff, and faulty drainage systems contribute to the toxic soup (Ashraf et al., 2022; Asia Water Wire, 2006; Gull & Gul, 2017; Shahid et al., 2019). This is not just limited to raw sewage; tests reveal high levels of metals and even pharmaceuticals in the water at the 72-kilometer (45 mi) stretch of the river from Lahore Siphon to Baloki headworks (Lai, 2022). The situation is further worsened by contaminated riverbed sediment acting as a secondary source of pollution (Asia Water Wire, 2006). Recognizing the seriousness of this issue, a UN-backed program is working to address the problem (Asia Water Wire, 2006).

PROJECT CONTEXT

Across the globe, a burgeoning initiative seeks to redefine our understanding of rivers. The "Rivers of Hope" project transcends the perception of rivers as mere bodies of water, instead highlighting their profound impact on the identity of the cities they traverse (*Rivers of the World | British Council*, 2021). This project unveils a compelling narrative through two distinct storylines. One narrative celebrates rivers that flourish due to dedicated conservation efforts. These success stories are powerful testaments to the harmonious coexistence of urban environments and the natural world. They illustrate how a thriving river can be the lifeblood of a vibrant city, fostering a unique cultural identity. The second narrative presents a stark contrast, highlighting rivers struggling due to a lack of maintenance and preservation resources. The ever-present threat of climate change further exacerbates these challenges. This stark reality underscores the urgent need for increased investment and innovative solutions to ensure the health of these vital ecosystems. The "Rivers of Hope" project serves as a global call to action, urging us to recognize the intrinsic value of rivers and foster a more sustainable future for our cities and the natural world (*Rivers of the World | British Council*, 2021).

This project aimed to raise awareness and was initiated with collaboration between Olomopolo Media (https://www.facebook.com/olomopolo/) and the British Council, Pakistan. The schools of Lahore were chosen due to the proximity of the students and the people associated with this vital resource. The selection of schools close to River Ravi was intentional and served a twofold purpose. First, their proximity to the impacted ecosystem allowed participants to form a concrete connection to the environmental issue at hand. Participants could relate well to the situation by directly experiencing the ecological crisis, which increased their initiatives' effectiveness. Second, the selection of schools close to River Ravi aligned with the broad goals of the program, which were community participation and localized environmental activism. As the project sought to galvanize community-based efforts toward environmental conservation, the project goals were:

- Environmental awareness
- Using art as an educational tool
- Empowerment and community engagement
- Cultivating critical inquiry and reflection
- Fostering collaborative creativity

- Promoting sustainable behavioral change
- Professional development for educators

Art for Action: Ravi River Crisis Through Printmaking

Art thrives as a potent force for environmental awareness, fostering emotional engagement and intellectual reflection (Jónsdóttir, 2017). Digital art platforms are popular, but traditional methods for sparking change remain as relevant as ever. As highlighted by the United Nations Educational, Scientific and Cultural Organization (UNESCO, 2023), visual art's capacity for subtle expression and thought-provoking engagement positions it as a powerful tool for fostering social change, particularly in shaping the environmental consciousness of younger generations. By integrating multidisciplinary approaches that leverage the power of art, educational institutions can empower students to become active participants in addressing the current environmental crisis. Aligned with UNESCO's emphasis on art's power to shape young minds, this project employed traditional printmaking to raise awareness about the ecological crisis plaguing the Ravi River.

Action Plan

The project execution unfolded over six days (two days for each school), encompassing the following key phases:

Phase 1—Team Formation: All team members were recruited via public calls for interviews disseminated on social media platforms or through an open invitation to the workshop. This inclusive approach attracted individuals from diverse backgrounds and life experiences, enriching the pool of potential candidates and fostering a collective sense of ownership over the project.

The program assistants were selected meticulously, considering their qualifications, prior experience, and alignment with the project's objectives and principles. The printmaking workshop team represented various professional backgrounds, i.e., software engineers, photographers, architects, performance artists, and filmmakers. Leveraging their diverse expertise and perspectives, each team member contributed to the project's depth and comprehensive execution. For instance, Author 2, an architect by profession, profoundly understood the built and natural environment. This expertise uniquely positioned her to offer insights into the impacts of human activities, such as urbanization and infrastructural development, on the ecosystem and water quality of the river.

Phase 2—School Selection and Time Allocation: The project engaged 20 to 24 students, aged 12 to 14, from three schools alongside 2 to 3 teachers from each institution. It lasted two weeks, including one week of printmaking workshops and the other for data compilation of the three public schools in the area—the initiative aimed to connect students and their immediate surroundings by employing the arts as a medium.

Phase 3—Printmaking Workshops Execution: Each school, overseen by Olomopolo Media's dedicated team, gathered resources and prepared workshop materials to conduct two-day printmaking workshops. Consequently, every school had the opportunity to participate in an extensive two-day workshop session. Author 2 collaborated with acclaimed printmaking artist Ms. Sarah Mumtaz to conduct the workshop.

Workshop Implementation and Student Engagement: The two-day workshops brimmed with excitement as the team introduced the art of printmaking to the students. Focusing on the Ravi River as a theme, presenters sparked attendees' curiosity as shown in Figure 1, by initiating a discussion through the questioning technique, i.e., why do you think we are here today? How are you connected to river Ravi? What is the earliest memory of the river? How do you remember it looked before how does it look now, and why is there so much difference between how it was and how it is now? Learning about the River Ravi resonated deeply with the attendees, as they had personal connections to the river. These connections often stemmed from cherished family stories passed down from mothers, fond recollections of childhood picnics by the riverbanks, or nostalgic memories of past visits. It allowed everyone to reminisce about their memories related to the river Ravi. All in all, young or old, everyone in the room had a happy memory related to the beautiful river that Ravi once was. Some students remembered the river as a vast, never-ending waterbody spreading for miles on end while traveling to some other city. On the other hand, the teachers also shared their boat rides and the pleasant monsoon weather at the Ravi. Using these positive associations, the presenters then transitioned into a discussion about the current decline of the Ravi River. To encourage students to explore the reasons behind the decline, they again initiated thought-provoking questions that helped them think about its causes, for example, where does our drinking water come from? Why is the river so smelly? Why the river has so much waste? Why are there so few visitors? What is the reason behind this decline? Sarah, author two, and the teachers explained how industrial waste and chemical pollution also threaten bird populations that once thrived along the river. They talked in detail about how the once-existent special turtle habitat (Hardella Thurjii) along the river Ravi has disappeared due to water and air pollution. Moreover, the increasing construction activity in the vicinity and the resulting rise in carbon emissions have threatened the surrounding forests' ecosystem. These expansive forests, once widespread across Ravi's surroundings, now exist in fragmented patches, requiring significant care and attention for preservation. Climate change has inflicted severe impacts on the river Ravi, altering its landscape to the extent that even culturally significant sites like “Kamran's Baradari” (twelve-door pavilion that a Mughal emperor constructed), originally designed to be situated in the river's center, now find themselves located on the riverbank, relegated to a corner due to the river's dwindling flow. The students listened intently, clearly concerned about the damage being done to a place they held dear.

Following the informative session, a Q&A session allowed them to understand their perspectives and what the Ravi River meant to them. This understanding became crucial as the teachers guided them to imagine and etch their vision for the future of Ravi, what they felt, and how they perceived the tragic truth about Ravi's current situation onto plexiglass sheets. Plexiglass sheet is a transparent, lightweight, and durable material made from acrylic. It is commonly used as a substitute for glass due to its clarity, shatter resistance, and versatility. Plexiglass sheets are utilized in various applications, including signage, protective barriers, displays, and artistic projects.

Figure 1. Artist Sarah Mumtaz Engaging Students in a Q&A Session about River Ravi

Source: https://www.facebook.com/media/set/?set=a.5454837377943538&type=3

Do it Yourself (DIY) Printmaking activity technique: Professional printmaking usually requires complex tools for the artwork, but the method used for this workshop was a simplified version that was easy for the students to understand. The “dry point” is a unique intaglio print technique focusing on small lines and delicate designs.

The artist scrapes into the plate by hand through an etching needle to produce rough patches (burrs) that hold the ink (Figure 2). When the printmaking ink is applied to the plexiglass, a rolling brush spreads it all on the sheet until the paint gets settled inside the incisions made by the dry point. A squeegee, a type of spongy material, is taken to help push the ink into the scratched areas. Since the team had only one squeegee, the team and the students also devised some DIY methods to spread the ink evenly, which involved using hard cards, SIM card boxes, etc. (Figure 3a, b). After spreading the ink and making sure the ink was settled in, newspapers were used to wipe off the excessive ink (Figure 4). After this, presenters took a piece of fine-grained textured paper dipped in water but in a wet-dry situation (Figure 5). The term "wet-dry" indicates that the paper should have the ideal moisture level—not overly wet nor completely dry—to achieve optimal results. Students put their paper on the plexiglass, took a clean rolling brush, and pressed it into the incisions they made. Sarah advised all the students to press it with full strength, and that is when the best and the most unpredictable part came out, which was the final painting. Figure 7 showcases students' artwork on the official website of the Thames Festival 2022.

Figure 2. Student Using a Pointy Needle for Etching

Source: https://www.facebook.com/media/set/?set=a.5454837377943538&type=3

Figure 3. Author 2 and Artist Guiding Students in Evenly Spreading Printing Ink with Hard Cards

Source: https://www.facebook.com/media/set/?set=a.5454609417966334&type=3

Figure 4. Artist Guiding Teachers and Students Through Hands-On Practice

Source: https://www.facebook.com/media/set/?set=a.5454609417966334&type=3

Figure 5. Students Using Newspaper to Remove Excess Ink and Incorporate it into Incisions

Source: https://www.facebook.com/media/set/?set=a.5454609417966334&type=3

Figure 6. The Wet-Dry Paper and Its Result

Source: https://www.facebook.com/media/set/?set=a.5454609417966334&type=3

Figure 7. Artwork Display at Thames Festival 2022

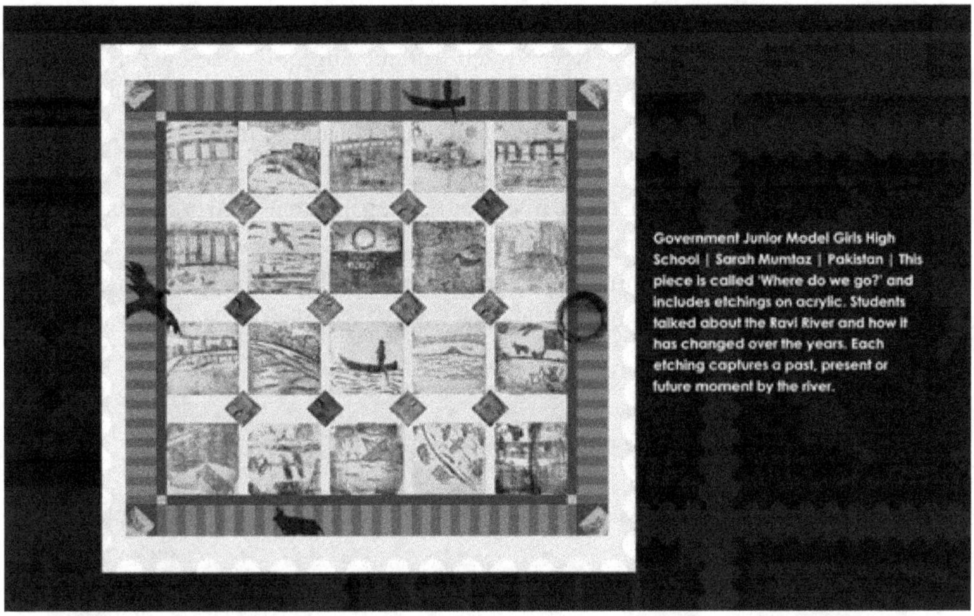

Source: https://thamesfestivaltrust.org/education-programme/river-of-hope/river-of-hope/roh22-d-h/

Reflection

This project served as a powerful example of how art education can be used to cultivate environmental awareness and empower young people. It successfully addressed the critical issue of River Ravi's decline through a two-pronged approach: fostering creative expression in students and equipping educators with effective teaching methods.

Beyond the artistic expression, the project yielded many benefits for all participants. Students learned firsthand about the critical situation faced by the Ravi River, one of Pakistan's natural treasures. They also delved into the fascinating world of printmaking, a traditional artistic technique. For the teachers, the project provided valuable insights into innovative and engaging pedagogical methods, specifically activity-based learning.

It was revealed that these schools typically focused on teaching drawing and painting as part of their academic curriculum. This was attributed to a shortage of skilled manpower, training opportunities, exposure to new ideas, awareness, and resources within the educational system. Consequently, printmaking emerged as a newly introduced art form that was easily manageable, cost-effective, and accessible for students and educators. The unpredictability of printmaking kept curiosity and excitement alive throughout the creative journey.

The culmination of this project was a vibrant exhibition showcasing diverse student perspectives on the Ravi River. Each student's unique vision was brought to life through their handcrafted prints. This exhibition, displayed during the Queen's Platinum Jubilee celebrations, was a powerful platform for environmental awareness (Figure 7).

This collaboration between the British Council, an international organization, and Olomopolo Media, a local entity, exemplifies the power of international partnerships. Such collaborations foster connections across borders, elevate global environmental concerns, and bring these issues to the forefront for collective consideration and solution-oriented action.

Figure 8. The Conclusive Outcome of the Printmaking Activity

Source: https://www.facebook.com/media/set/?set=a.545460941796633&type=3

Strengths of the Project

Engaging Format: Printmaking workshops provided a fun and engaging platform for students to learn about environmental issues. The children brought a variety of creative perspectives to the project. Some students depicted their current understanding of the Ravi River, while others aimed to portray a more hopeful vision. Still, others opted for symbolic expression, using black to represent the devastating impact of pollution on the river.

Personal Connection: By focusing on River Ravi, a local landmark with personal significance for the students, the project fostered a deeper emotional connection to the cause.

Dual Impact: The project not only educated students about the challenges faced by the river but also equipped teachers with new methods to integrate environmental education into their curriculum.

Empowering Students: Students became environmental activists by learning how art can address difficult topics.

Overall Impact

This project demonstrated the transformative potential of art education in environmental advocacy. By sparking creativity and fostering a sense of agency, it equipped young people with the knowledge, skills, and passion to become stewards of the environment.

CONCLUSION

This paper presented an environmental awareness endeavor through art in 3 public schools in Lahore-Pakistan by displaying the ability of creative efforts to teach environmental responsibility and awareness to everyone who can partake in the process of making a change in this society. Students, educators, and administrators developed intellectual, artistic, and social-emotional abilities while learning about the environmental problems affecting the Ravi River through Olomopolo Media's global "River of Hope" program. This initiative enhanced their understanding of local environmental issues and encouraged a sense of responsibility toward them.

Printmaking was used to draw attention to art's important role in advancing environmental concerns by inspiring action to restore the river or any other natural entity affected directly or indirectly by several factors. Steps were discussed through which students were introduced to an art form that wasn't known to them before. This was a powerful medium for expressing their concerns and hopes for the Ravi River. This hands-on approach allowed students to connect more deeply with the subject (the environment) and fostered a personal investment in environmental advocacy.

The final student-made printmaking art was exhibited during the Queen's Platinum Jubilee, highlighting the significance of international partnerships in addressing climate change. This event provided a platform for the student's creative expressions and underscored the global nature of environmental challenges and the need for collective action. The exhibition demonstrated how local efforts could resonate internationally, emphasizing the interconnectedness of communities worldwide.

Additionally, the project gave teachers innovative instructional strategies to integrate environmental education more successfully into their educational programs. By implementing creative techniques into conventional educational frameworks, the program fostered a comprehensive learning strategy that places equal emphasis on creativity and academic accomplishment. Thus, to ensure a sustainable future, this effort serves as a comprehensive approach to raising environmental awareness by fusing art, education, and global collaboration.

The success of this project opens doors for such initiatives through which students from diverse backgrounds and communities can get involved to create a more comprehensive understanding of the river's importance. Further, partnering with environmental organizations will provide students direct action and advocacy opportunities. Such initiatives also emphasize the crucial role of art as a medium of self-expression and incorporating them into the school curriculum or creating an ongoing art exchange program focused on environmental themes may be beneficial to generate awareness among all. Thus, by building upon the strengths of the "Rivers of Hope" project and scaling its impact, we can empower generations to protect our planet's vital ecosystems and create a more sustainable future.

REFERENCES

Ashraf, S., Ali, M., Shrestha, S., Hafeez, M. A., Moiz, A., & Sheikh, Z. A. (2022). Impacts of climate and land-use change on groundwater recharge in the semi-arid lower Ravi River basin, Pakistan. *Groundwater for Sustainable Development*, 17, 100743. 10.1016/j.gsd.2022.100743

Asia Water Wire. (2006, January 23). *Pakistan, India join hands to clean canal.* River Basin Initiative. Retrieved April 1, 2024, from https://web.archive.org/web/20110724140814/http://www.riverbasin.org/newsmaster.cfm?&menuid=45&action=view&retrieveid=152

Cervantes, B., Hemmer, L., & Kouzekanani, K. (2015). The impact of project-based learning on minority student achievement: Implications for school redesign. *Education Leadership Review of Doctoral Research*, 2(2), 50–66.

Dewey, J. (1976). *The middle works, 1899-1924* (Vol. 13). SIU press.

Ecosystems and Human Well-Being. Synthesis. (2005). In *Millennium Ecosystem Assessment*. Island Press, Washington, DC. Retrieved April 5, 2024, from https://www.millenniumassessment.org/documents/document.356.aspx.pdf

Flowers, A. A., Carroll, J. P., Green, G. T., & Larson, L. R. (2015). Using art to assess environmental education outcomes. *Environmental Education Research*, 21(6), 846–864. 10.1080/13504622.2014.959473

Gull, R., & Gul, R. (2017). In the surge of healing-the case of Indus basin.

IPCC. (2023). *Climate Change 2023: Synthesis Report* (Core Writing Team, H. Lee, & J. Romero, Eds.). https://www.ipcc.ch/report/ar6/syr/downloads/report/IPCC_AR6_SYR_LongerReport.pdf

Jónsdóttir, Á. B. (2017). *Artistic actions for sustainability: Potential of art in education for sustainability*. Lapland University Press.

Kumari, R., & Thakur, N. (2023). Role of youth in environmental preservation. *Role of Youth in Environmental Preservation, 10*(6).

Lai, O. (2022, May 11). *Pakistan's Ravi is most polluted river in the world from pharmaceutical pollution*. Earth.Org. Retrieved April 1, 2024, from https://earth.org/most-polluted-river-in-the-world-pharmaceutical-pollution/

Papavasileiou, V., Nikolaou, E., Andreadakis, N., Xanthacou, Y., & Kaila, M. (2021). The role of art in environmental education. *IJAEDU-International E-Journal of Advances in Education*, 6(18), 287–295.

Past climate. (2021, September 22). Climate.gov. Retrieved April 5, 2024, from https://www.climate.gov/maps-data/climate-data-primer/past-climate

Rivers of the World. (2021). *Rivers of the world*. British Council. Retrieved April 1, 2024, from https://england.britishcouncil.org/coventry2021/rivers-world

Shahid, M. J., Arslan, M., Siddique, M., Ali, S., Tahseen, R., & Afzal, M. (2019). Potentialities of floating wetlands for the treatment of polluted water of river Ravi, Pakistan. *Ecological Engineering*, 133, 167–176. 10.1016/j.ecoleng.2019.04.022

APPENDIX

List of Questions and Answers for Our Case

1) What is the overall problem presented in this case?

The overall problem presented in this case is the environmental degradation of the river Ravi in Lahore, Pakistan. This issue is compounded by a lack of awareness and engagement within the local community about the river's ecological problems and the broader importance of environmental sustainability. The initiative aims to address this problem by fostering environmental awareness and responsibility through an art-integrated educational program in public schools.

2) How does fostering environmental awareness in children contribute to a sustainable future?

Teaching children about environmental responsibility instills lifelong habits and values prioritizing sustainability. This early education helps build a more environmentally conscious society, ensuring future generations are equipped to address environmental challenges and promote ecological balance.

3) What factors affect the problem(s) related to this case?

The problems related to environmental degradation and awareness in this case are influenced by several factors. Pollution from industrial discharge, sewage, and waste dumping into the river Ravi harms aquatic life and degrades water quality. Limited community understanding of environmental issues, insufficient environmental education, and resource constraints hinder effective conservation efforts. Varying levels of engagement, rapid urbanization, weak environmental regulations, cultural attitudes, economic pressures, and climate change further compound these challenges. Addressing these factors through targeted interventions and comprehensive education is crucial for promoting environmental awareness and sustainable practices.

4) What are some of the emerging technologies that should be considered in solving the problem(s) related to the case?

Emerging technologies offer innovative solutions to address environmental challenges related to the river Ravi. These include IoT and sensor technology for real-time water quality monitoring, AI and machine learning for data analysis and prediction, and blockchain for transparent environmental data tracking. Other technologies like green infrastructure, advanced wastewater treatment, drones for aerial inspections, and bioremediation also contribute to sustainable management. Integrating renewable energy and citizen science through mobile apps further enhances monitoring and community engagement. These technologies collectively aim to improve water quality, mitigate pollution, and foster environmental sustainability along the river Ravi.

5) How did the initiative foster a sense of environmental responsibility in students, teachers, and administrators?

The initiative engaged participants in hands-on art projects focused on local environmental issues, fostering a deeper understanding and commitment to environmental stewardship. This collective involvement cultivated a shared responsibility for protecting and preserving the environment.

6) What key outcomes did the environmental awareness initiative achieve in Lahore's public schools?

The initiative raised awareness about river conservation, enhanced students' creative and academic skills, and fostered a sense of environmental responsibility among students, teachers, and administrators. It demonstrated the effectiveness of art-integrated education in promoting environmental awareness and action.

Epilogue

Looking back on the case of environmental awareness and river conservation along the river Ravi, the long-range effects are profound and encouraging. The initiatives raised local and global awareness about the river's environmental challenges and fostered a sense of community engagement and responsibility among students, teachers, and residents. Over time, sustained efforts in education, art integration, and community involvement have led to measurable improvements in water quality, ecosystem health, and environmental stewardship. The collaborative spirit and innovative use of emerging technologies continue to play crucial roles in ensuring the sustainability of the river Ravi for future generations.

List of Lessons Learned

1. **Integration of Art and Education:** Incorporating art into environmental education enhances engagement and creativity among students, fostering a deeper understanding and appreciation for environmental issues. This approach not only educates but also inspires action and advocacy within communities.
2. **Importance of Community Engagement:** Engaging local communities in environmental initiatives builds ownership and sustainability. Long-term environmental stewardship is fostered by involving residents in decision-making processes and conservation efforts, creating lasting positive impacts.
3. **Global Collaboration and Awareness:** Collaborative projects with international partners and participation in global events, like the Queen's Platinum Jubilee, amplify local efforts and raise global awareness about regional environmental issues. This global perspective fosters solidarity and shared responsibility in addressing global environmental challenges.
4. **Adaptability and Continuous Improvement:** Environmental conservation efforts require adaptability to changing conditions and continuous improvement in strategies and technologies. Learning from successes and challenges allows for the refinement and scalability of initiatives, ensuring sustainable outcomes over the long term.

List of Resources for Further Study

1) https://www.unesco.org/en/articles/advancing-sustainability-education-through-art-expression-and-culture
2) https://loksujag.com/story/pollution-destroyed-the-ravi-ecosystem-eng
3) https://thamesfestivaltrust.org/education-programme/river-of-hope/river-of-hope/roh22-d-h/
4) https://environment.ucdavis.edu/activism-through-art
5) https://www.neefusa.org/story/environmental-education/environmental-education-through-art
6) https://research.childrenandnature.org/research/arts-based-environmental-education-can-be-effective-in-promoting-nature-connectedness-and-pro-environmental-agency/
7) https://www.earthday.org/learning-through-art/
8) https://jerwoodvisualarts.org/art-education-and-methodologies-glossary/environmental-education-through-art/
9) https://www.learninginnature.com/art-based-environmental-education
10) https://files.eric.ed.gov/fulltext/EJ1104868.pdf

Chapter 5
Using the Integrative Learning Process to Enhance Competencies for Sustainable Development

Rungarun Rojrattanadamrong Chaisri
https://orcid.org/0000-0002-3926-6880
Srinakharinwirot University, Thailand

EXECUTIVE SUMMARY

This study aims to develop competencies for sustainable development and create corresponding indicators designed to address the current conditions, to study desirable conditions in accordance with the concept of the whole-school approach, to develop innovations and assess the effectiveness of the integrative learning process to enhance competencies for sustainable development. It also aims to propose education management strategies for sustainable development. The study's findings were developing competencies for sustainable development and indicators and dividing them into eight competencies. The study on the students' current conditions revealed average competencies ($x = 3.73$, $SD = 1.34$) regarding the desirable conditions of competencies. Students' overall wishes were at the highest level ($x = 4.68$, $SD = 0.67$). The integrative learning process consisted of four steps. The impoverished students had at least 70% of the competencies for sustainable development in each cycle. Recommendations on education management strategies for sustainable development consisted of four strategies.

DOI: 10.4018/979-8-3693-1710-5.ch005

1. INTRODUCTION

The General Assembly of the United Nations put a joint agenda in place to establish the Sustainable Development Goals (SDGs) for the next 10 years, until 2030, with 17 key goals, including Goal 4: Education for Sustainable Development "to ensure inclusive and equitable quality education and promote lifelong learning opportunities for all," in which the target 4.5 strives to eliminate disparities in education to all levels of education for the vulnerable" and the target 4.7 seeks to ensure that all learners acquire the knowledge and skills needed to promote sustainable development. Poverty is the most serious issue that must be addressed (United Nations, 2016; Office of Education Council, 2016). In formulating the Twelfth National Economic and Social Development Plan (2017–2021), the Office of the National Economic and Social Development Council (NESDB) adhered to the 20-year National Strategy Framework (2017–2036), the master plan for national development, the country's Sustainable Development Goals (SDGs), and the Thailand 4.0 Policy. The objectives of Strategy 1: Strengthening and Realizing the Potential of Human Capital are to prepare Thai people of all ages to acquire the skills needed for a quality life in the 21st Century world and apply a knowledge base and thinking system in a multidisciplinary manner by creating an integrative learning approach that places an emphasis on hands-on practice. Strategy 1 also aims to use research and innovation to drive education and provide educational opportunities to underserved communities. The impoverished, disabled, in particular, and destitute will have access to and the right to choose high-quality education services. This strategy begins with an understanding of the area's and community's true challenges (The 20-year national strategy [2018–2037]), 2018; Office of the National Economic and Social Development Board, Office of Prime Minister, 2016; National Strategy Secretariat Office, Office of the National Economic and Social Development Board, 2018; Office of the National Economic and Social Development Board, Office of Prime Minister, 2019). The National Education Plan 2017–2031 (Office of Education Council, 2017) has established five education development goals: access, equity, quality, efficiency, and relevancy.

According to an interview with Mrs. Wiratchani Suwan, director of the Amphasiriwong Temple Community School, Mitraparb 162 (Suwan, interviewee, 15 February 2021), a growing number of students in grades 1–6 who are now impoverished and destitute wished to sign up for the information system development project, the Office of the Equitable Education Fund, to ensure their learning opportunities.

Action Learning Process: 5Ts is the concept that focuses on "taking care of others as best as you can take care of yourself" through Think: reflect and revise, Touch: fieldwork encounters, Take Action, Tune-in: plan together, Talk: dialogue, mutual awareness via in-depth investigation, and a focus on resolving problems in learners' community and locality, inquiry learning based on the geographical context to develop action learning and sustainable behavior (O'Donoghue et al., 2020; WESSA, 2016; UNESCO, 2018b). The concept aligns with the Four-Noble-truth learning management method with its process: Ask a question, form hypotheses, conduct an experiment, and communicate a conclusion. It can be classified as Buddhist learning management owing to its emphasis on problem-solving along with practical approaches by using rational thinking processes to enable learners to solve problems intelligently on their own in current and future situations (Buasri, 1985; Khamkeaw, 2012; Kammanee, 2020; Sutthirat, 2016; Dhammakosachan, 2009; Kunaporn et al., 2013; Institute of Culture and Arts, 2010; Mulkam, 2002; Buasri, 2001; Sutthirat, 2009). The researcher synthesized the aforementioned concepts into the integrative learning process— raise a question, open inquiry, conduct an experiment, and communicate a conclusion—and holistically engaged in interdisciplinary activities. Competencies in terms of knowledge, skills, and attributes, while concentrating on the learning process from practice in solving various problems to enable learners to transfer learning, connect what they learn with the environment and actual conditions, and achieve competencies based on Education for Sustainable Development (ESD) (UNESCO, 2018a; Schreiber & Siege, 2016; O'Donoghue et al., 2020). To identify competencies for sustainable development and foster the ability of individuals to apply knowledge, skills, and attributes that are applicable, measurable, and evaluable, The researcher also synthesized data from the key competencies of learners from different countries and the necessary competencies (OECD, 2019; De Hann, 2010; Rieckmann, 2012; Wals, 2015; Wiek et al., 2015; UNESCO, 2018b). This process was in line with the global educational framework, which strives to prepare students for the labor market of the 21st century (Gardner, 2011; Brauer, 2010).

The whole-school approach is an interrelationship of the school's administrators, education personnel, teachers, students, parents, and communities. Only when all parties collaborate and methodically approach problems to achieve the goal of global citizenship education (GCED), along with learning by doing and a variety of assessments tailored to individual students and appropriate to each area, will the entire school development system be successful (Mills, 2003). The same curriculum shouldn't be implemented nationwide. Considering the backdrop of the SDGs, which are connected to the economy, society, and environment, Area-based education is one of the most effective approaches to enable community stakeholders to collaborate with the government. This is consistent with education for sustainable development (ESD), which promotes inclusivity and embraces diversity to foster world peace (Vanichsenee, 2015; Viphoouparakhot, 2021; Young & Commins, 2002; Oxfam Education, 2002; Education Improvement Commission, 2000; UNESCO, 2014; UNESCO, 2015a; UNESCO, 2015b; UNESCO, 2016; UNESCO, 2018b).

Guidelines for the whole-school approach may be broken down into four categories for sustainable school development: governance policy and capacity building, curriculum, teaching and learning, school facilities and operations, and community partnerships and relationships. These recommendations are combined with education for sustainable development to administer (UNESCO. 2018a) Southeast Asia teacher education management using participatory action research, a concept that blends action research with participatory research (Morales, 2016). This integrated method aims to benefit society, increase societal consciousness, understand the community, and prepare for conflict scenarios. The study is regarded as a dynamic effort at research for development and difficult issue solutions. It combines study and practice in keeping with the sufficiency economy philosophy and the context of Thai society (Jiawiwatkul, 2010; Office of the Education Council, 2009; Siribanpitak, 2010; Keahey, 2021).

Strategic planning played a crucial role in helping agencies keep pace with changes and plan in line with decentralization and government administration that require agencies at all levels to prepare strategic plans to be used as a tool for planning sustainable development in accordance with the long-range goals of the organization (Office of the National Primary Education Commission, 2000; Maneelek, 2002; Kesapat, 2000).

It is clear from the above that the learning of impoverished students varied from area to area and showed regional differences. One strategy to maximize the potential of human resources is the whole-school approach with a focus on the integrative learning process. With the collaboration of schools, communities, and areas, it is possible to attain the competencies for sustainable development required for the 21st century. To improve collaboration, stakeholders should foster a positive relationship between the school and the community. This relationship requires building trust, confidence, and social synergies. The best practices developed may be shared as a model for communities, schools, and other settings to address more significant policy issues by presenting education management strategies for sustainable development to adapt effectively to changes in Thai society and global society.

2. RESEARCH OBJECTIVES

1. To develop competencies for sustainable development and indicators
2. To study the current and desirable conditions of competencies for sustainable development based on the whole-school approach with impoverished students
3. To develop innovations using the integrative learning process to enhance sustainable development.

4. To assess the effectiveness of an integrative learning process to enhance competencies for Sustainable development based on the whole-school approach for impoverished students.
5. To suggest education management strategies for sustainable development.

3. THE LITERATURE REVIEW

According to the United Nations' Sustainable Development Education Framework Target 4.7, individuals should develop innovation competencies and focus on learning to meet the requirements of SDGs 4.7 in which they must have core competencies related to sustainable development in addition to general competencies as learning compass (The Learning Compass 2030) (Giangrande et al., 2019; OECD, 2019). The "Gestaltungskompetenz" model for sustainable development education presented core competencies, including future-oriented thinking, participation competencies, empathy, and compassion (De Haan, 2006). Many researchers have also advocated other competencies, such as interdisciplinary, cross-cultural understanding ability, empathy, compassion, solidarity, and self-motivation (Barth et al., 2007). Systemic, anticipatory, normative, strategic, and interpersonal competencies (Reid et al., 2011) as well as self-reflection, self-relevance, and mental wellness. Learners can reflect on their own development of these competencies (Wiek et al., 2011). Synergy and leadership and enlightened citizenship with a global awareness (Office of the Education Council, 2018; Young, 2018).

All areas of the most desirable core and sub-competencies comply with Rychen & Salganik's (2003) concept that outlined key competencies for a successful life and a better society that it is necessary to involve different sectors including problem solvers and change makers. The mechanisms required to conduct sustainable development studies with a focus on sustainability competencies necessitate a mixed research design that combines questionnaires and observation forms to answer research questions with a focus on sustainability competencies (Faham et al., 2017).

The development of sustainability competencies will require the use of innovation to develop learning. Colvin et al. (2014) also established a systematic innovation for sustainable development based on social learning and gained collaborative investment from researchers, stakeholders, and policymakers so that their design can be used as a framework for regulating and promoting systematic innovation for sustainable development. In addition, Clark et al. (2016) investigated ways to produce knowledge for sustainable development and concluded that knowledge, practice, innovation, and research community capacity building should be linked.

Effective instruction and learning necessitate collaboration from the stakeholders involved. Moreover, Vare et al. (2019) stated that putting competency development into action requires effective teaching. It is considered the most important for learners to understand the core values that drive and motivate the mind. Reed et al. (2006) discussed the adaptive learning approach for developing and applying sustainability indicators with local communities. The procedure may be viable if local indicators are developed and used to establish methodological frameworks and collect best practices.

Therefore, educators must collaborate, advocate for education that aligns with global trends, and formulate feasible policies. Gallwey (2016) discussed the challenges of reforming education toward SDG 4.7 goals. The necessities and the outcome must be balanced to create a practical policy. Additionally, Giangrande et al. (2019) urged education policymakers to be accountable for the successful implementation of the SDGs through a framework for curriculum planning and assessment so that future researchers and educators interested in Education for Sustainable Development (ESD) can investigate these problems and create more effective solutions. These concepts are line with those of Reed et al. (2006), who studied the paradigm of top-down experts and the use of community bases as bottom-up experts to address the process of adaptive learning to develop and deploy sustainability indicators with local communities.

4. RESEARCH METHODS

The researcher outlined the research in 4 phases:

4.1 Phase 1: Developing Competencies for Sustainable Development and Studying Current and Desirable Conditions

This phase involved developing performance indicators for sustainable development by inviting 8 experts in curriculum and teaching, assessment, evaluation, and educational research. According to the evaluation results, the index of consistency was 1.00 and relevant.

Current and desirable conditions concerning competencies for sustainable development according to the concept of the whole-school approach. According to 8 experts in curriculum and teaching, assessment, evaluation, and educational research, the consistency index of 1.00 is relevant. The informants were 68 impoverished students, 25 parents, and 14 educational staff. A total of 107 informants were chosen by purposive sampling.

4.2 Phase 2: Innovation Development and Integrative Learning Process

Develop Board games by involving 8 curricula and teaching, Thai, Chinese, English, elementary education, educational technology, children's literature, assessment, and evaluation. The overall mean for consistency, relevance, and appropriateness was ($\bar{x}$) 4.90, with a standard deviation (SD) of 0.21, indicating the highest level of applicability. The integrative learning management plan had the highest level of relevance with a total mean of ($\bar{x}$) 4.89 and an SD of 0.23.

Develop 17 fairy tales by involving 3 curriculum and teaching professionals. The overall mean ($\bar{x}$) ranged from 4.45 to 4.76, and the SD ranged from 0.32 to 0.72. The integrative learning management plan's total mean ($\bar{x}$) was between 4.14 and 4.58 with an SD between 0.24 and 0.58, indicating a high to very high degree of appropriateness.

An integrative learning process was designed with the assistance of 8 experts in curriculum and teaching, assessment, evaluation, and educational research with a total mean ($\bar{x}$) of 4.71 and an SD of 0.46, demonstrating the highest level of appropriateness.

4.3 Phase 3: Evaluation of the Effectiveness of the Integrative Learning Process to Enhance Competencies for Sustainable Development Based on the Whole-School Approach of Impoverished Students

The researcher used the tool to try out with impoverished students who were not part of the sample group. They were students from the Ban Khlong Yai School Khlong Yai District, Nakhon Nayok Province, Thailand in grades 1–6. The researcher then applied the research tools to the next target group.

The researcher implemented the integrative learning process that enhances competencies for sustainable development based on the concept of the whole-school approach to the target group that included students in grades 1–6 from the Ampha Siriwong Temple Community School, Mittraparb No. 162, Ongkharak District, Nakhon Nayok Province, Thailand. The researcher used purposive sampling to pick 44 students at the school based on their distinctiveness and geographical diversity.

Study variables

The independent variable is an integrative learning process.

Dependent variables are competencies for the sustainable development of impoverished students.

4.4 Phase 4: Development of Education Management Strategies for Sustainable Development

This phase included 8 experts in curriculum and teaching, assessment, evaluation, educational research educational equality, area-based studies, education administration, lifelong education, and sustainable development. The strategies had the highest level of applicability in terms of consistency, relevance, and appropriateness of the strategy with a total mean ($\bar{x}$) of 4.48 and an SD of 0.52.

The informants were 20 educational staff and government agencies by purposive sampling.

5. THE RESULTS

5.1 Competencies for Sustainable Development and Indicators

The researcher synthesized competencies for sustainable development and indicators from relevant documents and research from key competencies from different countries and OECD transformative competencies (OEDC, 2019), ESD Competencies (De Hann, 2010; Rieckmann, 2012), sustainability competencies (Wals, 2015; Wiek et al., 2015; UNESCO, 2018b), and life and career (Brolin, 1992) competencies to determine the competencies framework for sustainable development. The researcher then developed competence indicators for sustainable development and analyzed data, resulting in 30 indicators in 8 sustainable development competencies as shown in Table 1.

Table 1. Competencies for Sustainable Development and Indicators

Competencies	Indicators
1. Interdisciplinary competency	1. Ability to combine knowledge of related subjects
	2. Ability to integrate knowledge of diverse subjects into daily life to become a sustainable global citizen
	3. Ability to use information from a variety of areas to shape attitude and appreciation
2. Communication competency for sustainable peace	1. Use language for communication
	2. Verbal and nonverbal communication
	3. Use language to foster peace and tolerance for diversity in a multicultural society

continued on following page

Table 1. Continued

Competencies	Indicators
3. Higher-order thinking competency	1. Analytical and synthetical thinking
	2. Critical thinking
	3. Problem-solving thinking
	4. Creative thinking
	5. System thinking
	6. Anticipatory thinking
4. Collaborative competency	1. Ability to collaborate with others
	2. Share, exchange knowledge, organize ideas, and assistance in a collaborative manner
	3. Have self-respect and respect for others
	4. Be a good leader and follower to achieve goals.
5. Digital literacy competency	1. Access, comprehend, use information, use digital and media technologies
	2. Information, technology, and digital literacy
	3. Use digital and information media technology to develop oneself and society
6. Creative innovation competency	1. Ability to design, develop innovations
	2. Improve and enhance to achieve something fresh and new
	3. Recognize the advantages and value of innovation
7. Self and social awareness competency	1. Recognize the importance of controlling one's own emotions and having empathy for others
	2. Responsibility to oneself and society
	3. Respond to varied situations in a suitable manner
8. Life and career competency	1. Learn and live properly
	2. Adapt and recover to equilibrium from various situations appropriately (resilience)
	3. Protect oneself and others against a variety of risks in life
	4. Choose a career according to one's aptitude and interest
	5. Live by the sufficiency economy philosophy

The competencies for sustainable development and indicators listed in Table 1 provide a competency framework used to evaluate behavior to represent students' performance. To build a behavioral observation model for further studies, the researcher employed the aforementioned standards as a tool.

5.2 The Current and Desirable Conditions of Competencies for Sustainable Development Based on the Whole-School Approach With Impoverished Students

The findings of the analysis of the current and desirable conditions for competencies for sustainable development are based on the whole-school approach of impoverished students from the interviews with impoverished students, education staff, and parents can be divided into two parts:

1. The current and desirable conditions of competencies for sustainable development based on the whole-school approach of impoverished students from interviews with impoverished students.

Table 2. The Current and Desirable Conditions of Competencies for Sustainable Development Based on the Whole-School Approach of Impoverished Students from interviews with impoverished students

Competencies for Sustainable Development	Current Condition		Analysis	Position	Desirable Condition		Analysis	Position	PNI modified
	$\bar{x}$	SD			$\bar{x}$	SD			
1. Apply knowledge of many disciplines in everyday life	3.81	1.26	Moderate	4	4.74	0.56	Highest	5	0.18
2. Willing to learn and practice	4.04	1.27	High	2	4.75	0.58	Highest	3	0.18
3.Communication etiquette	3.74	1.28	Moderate	6	4.63	0.69	Highest	6	0.24
4. Experiment with alternative ways of thinking	3.46	1.39	Moderate	8	4.74	0.64	Highest	4	0.37
5. Enjoy collaborating with others	3.69	1.44	Moderate	7	4.53	0.89	Highest	10	0.23
6. Understand how to use technology	3.38	1.53	Low	10	4.56	0.80	Highest	9	0.35
7. Enjoy creating and designing new items	3.82	1.28	Moderate	3	4.62	0.77	Highest	8	0.21

continued on following page

Table 2. Continued

Competencies for Sustainable Development	Current Condition		Analysis	Position	Desirable Condition		Analysis	Position	PNI modified
	$\bar{x}$	SD			$\bar{x}$	SD			
8. Take responsibility for oneself and others.	3.74	1.31	Moderate	5	4.76	0.52	Highest	2	0.27
9. Adaptability to the environment	3.44	1.34	Low	9	4.63	0.69	Highest	7	0.35
10. Choose a desired profession	4.18	1.28	High	1	4.82	0.54	Highest	1	0.15
Total Average	3.73	1.34	Moderate	-	4.68	0.67	Highest	-	0.25

According to the study of the current condition, students' overall competencies for sustainable development were at a moderate level ($\bar{x}$ = 3.73, SD = 1.34). The top three areas which students assessed themselves the highest were as follows: choose a desired profession ($\bar{x}$ = 4.18, SD = 1.28) willing to learn and practice ($\bar{x}$ = 4.04, SD = 1.27), enjoy creating and designing new items ($\bar{x}$ = 3.82, SD = 1.28) Regarding the desirable conditions of competencies for sustainable development that students wanted to acquire in the future, Students' overall wishes were at the highest level ($\bar{x}$ = 4.68, SD = 0.67). Ability to choose a desired profession was the most desired competency ($\bar{x}$ = 4.82, SD = 0.54), followed by to take responsibility for oneself and others ($\bar{x}$ = 4.76, SD = 0.52) and willing to learn and practice ($\bar{x}$ = 4.75, SD = 0.58). Students rated the other areas as follows: experiment with alternative ways of thinking ($\bar{x}$ = 4.74, SD = 0.64), apply knowledge of many disciplines in everyday life ($\bar{x}$ = 4.74, SD = 0.56), understand how to use technology ($\bar{x}$ = 4.56, SD = 0.80), adaptability to the environment ($\bar{x}$ = 4.63, SD = 0.69), enjoy creating and designing new item ($\bar{x}$ = 4.62, SD = 0.77), understand how to use technology ($\bar{x}$ = 4.56, SD = 0.80), and enjoy collaborating with others ($\bar{x}$ = 4.53, SD = 0.89), respectively.

2. The current and desirable conditions of competencies for sustainable development based on the whole-school approach of impoverished students from interviews with education personnel and parents.

Table 3. The Current and Desirable Conditions of Competencies for Sustainable Development Based on the Whole-School Approach of Impoverished Students from interviews with education personnel and parents

Sustainability Development Competency	Current Condition		Analysis	Position	Desirable Condition		Analysis	Position	PNI modified
	$\bar{x}$	SD			$\bar{x}$	SD			
1. Interdisciplinary Competency									
1. Combine knowledge of related subjects	3.69	0.77	Moderate	3	4.51	0.60	Highest	4	0.22
2. Integrate knowledge of diverse subjects into daily life	3.79	0.80	Moderate	1	4.72	0.46	Highest	1	0.25
3. Use knowledge to become a global citizen	3.64	0.71	Moderate	4	4.56	0.64	Highest	3	0.25
1.4 Use information from a variety of areas to shape attitude and appreciation	3.72	0.72	Moderate	2	4.62	0.54	Highest	2	0.24
Total	**3.72**	**0.74**	**Moderate**	**6**	**4.63**	**0.55**	**Highest**	**6**	**0.25**
2. Communication Competency for Sustainable Peace									
1. Use language for communication	3.77	0.78	Moderate	1	4.74	0.44	Highest	2	0.26
2. Verbal and non-verbal communication	3.72	0.76	Moderate	2	4.62	0.49	Highest	4	0.24
2.3 Language skills to alleviate conflicts and foster peaceful coexistence	3.56	0.82	Moderate	3	4.77	0.43	Highest	1	0.34
2.4 Consider societal and cultural differences in choosing the language for communication	3.56	0.75	Moderate	4	4.67	0.48	Highest	3	0.31
Total	**3.65**	**0.78**	**Moderate**	**7**	**4.70**	**0.46**	**Highest**	**2**	**0.29**
3. Higher-Order Thinking Competency									

continued on following page

Table 3. Continued

Sustainability Development Competency	Current Condition		Analysis	Position	Desirable Condition		Analysis	Position	PNI modified
	$\bar{x}$	SD			$\bar{x}$	SD			
1. Analytical and Synthetical Thinking	3.64	0.81	Moderate	1	4.72	0.46	Highest	3	0.30
2. Critical Thinking	3.49	0.85	Moderate	4	4.82	0.39	Highest	1	0.38
3. Problem - Solving Thinking	3.58	0.83	Moderate	2	4.69	0.52	Highest	4	0.31
4. Creative Thinking	3.54	0.85	Moderate	3	4.72	0.51	Highest	2	0.33
5. System Thinking	3.33	0.96	Low	6	4.59	0.64	Highest	6	0.38
6. Anticipatory Thinking	3.46	0.76	Low	5	4.64	0.67	Highest	5	0.34
Total	**3.51**	**0.84**	**Moderate**	**8**	**4.70**	**0.53**	**Highest**	**1**	**0.34**
4. Collaborative Competency									
1. Ability to collaborate with others	3.90	0.85	Moderate	4	4.74	0.44	Highest	1	0.22
2. Share, exchange knowledge	4.08	0.66	High	1	4.72	0.46	Highest	3	0.16
3. Brainstorm and promote collaboratively	3.95	0.72	Moderate	5	4.72	0.51	Highest	2	0.19
4. Have self-respect and respect for others	4.00	0.83	High	2	4.62	0.54	Highest	4	0.16
5. Be a good leader and follower to achieve goals.	4.00	0.69	High	3	4.59	0.55	Highest	5	0.15
Total	**3.99**	**0.75**	**Moderate**	**1**	**4.68**	**0.50**	**Highest**	**3**	**0.17**
5. Digital Literacy Competency									
5.1Comprehend information, digital and media technologies	3.95	0.69	Moderate	1	4.59	0.55	Highest	4	0.16

continued on following page

Table 3. Continued

Sustainability Development Competency	Current Condition		Analysis	Position	Desirable Condition		Analysis	Position	PNI modified
	$\bar{x}$	SD			$\bar{x}$	SD			
5.2 Access IT and digital technology	3.87	0.80	Moderate	4	4.69	0.47	Highest	3	0.21
5.3 IT and digital literacy	3.90	0.64	Moderate	3	4.67	0.48	Highest	2	0.20
5.4 Use digital and information media technology to develop oneself and society	3.92	0.66	Moderate	2	4.74	0.44	Highest	1	0.21
Total	**3.91**	**0.70**	**Moderate**	**2**	**4.67**	**0.49**	**Highest**	**4**	**0.20**
6.Creative Innovation Competency									
6.1 Ability to design, develop innovations	3.72	0.79	Moderate	3	4.62	0.54	Highest	3	0.24
6.2 Improve and enhance to achieve something fresh and new.	3.77	0.74	Moderate	2	4.69	0.47	Highest	2	0.24
6.3 Recognize the advantages and value of innovation.	3.90	0.68	Moderate	1	4.54	0.55	Highest	1	0.16
Total	**3.80**	**0.74**	**Moderate**	**5**	**4.62**	**0.52**	**Highest**	**8**	**0.22**
7. Self and Social awareness Competency									
7.1. Control one's own emotions and have empathy for others.	3.92	0.70	Moderate	1	4.72	0.51	Highest	1	0.20
7.2. Responsibility to oneself and society	3.85	0.74	Moderate	2	4.69	0.52	Highest	2	0.22
7.3. Respond to varied situations	3.82	0.76	Moderate	3	4.58	0.60	Highest	3	0.20
Total	**3.86**	**0.73**	**Moderate**	**4**	**4.66**	**0.54**	**Highest**	**5**	**0.21**
8. Life and Career Competency									
8.1 Learn and live properly	3.97	0.63	Moderate	1	4.67	0.53	Highest	1	0.18

continued on following page

Table 3. Continued

Sustainability Development Competency	Current Condition		Analysis	Position	Desirable Condition		Analysis	Position	PNI modified
	$\bar{x}$	SD			$\bar{x}$	SD			
8.2 Resilience	3.87	0.66	Moderate	3	4.54	0.55	Highest	5	0.17
8.3 Self-protection against risks	3.79	0.66	Moderate	5	4.64	0.49	Highest	3	0.22
8.4 Choose a career according to one's aptitude and interest.	3.90	0.72	Moderate	2	4.62	0.49	Highest	4	0.18
8.5 Live by the Sufficiency Economy Philosophy	3.85	0.71	Moderate	4	4.67	0.48	Highest	2	0.21
Total	**3.88**	**0.68**	**Moderate**	**3**	**4.63**	**0.51**	**Highest**	**7**	**0.19**
Total average	**3.78**	**0.75**	**Moderate**	**-**	**4.66**	**0.51**	**Highest**	**-**	**0.23**

According to the interview with education personnel and parents on the current conditions of competencies for sustainable development based on the whole-school approach of impoverished students, most students performed moderately ($\bar{x}$ = 3.78, SD = 0.75). From high to low, students demonstrated collaborative competency ($\bar{x}$ = 3.99, SD = 0.75), digital literacy competency ($\bar{x}$ = 3.91, SD = 0.70), life and career competency ($\bar{x}$ = 3.88, SD = 0.68). Self- and social awareness competency ($\bar{x}$ = 3.86, SD = 0.73), innovation competency ($\bar{x}$ = 3.80, SD = 0.74), interdisciplinary competency ($\bar{x}$ = 3.72, SD = 0.74), communication competency for sustainable peace ($\bar{x}$ = 3.65, SD = 0.78) higher-order thinking competency ($\bar{x}$ = 3.51, SD = 0.84). Students performed on average in all areas of key competencies. The desirable conditions of competencies for sustainable development were rated the highest ($\bar{x}$ = 4.66, SD = 0.51, PNI modified = 0.23). From high to low, the desirable competencies included higher-order thinking competency ($\bar{x}$ = 4.70, SD = 0.53, PNI modified = 0.34), communication competency for sustainable peace ($\bar{x}$ = 4.70, SD = 0.46, PNI modified = 0.29), collaborative competency ($\bar{x}$ = 4.68, SD = 0.50, PNI modified = 0.17), digital Literacy Competency, self-and social-awareness competency ($\bar{x}$ = 4.66, SD = 0.54, PNI modified = 0.21), interdisciplinary competency ($\bar{x}$ = 4.63, SD = 0.55, PNI modified = 0.25), life and career competency ($\bar{x}$ = 4.63, SD = 0.51, PNI modified = 0.19) and innovation competency ($\bar{x}$ = 4.62, SD = 0.52, PNI modified = 0.22).

5.3 Innovation Development Using the Integrative Learning Process to Enhance Competencies for Sustainable Development

To produce board games and stories, Themes have been established with the intention of promoting the 17 SDGs. Experts were also brought in experts to evaluate the consistency, relevance, and appropriateness. The ratings were favorable, extremely favorable, and applicable.

The Integrative Learning Process was synthesized into 4 steps.

1. Raise a question: This step involves posing a question and solving problems through reflection and consideration by integrating interdisciplinary knowledge through simple questions about real-life conditions and everyday scenarios.
2. Open inquiry: This step involves investigating a problem to find the root cause of the problem and sharing opinions from real experiences about the problem.
3. Conduct an experiment: This step involves experimenting and collecting data from actual operations.
4. Communicate a conclusion: This step involves conversing, communicating, planning collaboratively, and analyzing data from experience, knowledge, and evaluations for problem solutions.

Innovations development is the integration of interdisciplinary knowledge subjected to the Basic Education Core Curriculum, 2011, which is related to one another in a holistic manner. Innovations also promote all 17 sustainable development goals. Goal 4 (Quality Education) is the major goal that enables learners to connect stories with real-life conditions, acquire practical knowledge, skills, and attributes in daily life to become sustainable global citizens.

5.4 The Effectiveness of the Integrative Learning Process to Enhance Competencies for Sustainable Development Based on the Whole-School Approach of Impoverished Students

The effectiveness of the integrative learning process to enhance competencies for sustainable development based on the whole-school approach of impoverished students was assessed. Participatory action research was employed in the experiment (Crane & Richardson, 2000) with the target group of impoverished students in grades 1-6, Ampha Siriwong Temple Community School, Mittraphap 162, Ongkharak District, Nakhon Nayok Province, to create innovation and carry out learning management: 1. Plan 2. Act 3. Observe 4. Reflection. After four weeks, the 10 study cycles were repeated, and the results were as follows:

1. Plan: learners discuss the difficulties of designing integrated activities, the integrative learning process, innovations including media, activity duration, and evaluation criteria to plan and develop.
2. Act: the plan implementation and the experiment with students. Activities were carried out to tackle learning problems and promote competencies for sustainable development using the integrative learning process, innovative media, and learning resources. Each cycle included four steps of the integrative learning process: Raise a question, Open Inquiry, Conduct an experiment, and Communicate a conclusion. Learners were encouraged to practice improving competencies and were given opportunities to design innovations that matched their aptitudes and interests. Examples of works of students in grades 1-6 are shown in the following figure.

Figure 1. Artworks by Students in Grades 1-6

3. Observe: Learners' behaviors were observed and documented. The outcomes of activities were tracked while teachers gathered data on problem-solving and learning outcomes. Each cycle featured records of problem obstacles, solutions, suggestions at the end of the learning management plan, and guidelines for

completing each cycle, such as splitting the number of students into smaller groups for better learning activity implementation.

4. Reflect: the practice of reflecting on ideas produced from the value of outcomes to increase competencies, establish an integrative learning process, and improve innovation quality. With the implementation and expansion of the outcomes, everyone would be allowed to play a role as a member of the learning community based on the whole school approach within the operation cycle.

The effectiveness of the integrative learning process for competencies for sustainable development based on the whole-school approach of impoverished Students. After four weeks, competencies for sustainable development Cycle 1-10 of impoverished students were repeated, and the outcomes were shown in the following table:

Table 4. Results for Competencies for Sustainable Development Cycle 1-10

Operation Cycle	Percentage of Students with Overall Competencies for Sustainable Development			
	High	**Moderate**	**Low**	**Not pass**
1	20	57.78	22.22	0
2	55.55	37.78	6.67	0
3	77.78	22.22	0	0
4	93.33	6.67	0	0
5	100	0	0	0
6	100	0	0	0
7	100	0	0	0
8	100	0	0	0
9	100	0	0	0
10	100	0	0	0

According to the table above, it was found that all impoverished students had 70% or more of the competencies for sustainable development in each cycle. The overall competencies can be rated as high, moderate, and low. For Cycle 1, 20% of students demonstrated a high level of overall competencies for sustainable development, 57.78% had a moderate level of competencies, and 22.22% possessed a low level of competencies. For Cycle 2, 55.55% of the students had a high level of the overall competencies for sustainable development, 37.78% demonstrated a moderate level of competencies, and 6.67% showed a low level of competencies. For Cycle 3, 77.78% of students had a high level of overall competencies for sustainable development, and 22.22% demonstrated a moderate level. For Cycle 4, 93.33% of students had a high level of competencies for sustainable development, whereas 6.67

percent had a moderate level of competencies. In Cycles 5–10, 100% of students demonstrated a high level of competencies for sustainable development. Learners demonstrated applicable and measurable competencies for sustainable development in knowledge, skills, and attributes.

5.5 Education Management Strategies for Sustainable Development

This research presents the analysis results and proposes education management strategies for sustainable development using Delphi techniques. (Giannarou & Zervas, 2014; Fink-Hafner et al., 2019). 20 teachers, educational staff, and government entities responded to open-ended and closed-ended questionnaires on a rating scale.

In the open-ended questionnaires, experts provided information on education management strategies for sustainable development in terms of school management and capacity building, curriculum, teaching and learning development, coordinate collaboration and community relationships, supporting facilities and operations, and education management implementation, and a form of education management planning that is decentralized to encourage participation of all stakeholders. In the second round, the researcher used experts' responses from the questionnaire on education management strategies for sustainable development from the first round to produce a closed-ended questionnaire with a rating scale.

An analysis of experts' recommendations for education management strategies for sustainable development. According to the results of the Delphi questionnaire, education management strategies for sustainable development include four strategies:

1. School management and capacity building
2. Curriculum, teaching and learning development
3. Coordinate collaboration and community relationships
4. Support facilities and operations

The median and interquartile range as shown in Table 5.

Table 5. Median and Interquartile Ranges on Educational Management Strategies for Sustainable Development

Recommendations	**Median**	**Interquartile Range**
Strategy 1: School management and capacity building		

continued on following page

Table 5. Continued

Recommendations	Median	Interquartile Range
1. Manage educational institutions to foster sustainable development in response to global changes.	5	1.0
2. Manage educational institutions by connecting societal, economic, and environmental relationships for sustainable development.	5	1.0
3. Promote the sufficiency economy philosophy and cultivate awareness of the environment in daily life.	5	0.25
4. Define action plans and prepare action plans, projects, and innovation activities to promote sustainable development with participation from all sectors.	5	1.0
5. Supervise, follow up, evaluate, and improve the management of education in the area together with stakeholders.	4	1.0
6. Create a sustainable economy, society, and environment to prepare people for lifelong learning and to become global citizens.	5	1.0
7. Promote equality and equality in education to improve the quality of educational institutions with special needs and that require immediate support.	5	1.0
8. Plan budget allocation in accordance with the administration of policies and plans required for long-term development.	4	1.0
Strategy 2 Curriculum, teaching and learning development		
9. Create curriculum, teaching, and learning that promote long-term growth in schools and communities.	5	1.0
10. Develop curriculum for educational institutions in accordance with the cultural context and environment of the school and community.	5	1.0
11. Manage learning in accordance with the school curriculum to achieve sustainable development.	4	1.0
12. Manage learning by emphasizing the integration of information, emotional, and social abilities, and effectiveness in learners. Developing competencies for sustainable development that may be used in everyday life.	5	1.0
13. Encourage the organization of the integrative learning process, which includes measuring and evaluating learning outcomes in accordance with the curriculum.	5	1.0
14. Integrate multidisciplinary knowledge to organize holistic learning.	4.5	1.0
15. Encourage knowledge application until attitudes and appreciation improve.	4.5	1.0
16. Gain practical skills that enable sustainable environmental quality enhancement.	5	1.0
17. Promote peace communication skills to reduce conflicts.	5	1.0
18. Improve diverse thinking processes by logical problem-solving.	5	1.0

continued on following page

Table 5. Continued

Recommendations	Median	Interquartile Range
19. Promote effective leadership and follower cooperation to produce practical guidelines for accomplishing education goals for sustainable development.	5	1.0
20. Information and digital technology literacy	5	0.25
21. Education to create environmentally friendly innovation	5	1.0
22. increased awareness of oneself and society. In a multicultural society, coexistence is respected.	5	1.0
23. Promote career path based on aptitudes and interests.	5	0.25
24. Learn to adapt, protect oneself, and resilience from various forms of risk.	5	1.0
25. Promote action research interwoven with education for sustainable development.	5	1.0
Strategy 3: Coordinate collaboration and community relationships		
26. To support sustainable development, establish educational networks among personnel, schools, homes, communities, and relevant agencies.	5	1.0
27. Encourage all sectors to participate in area-based education management for sustainable growth.	5	1.0
28. Create a research network to strengthen network collaboration.	5	1.0
Strategy 4 Supporting facilities and operations		
29. Budget support and environmental allocation Learning resources and materials to improve one's quality of life.	5	1.0
30. Manage natural resources in collaboration with all sectors for long-term development.	5	1.0
31. Allocate a variety of educational learning sources both within and outside the system to develop realistic guidelines that are consistent with cultural and environmental circumstances.	5	1.0

According to the analysis of experts on education management strategies for sustainable development, the median and interquartile range were consensus with the majority, strategies, and indicators with an interquartile range of 0.25–1.0, which did not exceed 1.50 as defined. The researcher then verified that all strategies and indicators had the distribution coefficient larger than 0, but not more than 0.5. As a result, the Delphi research process was terminated in the second round, with the median value being 4–5 higher than the stipulated requirements, showing that such indications were consensus, and expert opinions are compatible with the median value of 1 item at level 4. In other words, the expert group agreed that establishing education management strategies for sustainable development in the components required strategies and indicators in governance, monitoring, evaluation, and improvement of area-based education management in collaboration with stakeholders. Concerning the remaining 31 items, the expert group unanimously agreed that the strategies and indicators were suitable for developing education management strategies for sustainable development in that component.

6. DISCUSSION

6.1 Competencies for Sustainable Development and Indicators

The eight competencies for sustainable development include interdisciplinary competency, communication competency for sustainable peach, higher-order thinking (HOT) competency, collaborative competency, digital literacy competency, creative innovation competency, self and social awareness competency, and life and career competency. Each competency has indicators to assess behaviors, especially those related to education for sustainable development. Competencies are an individual's foundation in terms of knowledge, skills, and attributes that motivate people to work toward the desired results, preparing them for changes in the world in the 21st century and moving them toward being a "sustainable citizen" (Hay Group, 2003; UNESCO, 2010; Ministry of Education, Singapore. 2010; Tangprapha, 2013; Schreiber & Siege, 2016; O'Donoghue et al., 2020). The criteria should be in line with competencies with a focus on improving teaching and learning, reflecting democratic values, and involving stakeholders (Mogensen & Schnack, 2010). Furthermore, educational indicators for sustainable development should be developed with quality criteria and standards that lead to a practical instrument that can be tested and evaluated. Education indicators and competencies for sustainable development are critical for achieving the SDGs (Rode & Michelsen, 2008) so that assessment and monitoring indicators can be used in the research process by educational policymakers, curriculum developers, teachers, school leaders, and researchers (Radovanović & Noll, 2020). According to the United Nations' Sustainable Development Education Framework Target 4.7, individuals should develop innovation competencies and focus on learning to meet the requirements of SDGs 4.7 in which they must have core competencies related to sustainable development in addition to general competencies as a learning compass. (The Learning Compass 2030) (Giangrande et al., 2019; OECD, 2019).

6.2 The Current and Desirable Conditions of Competencies for Sustainable Development Based on the Whole-School Approach With Impoverished Students

The study of the current condition of the students revealed that their overall competencies for sustainable development were moderate. From high to low, competencies for sustainable development demonstrated in learners included choosing a desired profession, being willing to learn and practice, and enjoying creating and designing new items. However, the learners wanted to develop their own skills in the future. The overall wishes, as the report suggested, were at the highest level. Choosing a career that best matches their own self, responsibility for oneself and others, eagerness to learn and practice, practicing thinking in various situations, applying knowledge in various matters in everyday life, knowing how to use technology, adaptability to the environment, responsibility for yourself and others, understanding how to use technology, and collaboration with others were among the desirable competencies, ranging from high to low. Most students' competencies were rated as moderate by education personnel and parents. On a scale from high to low, the core competencies that students demonstrated were collaborative competency, digital literacy Competency, life and career competency, Self and social awareness competency, creative innovation Competency, Interdisciplinary competency, communication competency for sustainable peace, and higher-order thinking competency. Learners demonstrated all core competencies at a moderate level. The desirable condition for students to have overall competency in sustainable development was at the greatest level. From high to low, students desired higher-order thinking competency, communication competency for sustainable peace, collaborative competency, digital literacy competency, self and social awareness competencies, Interdisciplinary competency, life and career competency, and creative innovation competency. By prioritizing data from questionnaires with students and teachers about competencies for sustainable development, Cebrián & Junyent (2015) also added knowledge and practical skills related to nature and science, promoting ethical values, a positive attitude toward sustainability, and emotional management of primary school students to a list of the future sustainable development competencies. Saqib et al. (2020) conducted a survey on students' and teachers' attitudes of sustainability in Pakistan. According to the findings, education for sustainable development in Pakistan was at a poor level. Teachers lacked awareness of sustainability and holistic approaches, while students needed greater education on sustainable development. Therefore, effective student development requires an inquiry into the current and desired conditions to develop learning that meets the needs of students.

6.3 Innovation Development Using the Integrative Learning Process to Enhance Competencies for Sustainable Development

In developing innovation and integrative learning processes, board games, and stories were developed under "themes." The outcomes were expanded to other local schools in Nakhon Nayok Province in addition to innovations and integrative learning processes for use in target schools. The content and learners' skills and mental development were integrated (Pang & Good, 2000; Hurley, 2001; Lenoir & Hasni, 2010; Kaemmanee, 2020). Goal 4 is to promote learning opportunities, create an equitable and inclusive education, and promote lifelong learning for all. This goal was considered as a main goal of developing innovation and organizing integrated activities that were related to 17 SDGs (UNESCO, 2016; UNESCO, 2017; UNESCO, 2018a; UNESCO, 2018b). In conjunction with area-based education management, the conceptual framework of the whole-school approach to sustainable development (UNESCO, 2015b; UNESCO, 2018b) was merged with education for sustainable development (ESD) (UNESCO, 2018a). Hence, incorporating an integrative learning process is a crucial framework for equipping learners with the skills to address and resolve challenges effectively until pupils possess the knowledge, skills, and attributes required competencies for sustainable development.

6.4 Effectiveness of the Integrative Learning Process to Enhance Competencies for Sustainable Development Based on the Whole-School Approach of Impoverished Students

The competencies for sustainable development of impoverished students were enhanced through the integrative learning process. The plan started with raise a question, open inquiry, conduct an experiment, and communicate a conclusion. Following behavior observation and reflecting on the learning results, each time following an operating cycle. In terms of effectiveness of the integrative learning process to enhance competencies for sustainable development based on the whole-school approach of impoverished students, The researcher discovered that all impoverished students demonstrated the competencies for sustainable development in each cycle, passing the requirement of 70% or higher. In each cycle, learners' overall competencies could be assessed as high, moderate, and low. The competencies for sustainable development are also the abilities of learners to apply knowledge, skills, and attributes which are applicable and measurable. The eight competencies include interdisciplinary competency, communication competency for sustainable peach, higher-order thinking (HOT) competency, collaborative competency, digital literacy competency, creative innovation competency, self and social awareness competency, and life and career competency. According to White & Harder (2013), the links between institutions, particularly universities in terms of teaching, research, and community-related projects, will create a broader engagement with society, resulting in the exchange of knowledge, sharing, and developing skills through various activities. Mader (2013) discussed the evaluation of the sustainability process in terms of changeability based on the model "Graz Model for Integrative Development." The integrative development by modelling the leadership, social networks, participatory education, and research integration is a foundation of development to create change.

6.5 Educational Management Strategies for Sustainable Development

The education management strategies for sustainable development serve as a set of recommendations or feasible methods that may be implemented and satisfy the needs of stakeholders according to the area-based education management. There are four strategies: school management and capacity building, curriculum teaching and learning development, coordinate collaboration and community relationships, and supporting facilities and operations. Strategies have a total of 31 indicators that are aligned with SDG Goal 4: ensure inclusive and equitable quality education and promote lifelong learning opportunities for all. Target 4.7 ensures that by 2030, all learners acquire the knowledge, skills, and attributes needed to promote sustainable development, including among others through education for sustainable development and sustainable lifestyles, human rights, gender equality, promotion of a culture of peace and nonviolence, global citizenship, and appreciation of cultural diversity and of culture's contribution to sustainable development (UNESCO: 2018b). The concepts are line with Gallwey (2016), who discussed the challenges of reforming education toward SDG 4.7 goals. The necessities and the outcome must be balance in order to create a practical policy.

Additionally, it calls for collaboration, exchange, and communication between practitioners and policymakers. In addition, education should meet local needs and encourage the use of cutting-edge media and technology that are appropriate for small schools, according to the strategies for success in education management based on the educational reform in basic education institutions in terms of curriculum management, student development, teaching, and teacher quality (Jumpathip et al., 2014; Phochanatharn & Sudsawat, 2016; Rosenström & Kyllönen, 2007; Comyaoun et al., 2019; and Suthaso et al., 2022) Administration basic education by conducive to the development of desirable characteristics of learners according to the philosophy of sufficiency economy (Wechayanan et al., 2016). According to the study of the impact of a participatory approach on the national sustainable development indicators in Finland, strong and widespread engagement of experts and civil officials would help increase the capacity of outcomes and improve the effectiveness of working methods. Wahyudin (2018) discussed the effort made by Indonesia's Ministry of Education and Culture to develop a new peace education curriculum for use in elementary and secondary schools in accordance with a sustainable development goal. In the 21st century, amid advances in technology, society, economy, national, and global environment, indigenous teaching and learning activities were encouraged as a means of resolving disputes and addressing the issues of global citizenship. Additionally, best practices for evaluating local sustainability were chosen as indicators, which stakeholders might combine to evaluate the process, and a comprehensive strategy was suggested, putting special emphasis on the value of involvement in local contextualization and the function of specialists in the evaluation and dissemination of indicators. International studies have emphasized how these steps can be used to create both quantitative and qualitative indicators. The government also engages a part in establishing policies and improving practices that lead to better outcomes for learners (Crane & Richardson, 2000; Wongwanich, 2006; Suratruangchai, 2003).

7. CONCLUSION

Research and development through brainstorming, analyzing the problem, arranging learning activities, and employing the integrative learning process to enhance competencies for sustainable development in accordance with the concept of the whole-school approach. All of this leads to the exchange of ideas, awareness, and problem-solving together through adjusting one's thinking from solving a problem. It is a way to encourage common thinking and solving problems in all dimensions. The best practice can be used as a model for schools and other areas to use as a guideline for successful development through the participation of schools and communities, as well as coordinating external networks to participate in the development of the integrative learning process to strengthen competencies for sustainable development for students to have the ability to learn effectively in an interdisciplinary manner. Furthermore, creative innovations can be employed for commercial purposes, expanding industrial capacity, generating economic growth, and addressing long-term education issues. The education management strategies for sustainable development can be adopted at the school, community, societal, national, and global levels to raise awareness of all obligations shared by all sectors.

8. RECOMMENDATIONS

According to the study on using the integrative learning process to enhance competencies for sustainable development, the researchers suggested the following recommendations:

1. Competencies for sustainable development should be used to measure and evaluate other learning outcomes.
2. The current and desired conditions related to learning outcomes should be studied to see the problems and guidelines for learning development.
3. The development of various forms of innovation should be promoted for use in teaching and learning.

4. Suggest the implementation of integrative learning processes in learning management across a range of academic disciplines.
5. Advocate for the adoption of educational management strategies that contribute to sustainable development and advance the field of educational management.

REFERENCES

Barth, M., Godemann, J., Rieckmann, M., & Stoltenberg, U. (2007). Developing key competencies for sustainable development in higher education. *International Journal of Sustainability in Higher Education*, 8(4), 416–430. 10.1108/14676370710823582

Brauer, B. T. (2010). Twenty-first century skills: Learning for life in our times. *The Journal of General Education*, 59(1), 82–84. 10.5325/jgeneeduc.59.1.0082

Brolin, D. E. (1992). *Life centered career education: Competency units for personal-social skills*. Council for Exceptional Children.

Buasri, S. (1985). *Study and research in educational philosophy*. Srinakharinwirot University.

Buasri, S. (2001). *Education, philosophy of education, democracy, and ethics*. Suweeriyasart.

Cebrián, G., & Junyent, M. (2015). Competencies in education for sustainable development: Exploring the student teachers' views. *Sustainability (Basel)*, 7(3), 2768–2786. 10.3390/su7032768

Clark, W. C., van Kerkhoff, L., Lebel, L., & Gallopin, G. C. (2016). Crafting usable knowledge for sustainable development. *Proceedings of the National Academy of Sciences of the United States of America*, 113(17), 4570–4578. 10.1073/pnas.160126611327091979

Colvin, J., Blackmore, C., Chimbuya, S., Collins, K., Dent, M., Goss, J., Ison, R., Roggero, P. P., & Seddaiu, G. (2014). In search of systemic innovation for sustainable development: A design praxis emerging from a decade of social learning inquiry. *Research Policy*, 43(4), 760–771. 10.1016/j.respol.2013.12.010

Comyaoun, C., Yawirach, P., Tunkeaw, S., & Rattanachuwang, P. (2019). Strategies for management of collaborative networks to enhance the quality of education management in small primary schools. *The Journal of Academic Social Science*, 12(2), 30–36.

Crane, P., & Richardson, L. (2000). *Reconnect action research kit*. Australia Department of Family and Community Services.

De Haan, G. (2006). The BLK '21'programme in Germany: A 'Gestaltungskompetenz'-based model for education for sustainable development. *Environmental Education Research*, 12(1), 19–32. 10.1080/13504620500526362

De Haan, G. (2010). The development of ESD-related competencies in supportive institutional frameworks. *International Review of Education*, 56(2), 315–328. 10.1007/s11159-010-9157-9

Dhammakosachan, P. (2009). *Buddhism and science create a culture of wisdom*. Pithak Printing.

Education Improvement Commission. (2000). *School improvement planning for principal, teachers and school councils*. Education Improvement Commission.

Faham, E., Rezvanfar, A., Mohammadi, S. H. M., & Nohooji, M. R. (2017). Using system dynamics to develop education for sustainable development in higher education with the emphasis on the sustainability competencies of students. *Technological Forecasting and Social Change*, 123, 307–326. 10.1016/j.techfore.2016.03.023

Fink-Hafner, D., Dagen, T., Doušak, M., Novak, M., & Hafner-Fink, M. (2019). Delphi method: Strengths and weaknesses. *Advances in Methodology and Statistics*, 16(2), 1–19. 10.51936/fcfm6982

Gallwey, S. (2016). Capturing transformative change in education: The challenge of tracking progress towards SDG Target 4.7. *Policy & Practice: A Development. Education Review*, 23, 124–138.

Gardner, H. (2008). The five minds for the future. *Schools*, 5(1/2), 17–24. 10.1086/591814

Giangrande, N., White, R. M., East, M., Jackson, R., Clarke, T., Saloff Coste, M., & Penha-Lopes, G. (2019). A competency framework to assess and activate education for sustainable development: Addressing the UN sustainable development goals 4.7 challenge. *Sustainability (Basel)*, 11(10), 2832. 10.3390/su11102832

Giannarou, L., & Zervas, E. (2014). Using Delphi technique to build consensus in practice. *International Journal of Business Science and Applied Management*, 9(2), 65–82.

Hay Group. (2003). *Using competencies to identify high performers: An overview of the basics*. Hay Group, Inc.

Hurley, M. M. (2001). Reviewing Integrated Science and Mathematics: The Search for Evidence and Definitions From New Perspectives. *School Science and Mathematics*, 101(5), 259–268. 10.1111/j.1949-8594.2001.tb18028.x

Institute of Culture and Arts. (2010). *In remembrance of Professor Dr. Saroj Buasri's contributions*. Srinakharinwirot University.

Jiawiwatkul, U. (2010). *Participatory action research: Concepts, principles, and lessons*. P.A. Leaving.

Jumpatip, W., Asavapoom, S., & Pakothang, J. (2014). Strategies for successful education management based on the educational reform in basic education institutions. *Journal Buriram Rajabhat University*, 9(2), 163–169.

Kammanee, T. (2020). *Teaching science: Body of knowledge for an effective learning process* (24th ed.). Chulalongkorn University Press.

Keahey, J. (2021). Sustainable development and participatory action research: A systematic review. *Systemic Practice and Action Research*, 34(3), 291–306. 10.1007/s11213-020-09535-8

Kesapat, P. (2000). *Strategic planning for educational management*. Suweeriyasat.

Khamkeaw, W. (2012). *Creating world peace based on the concept of 4 noble truths*. Faculty of humanities. Srinakharinwirote University.

Kunaporn, P. P. (2013). *Thailand's philosophy of education, Buddhadharma section: Educational leader.* Wat Nyanavesakavan.

Lenoir, Y., & Hasni, A. (2010). Interdisciplinarity in Quebec schools: 40 years of problematic implementation. *Issues in Interdisciplinary Studies*, 28, 238–294.

Mader, C. (2013). Sustainability process assessment on transformative potentials: The Graz model for integrative development. *Journal of Cleaner Production*, 49, 54–63. 10.1016/j.jclepro.2012.08.028

Maneelek, R. (2002). *Strategic planning in educational institutions*. Metheetips.

Mills, G. (2003). *Action research: A guide for the teacher researcher*. Merrill Prentice Hall.

Ministry of Education Singapore. (2010). *Nurturing our young for the future: Competencies for the 21st Century.* Ministry of Education, Singapore.

Mogensen, F., & Schnack, K. (2010). The action competence approach and the 'new' discourses of education for sustainable development, competence and quality criteria. *Environmental Education Research*, 16(1), 59–74. 10.1080/13504620903504032

Morales, M. P. E. (2016). Participatory action research (PAR) cum action research (AR) in teacher professional development: A literature review. [IJRES]. *International Journal of Research in Education and Science*, 2(1), 156–165. 10.21890/ijres.01395

Mulkam, S. (2002). *20 Methods of learning management to develop morality, ethics, values and autonomous learning management*. Printmaking.

National Strategy Secretariat Office, Office of the National Economic and Social Development Board. (2018). *National strategy (2018–2037).* Office of the National Economic and Social Development Board, Office of the Prime Minister.

O'Donoghue, R., Henze, C., Shimray, C., Sarabhai, K. V., & Rivera, J. C. A. S. (2020). Hand-print CARE: Towards ethics-led action learning for ESD in school subject disciplines. *Journal of Education for Sustainable Development*, 14(1), 41–60. 10.1177/0973408220934647

OECD. (2019). *OECD future of education and skills 2030: OECD Learning Compass 2030.* Organization for Economic Co-operation and Development. www.oecd.org

Office of the Education Council. (2009). *The situation of Thai education on the world stage*. Prikwan Graphic. (Original work published 2007)

Office of the Education Council. (2016). *Directions and research issues of Thai education to achieve the world's sustainable development goals* (2nd ed.). Office of the Education Council.

Office of the Education Council. (2017). *Directional framework of the national education plan 2017–2031* [Draft]. Ministry of Education.

Office of the Education Council. (2018). *Report on the research results and development of the competency framework for grade 1–3 students for basic education courses*. Office of the Education Council.

Office of the National Economic and Social Development Board, Office of the Prime Minister. (2016). *Strategies for national development according to the 12th national economic and social development plan (2017–2021).* Office of the National Economic and Social Development Board, Office of the Prime Minister.

Office of the National Economic and Social Development Board, Office of the Prime Minister. (2019). *Brief master plan under the national strategy (2018–2037*). (2nd Ed.). Office of the National Economic and Social Development Board, Office of the Prime Minister.

Office of the National Primary Education Commission. (2000). *9-year strategic planning for compulsory education management.* Department of Religious Affairs.

Oxfam Education. (2002). *Whole School.* Oxfam Education. https://www.oxfam.org.uk/education/whole-school

Pang, J., & Good, R. (2000). A review of the integration of science and mathematics: Implications for further research. *School Science and Mathematics*, 100(2), 73–82. 10.1111/j.1949-8594.2000.tb17239.x

Phochanathan, S., & Sudsawat, S. (2016). Local administrative organizations' strategies for educational management of Phuket Province. *Rajabhat University Academic Journal, 12*(1).

Radovanović, D., Holst, C., Belur, S. B., Srivastava, R., Houngbonon, G. V., Le Quentrec, E., Miliza, J., Winkler, A. S., & Noll, J. (2020). Digital literacy key performance indicators for sustainable development. *Social Inclusion (Lisboa)*, 8(2), 151–167. 10.17645/si.v8i2.2587

Reed, M. S., Fraser, E. D. G., & Dougill, A. J. (2006). An adaptive learning process for developing and applying sustainability indicators with local communities. *Ecological Economics*, 59(4), 406–418. 10.1016/j.ecolecon.2005.11.008

Reid, A., Dahlgren, M. A., Dahlgren, L. O., & Petocz, P. (2011). *From expert student to novice professional* (Vol. 99). Springer Science & Business Media. 10.1007/978-94-007-0250-9

Rieckmann, M. (2012). Future-oriented higher education: Which key competencies should be fostered through university teaching and learning? *Futures*, 44(2), 127–135. 10.1016/j.futures.2011.09.005

Rode, H., & Michelsen, G. (2008). Levels of indicator development for education for sustainable development. *Environmental Education Research*, 14(1), 19–33. 10.1080/13504620701843327

Rosenström, U., & Kyllönen, S. (2007). Impacts of a participatory approach to developing national level sustainable development indicators in Finland. *Journal of Environmental Management*, 84(3), 282–298. 10.1016/j.jenvman.2006.06.00816901619

Rychen, D. S., & Salganik, L. H. (Eds.). (2003). *Key competencies for a successful life and well-functioning society*. Hogrefe & Huber Publishers.

Saqib, Z. A., Zhang, Q., Ou, J., Saqib, K. A., Majeed, S., & Razzaq, A. (2020). Education for sustainable development in Pakistani higher education institutions: An exploratory study of students' and teachers' perceptions. *International Journal of Sustainability in Higher Education*, 21(6), 1249–1267. 10.1108/IJSHE-01-2020-0036

Schreiber, J.-R., & Siege, H. (2016). *Curriculum framework: Education for sustainable development*. Cornelsen.

Siribanpitak, P. (2010). *Educational management for sustainable development: Basic education in economic, social and environmental aspects* (2nd ed.). Thai Samphan.

Suratruangchai, W. (2003). Research for teaching and learning development. *Journal of Education Science*, 14(2), 31–42.

Sutthaso, W., Sutthaso, R., & Piantham, C. (2022). Strategies for the development of education management of Pak Kret Municipality Pak Kret District Nonthaburi Province under the 20-year national strategy. *Academic Journal Faculty of Humanities and Social Sciences Phranakhon Rajabhat University*, 6(1), 29–41.

Sutthirat, C. (2009). *80 Innovations: Learner-centered learning management*. P Balans Design and Printing.

Sutthirat, C. (2016). *Techniques for using questions to develop thinking* (4th ed.). Chulalongkorn University Book Center.

Tangprapha, T. (2013). Educational management for student competency development: A case study of elevating education management for student competency and outcome development in the 21st century in Singapore. *Journal of Education*, 41(3), 203–227.

The 20-year national strategy (2018–2037). (2018, 13 October). Vol. 135 Chapter 82, Kor. 1–63.

UNESCO. (2010). *World data on education 7th edition 2010/11*.http://www.ibe.unesco.org

UNESCO. (2014). *Shaping the future we want. UN decade of education for sustainable development (2005–2014)*. UNESCO. https://unesdoc.unesco.org/images/0023/002301/230171e.pdf (accessed 17 May 2017).

UNESCO. (2015a). *Getting climate-ready: A guide for schools on climate action*. UNESCO.

UNESCO. (2015b). *Global citizenship education: UNESCO's approach*. UNESCO.

UNESCO. (2016). *School in Action global citizens for sustainable development: A guide for students*. UNESCO.

UNESCO. (2017). *Education for sustainable development goals: Learning objectives*. UNESCO.

UNESCO. (2018a). *Integrating education for sustainable development (ESD) in teacher education in South-East Asia: A guide for teacher educators*. UNESCO.

UNESCO. (2018b). *Issue and trends in education for sustainable development*. UNESCO Publishing.

United Nations. (2016). Transforming our world: The 2030 agenda for sustainable development. *Civil Engineering (Johannesburg, South Africa)*, 2016(1), 26–30.

Vanichsenee, S. (2015). Factors and success conditions for basic education management. Lessons learnt from the past decades and future possibilities in looking backward and forward of basic education reforms. Office of the Basic Education Commission. Ministry of Education.

Vare, P., Arro, G., de Hamer, A., Del Gobbo, G., de Vries, G., Farioli, F., Kadji-Beltran, C., Kangur, M., Mayer, M., Millican, R., Nijdam, C., Reti, M., & Zachariou, A. (2019). Devising a competence-based training program for educators of sustainable development: Lessons learned. *Sustainability (Basel)*, 11(7), 1890. 10.3390/su11071890

Viphoouparakhot, V. (2021). Area-based education management in participatory administration of basic school administrators. *Panyapiwat Journal*, 13(3), 320–332.

Wahyudin, D. (2018). Peace education curriculum in the context of education sustainable development (ESD). *Journal of Sustainable Development Education and Research*, 2(1), 21–32. 10.17509/jsder.v2i1.12354

Wals, A. E. J. (2015). *Beyond unreasonable doubt. Education and learning for socio-ecological sustainability in the anthropocene*. Wageningen University.

Wechayanan, P., Siribanpitak, P., Sriumpai, A., & Intharamaphan, W. (2016). Family's basic education management towards learners' desirable characteristics based on sufficiency economy. *Journal of Educational Administration Khon Kaen University*, 12(1), 145–155.

WESSA. (2016). *Stepping up to the sustainability goals: A practical guide to integrating the SDGs in our daily lives*. WESSA.

White, R. M., & Harder, M. K. (2013). The journey towards sustainability via community: Lessons from two UK universities. In Sterling, S., Maxey, L., & Luna, H. (Eds.), *The Sustainable University: Progress and Prospects* (pp. 132–150). Routledge.

Wiek, A., Bernstein, M., Foley, R. W., Cohen, M., Forrest, N., Kuzdas, C., & Keeler, L. W. (2015). Operationalising competencies in higher education for sustainable development. In *Routledge handbook of higher education for sustainable development* (pp. 241–260). Taylor and Francis.

Wiek, A., Withycombe, L., & Redman, C. L. (2011). Key competencies in sustainability: A reference framework for academic program development. *Sustainability Science*, 6(2), 203–218. 10.1007/s11625-011-0132-6

Wongwanich, S. (2006). *Action research in classrooms* (9th ed.). Chulalongkorn University Press.

Young, M., & Commins, E. (2002). *Global citizenship: The handbook for primary teaching*. Oxfam.

Youngs, H. (2018). *Developing 3R8C skills towards Thailand 4.0 innovative leaders*. Thailand Tribune.

Section 2
Exploring Pedagogical Aspects in Environmental Education Teaching

The following section broadens the scope of environmental education by exploring some educational dimensions of environmental awareness practices that can enhance students' learning across all subjects. These dimensions include interdisciplinary connections, fostering critical thinking and problem-solving skills, nurturing environmental stewardship, encouraging community engagement, and employing creative teaching methodologies and evaluation tools.

Chapter 6
Fostering Environmental Consciousness Through Multimodal Strategies in Undergraduate English Literature and Composition Courses in Algeria and the USA:
A Reflective Approach

Nada Tayem
Indiana University of Pennsylvania, USA

Bessedik Fatima Zahra
University of Oran 2, Algeria

EXECUTIVE SUMMARY

This chapter is a reflective teaching experience of incorporating multimodal strategies in English literature and composition writing courses to undergraduate students at Algerian and American universities. What this chapter primarily aspires to do, particularly, is to outline an account on how the teaching of different fiction, non-fiction, and poetry, such as Romantic poetry by William Wordsworth and Modernist poetry by by T.S. Eliot, fosters a conscious interest in the natural environment and its crises. The instructors propose the inclusion of a variety of multimodal texts on the environment to allow students to better understand environmental related issues

DOI: 10.4018/979-8-3693-1710-5.ch006

such as climate change and drought. This chapter will reflect on the teaching experiences of two instructors from different locations and include the course design employed in the teaching of different forms of literature. The chapter also provides the selected texts and the instructors' teaching recommendations.

INTRODUCTION

At colleges, environmental studies are a hotly contested and important topic. Teachers at all educational levels assign writing assignments with an eye toward many parts of nature and environmental concerns. English literature works emphasizing on nature included in the first- and second-year college courses in the Department of English provide a different viewpoint on environmental problems, therefore encouraging critical thinking and empathy in the students. Environmental problems span many challenges with both academic and pragmatic answers needed. Blibech addresses "emerging approaches to environmental thinking in higher education, perceived as a need but also as a "trend" that leads to versatile orientations impacting both educational goals and curriculum design," in his paper "Environmental Thinking in Higher Education: Towards a Systemic Integration." Growing awareness of the socio-physical surroundings has turned into a challenging process stressing the need of innovative solutions for environmental preservation and student education about environmental damage. Growing knowledge of the socio-physical surroundings has evolved into a difficult process that emphasizes the requirement of creative solutions for controlling and teaching students about environmental protection. One answer to this challenge is to incorporate environmental elements into the curriculum for teaching English literature and composition writing at the university level while also fostering collaboration between educational institutions from diverse cultural backgrounds. In this context, lecturers at different levels of study give students writing assignments on topics related to nature and the environment. This was initiated in "Place-Based Education: Learning to Be Where We Are" by Smith (2002).

Another crucial aspect of this process involves adequately equipping curriculum designers and instructors with the information and resources required to effectively encompass environmental issues in their courses. When instructors and students reflect on the selected literary works that address environmental themes, students can gain a deeper understanding of the interconnectedness between humans and the natural world with exposure to real-life scenarios. Therefore, to achieve environmental education, as well as individual and social responsibility, this chapter is a reflection on the teaching experience on environmental education using selected English fiction, non-fiction and poetry literary texts, and the collaboration between two professors from the University of Oran 2 Mohamed Ben Ahmed, in Algeria, and Indiana University of Pennsylvania in the USA. Given that this is a co-authored reflection, faculty members have come together remotely to collaborate and discuss the methods and selections of texts integrated in teaching and learning in these different educational settings.

METHOD

Current academic research on environmental education in higher education emphasizes a range of methods for promoting environmental consciousness. (Jane, Salvador, Castillo., Bryan, Mallillin, and Nozaleda, 2022). In an article titled "Environmental Education and English Language: In Search of a Sustainable and Liberating School," the authors have emphasized "the need to build a teaching practice that addresses the theme of the environment, considering the inseparability between education and social practice to face issues related to socio-environmental issues" (9). They argue that incorporating environmental education into the curriculum can help students develop critical thinking skills and a sense of responsibility towards the environment. By integrating environmental issues into language learning, educators can empower students to become agents of change in their communities.

As educators worked together in this chapter, we used reflective writing practices as our main research approach. Reflective writing facilitates a thorough examination of personal experiences, thoughts, and emotions, providing educators with useful insights into their teaching practices and professional development (Bolton 2010). This method allowed educators to critically analyze their teaching methods and make necessary adjustments to better address environmental themes in their classrooms. Through reflective writing, educators are able to enhance their awareness of the interconnectedness between education and social practice in addressing socio-environmental issues.

In *Reflecting on Reflective Practice*, Farrell (2012) asserts that reflective writing is a crucial component of reflective practice, allowing educators to gain insight into their experiences and enhance their pedagogical abilities (p. 14). The process of reflection fosters educators' involvement in continual self-assessment and ongoing enhancement, ultimately resulting in more effective and influential teaching methods. By integrating reflective writing into their professional development, educators can establish a more dynamic and captivating learning environment for their students.

Our objective is to engage in reflective writing to uncover meaningful connections and insights that will enhance our comprehension of our experiences as educators. The essay "The Appearance of Voice: EAP and Academic Literacies, Approaches to Teaching Reflective Writing" argues that voice in reflective writing can be understood as a construct. Contrary to other types of academic writing, reflective writing requires the author to convey both their personal and public identity simultaneously. In order to enhance our collaborative introspective writing approach, we devised a shared digital document where every instructor may provide their reflections over the course of a month. We have created a set of questions that are specifically tailored to promote self-examination and contemplation over teaching experiences, challenges, and achievements. Zwozdiak-Myers (2012) asserts that collaborative reflective writing offers instructors a supportive setting to analyze their teaching techniques, exchange ideas, and acquire information from each other (p. 27). By writing together and sharing our thoughts, we build a sense of unity and encouragement that expands our personal and collective knowledge. During the 3 month-long reflective writing period, each instructor committed to writing for at least 45 minutes per week. Moon (2006) proposes that engaging in regular reflective writing fosters the development of a reflective habit and has the potential to facilitate transformative learning (p. 43).

At the University of Oran 2 Mohamed Ben Ahmed in Algeria and Indiana University of Pennsylvania in the United States, we effectively integrated environmental concerns into our curriculum by broadening our teaching methods and viewpoints. Through the exchange of ideas and experiences, we acknowledge its significance. By engaging in this cooperative endeavor, our aim is to motivate fellow educators to actively encourage environmental education inside their own classrooms. Through the exchange of our achievements and difficulties in incorporating environmental concerns into education, our aim is to establish a community of educators committed to equipping students with the knowledge and values to become environmentally aware global citizens. In the subsequent parts, we will analyze the perceptions we acquired from our reflective writing exercises, as well as how they have enhanced our understanding of our collective experiences as educators. Our objective was to document the evolution of our viewpoints and comprehension as educators from different institutes and locations by engaging ourselves in regular reflective writing. At the conclusion of the meetings we had over months, we gathered our reflection submissions and arranged virtual meetings to discuss our views.

LITERATURE REVIEW

A recent study on environmental education in higher education identified numerous techniques for fostering environmental awareness. Higher education institutions have placed growing emphasis on environmental education in response to increased awareness among educators. Given this urgent need, it is essential to acquaint learners with current concepts and address the environmental concerns, such as climate change and biodiversity loss, that are relevant in the 21st century.

(Kumar, 2020). We must equip learners with the knowledge and skills that will enable them to become responsible global citizens and make informed decisions, thereby fostering a sustainable future for everyone.

Contemporary scholarly research in this field presents various approaches to foster ecological awareness and engagement among university students. An important field of research examines the importance of place-based education in cultivating students' sense of connection and responsibility towards their immediate environment. Gruenewald (2003) introduced a teaching framework known as "critical pedagogy of place," which investigates the social and ecological dimensions of students' authentic experiences. As he remarks in the article "The Best of Both Worlds: A Critical Pedagogy of Place," a critical pedagogy of place challenges all educators to reflect on the relationship between the kind of education they pursue and the kind of places we inhabit and leave behind for future generations (1).

Likewise, in the first chapter of David Sobel's "Place-Based Education: Connecting Classrooms and Communities"' (2004), Sobel argued for the application of location-based education to support environmental responsibility and community involvement. Robert Brooke's book "Rural Voices: Place-Conscious Education and the Teaching of Writing" (2003). Brooke explores the use of place-based approaches in English instruction, emphasizing how connecting literature to students' local surroundings and communities can improve their knowledge and engagement with the work. A subset of researchers has focused on integrating environmental subjects into the curriculum David Orr (2004), "Earth in Mind: On Education, Environment, and the Human Prospect", highlights the importance of incorporating environmental principles into all areas of education, something he refers to as "ecological literacy." He contends that comprehending ecological principles is critical for all disciplines, scientific or humanistic, in order to promote a more sustainable future. Garrard (2007) promotes ecocriticism, which uses ecological themes to study literature and culture. While in "Teaching for Joy and Justice: Re-imagining the Language Arts Classroom," Christensen presents doable strategies for including environmental justice issues into the instruction of rhetoric and writing. Through their writing and classroom activities, she offers techniques for involving students in debates on social and environmental concerns.

An increasing number of studies have investigated the importance of environmental education in acknowledging the influence of ethics on students' attitudes and behaviors toward the natural world. Gruenewald's (2009) book "Place-Based Education in the Global Age: Local Diversity" investigates the idea of place-based education, stressing learning anchored in the local surroundings and community. Gruenewald contends, along with other writers, that by tying students' life to the locations they live in, education should be relevant. The book explores how place-based education could boost community involvement, encourage ecological sustainability, and help to create local identity. Key important themes are the value of local diversity, the link between global and local settings, and the part education plays in tackling environmental and social concerns. Also, Macy and Brown's (1998) publication "Coming Back to Life: Practices to Reconnect Our Lives, Our World" offers a variety of activities and ideas meant to enable people and groups to reawaken themselves, and the natural surroundings. Introduced by Macy and Brown as a paradigm for both personal and social transformation is the "work that connects." Emphasizing the need of empathy and compassion for all living entities, the book It provides ways for building resilience, hope, and active participation as well as for handling emotions of hopelessness and alienation in the face of world ecological and catastrophes.

On the other hand, hooks (2009) argues for an "engaged pedagogy" that recognizes the complete character of students' identities and promotes a feeling of interconnectivity. Kaza (2008) in *Mindfully Green: A Personal and Spiritual Guide to Whole Earth Thinking,* proposes that "While practical approaches to an eco-responsible lifestyle offer important first steps, it is critical that we ground these actions in broader understanding so that we can effect real change in the world" (161). Reflective practices and environmental problems solutions essays as examples of real-world experiences, enhance students' relationships with the environment.

Teaching Context:

In this section, we provide an illustration of our selection of reading materials and assignments, as well as our educational approach, which combines the study of English literature with writing. Initially, we provided pertinent information about the context in which our approach operates.

A) Algerian Context and Course Design

The English undergraduate program at Algerian University requires a satisfactory completion of 10 semester courses within five years. These courses encompass several areas such as literary text analysis, cultural studies, linguistics, research techniques, and other issues pertaining to the English language. Algerian universities require students to successfully complete a three-year curriculum to receive a bachelor's degree. Students then have the option to continue their studies for an additional two years to earn a master's degree, culminating in the presentation and defense of a thesis. Each semester consisted of 13 courses, excluding the evaluation sessions. Therefore, we can infer that Algerian students learning English experience a reduction in their understanding of English literature as they begin their university education. This is a result of a shortcoming in the high school system, which neglects the inclusion of literary education in the English language curriculum.

As far as environmental awareness is concerned, it is obvious that students studying English literature in the English Department have little interest in environmental issues, and that ecological aspects are not considered in bachelor's or master's programs. Students showed a lack of awareness regarding environmental hygiene and the safe disposal of waste products. Unfortunately, the term 'city cleaner' is considered offensive in the community. Unfortunately, few Algerian young people actively volunteer for environmental organizations whose goal is to clean forests, beaches, streets, and other public spaces. Furthermore, certain people do not prioritize environmental concerns. This emphasizes the urgent need for environmental education. Algerian youth require the immediate acquisition of essential knowledge and fundamental skills to enable them to make informed decisions and cultivate positive attitudes towards the natural environment. However, it is important to recognize that cultural attitudes towards the environment in Algeria are currently undergoing positive and progressive changes.

In Algerian universities, the classroom is characterized by a significant student population, with two groups of 30 students each attending a literature session every week. Our second-year literature course consisted of two cohorts of 30 students each, who attended weekly. The students' ages ranged between 20 and 30 years. The administration supports the groups throughout the three-year program and promotes a conducive learning environment in which students can easily participate in group work and debates based on the rapport built in the previous year.

Course Outline

The second-year undergraduate course centers on the development of English literature from the Restoration of King Charles II in 1660 to the 1900s, which includes the age of world wars. Course design promotes the analysis of diverse poetry that establishes links with nature, thus promoting a thorough exploration of English literature, history, and literary movements. The course incorporates virtual showcases of prominent paintings and artworks accompanied by documentaries and memoirs to provide concrete examples that facilitate the understanding of literary revolutions. These resources improve students' understanding of the current cultural environment and give them a glimpse of the modern world. An essential resource that would provide the course's framework is the documentary film *Before the Flood* (2016). The Algerian program exhibits a varied assortment of neoclassical and Romantic paintings. The literary texts incorporated into the environmental education curriculum consist of selected passages from Alexander Pope's "An Essay on Man," William Wordsworth's "The World is Too Much with Us," "Expostulation and Reply," and "It is a beauteous evening, calm, and free," William Butler Yeats' "The Second Coming," and T.S. Eliot's "The Waste Land." These writings serve the purpose of examining the distinct attributes of each time period and analyzing the parallels and disparities in thoughts. In addition, we analyze the reoccurring motif of nature throughout history and its significance in relation to present-day concerns, such as industry, companionship, conflicts, and calamities. The purpose of this course is to empower students to analyze different perspectives on the natural environment and its historical changes.

B) The American Context and Course Design

In contrast to Algerian universities, the English composition course at Indiana University of Pennsylvania has a smaller student population. The class often consists of a group of 20–25 individuals, allowing for a greater level of personalized attention and interaction between the lecturer and students. The students' ages ranged from 18 to 22 years old, and they included a diverse group of individuals in their first, second, third, and fourth years of college.

Course Outline

In the Spring semester of 2024, Indiana University of Pennsylvania offered the Composition Writing 101 course. To design the course, the instructor has integrated a range of fiction and non-fiction literary readings, poems, documentaries, and assigned projects that specifically address environmental themes. The syllabus incorporates various essential readings pertaining to the environment, including the memoir "Ecology of a Cracker Childhood" by Jane Ray, "Yellow Woman and a Beauty of the Spirit: Essays on Native American Life Today" by Leslie Marmon Silko, and "Everything I Need to Know I Learned in the Forest" by Vandana Shiva. These texts explore the interconnectedness between geography, identity, and the natural world, encouraging students to reflect on their personal interactions with their environment. In addition, the course syllabus centers on the concept of place, examining the influence of our surroundings on our identity and the reciprocal impact of our actions on our environment. Supplementary readings encompass writings by Chimamanda Ngozi Adichie (2009), Leslie Marmon Silko (1997), Barry Lopez (1997), Vandana Shiva (2020), Carolyn Finney (2010), and other authors. Students participate in discussions and activities specifically designed to engage them in reading and writing assignments. Also, the syllabus includes the film documentary of David Attenborough's A Life on Our Planet", the poem "The Earth is a Living Thing," which reinforces the significance of comprehending and safeguarding the environment. Other readings also cover current environmental concerns, such as deforestation in the Brazilian Amazon, such as "The Brazilian Amazon Deforestation Rate in 2020 Is the Greatest of the Decade" by Ceslo H.L. Silva, Jr. et al. These texts urge students to actively participate in addressing contemporary environmental issues and to contemplate possible solutions The syllabus demonstrates a resolute commitment to educating students about the environment by including a wide range of fiction and nonfiction readings, multimedia materials.

Different Writing Projects Undertaken in the Composition Writing Course of the Spring Semester of 2024 at Indiana University of Pennsylvania

This ENG 101 Writing Course's design allows students to develop and participate in projects that address specific rhetorical contexts. Specifically, it focuses on the concept of place. The term "place" can have a variety of meanings, ranging from a very precise location, such as one's own bedroom at home or place of residence, to a more expansive concept, such as a homeland. The diverse range of locations students have visited and lived in shapes their individual experiences and perspectives. Locations they have not physically visited but have encountered through written narratives in books and articles, or visual representations in life stories, memoirs, news, reports, and films, also influence individuals.

In this English 101 Writing Course, students engage in simultaneous reading and writing, often expected to critically think, analyze, and evaluate their own and the world's relationships within the larger social frameworks presented by the reading material. Students will explore fiction, short films, poetry, and articles concerned with place, environment, and justice. The course guides students to achieve writing growth in various genres, such as personal narratives, writing reviews, annotated bibliography writings, researched problem-solution essays, and research posters. From drafting a thesis statement to formatting a well-polished paper, the course anticipates that our efforts in English 101 writing will not only equip students for college-level writing but also foster a sense of responsibility and interesting knowledge among all.

Projects Description

Throughout the duration of the course, the instructor has designed a total of three significant writing projects, in addition to weekly questions and responses, one presentation, one digital poster, and a final reflective self-assessment essay. Each assignment is an example of a certain type of writing that has its own set of conventions and rhetorical strategies. For a more detailed explanation of each project, the instructor provides specific instructions for each one. The instructor will provide a brief overview of what each project addresses.

Project One: Hometown Presentation

Each student will present information about their hometown in order to build a class community, get to know each other better, learn from one another's diverse backgrounds, and emphasize the course theme of "place."

Project Two: Narrative Essay

A narrative essay is a piece that tells one consistent, cohesive story. In academic writing, a narrative essay will also convey a lesson, a moral, or a point that the writer wants the reader to take. For this narrative essay, students will write about a place that had a significant effect on them and describe at least one specific event, lesson, or moral that took place there. Students may relate this to a place that evokes feelings of comfort, joy, beautiful memories, or discomfort, a place they are familiar with, or a place they have visited only once or perhaps numerous times. Students will get to practice writing detailed descriptions, engaging in the narrative writing method, and crafting a subtle thesis. This project also engages in "light research" using non-scholarly sources. It also includes a narrative essay draft, peer review, and a final paper.

Project Three: Writing Reports

In this assignment, students can write a place profile report and establish basic facts about a certain location on campus or in the local community. This report can cover any type of place, including buildings, monuments, dining halls, dorm rooms, favorite study spaces, parks, forests, lakes, and vacation spots, among others. It could also be a local restaurant, coffee shop, museum, theater, bank, or public library. The writing report genre typically allows the writer to collect and present information about the place of interest and provide an objective perspective based on facts rather than opinion. A useful place profile report has both the unique characteristics that make a place intriguing and the broader historical concepts that connect it to a certain location and era. If students are encountering difficulties in the beginning, consider suggesting that they contemplate the location from the perspective of a guest or a tourist to this place. Here are a few examples of useful questions:

1. What are the essential aspects of this place or location that people should be aware of?
2. How could you improve the traditional account and add any unexpected elements? What type of data is available at this location? Make sure to maintain precision and accuracy, especially when utilizing data.
3. What are the evaluative criteria that form the foundation of this location report? In other words, is it research-based?

Project Four: Writing Problem Solution Essays and Creating Digital Posters

The focus of this project should be on a problem that is either location- or demographic-based, and how we can solve it. Students need to work on preparing resources for this research (4-6). To do so, please consider the following steps:

1. Choose a topic: Select a topic that intersects with the student's major or future career. (Students' choice).
2. Research the problem: 1. Have students describe the problem's nature and discuss its severity.
3. Identify the affected groups:
4. Ask students to specify the place and/or demographic affected by the problem.
5. Explain the Importance: Students need to explain why the audience should care about this problem.
6. Propose Solutions: Students should investigate potential fixes or partial remedies for this problem. After that, they need to present the research findings to a general academic audience.
7. Present Solutions and Recommendations: Here, students need to offer solutions or recommendations based on their research.
8. Make a call to action:
9. Ask students to conclude by encouraging the audience to engage with the issue.
10. Write the Essay: This step includes writing a draft and providing peer feedback.
11. Create a Digital Poster: After writing the essay, ask students to create a visually appealing digital poster that summarizes the research topic in a way that is most suitable for an academic conference. Students should concentrate on crafting a visual design that strikes a balance between text and images.

Project Five: Reflective Writing

The final assignment in this course will entail composing a reflective essay wherein students discuss their development as writers throughout this course. Students choose one of the writing projects proposed in this class and write about their writing experience on the place-based topic and the specific writing project of their choice. The instructor may ask students to consider providing answers to the following questions:

1. How do you view yourself in terms of your abilities as a writer about place and the environment?

2. What aspects of your writing do you appreciate or find unfavorable?
3. Describe your writing project. What was it? What are your areas of proficiency or weakness in writing?
4. Does writing about this topic hold significance for you? Explain. Do you expect it to happen in the future? What empirical data and examples can you provide to support your answers?
5. What were your initial sentiments towards this assignment before completing it? Did you anticipate deriving pleasure from it? Explain. Did you anticipate that it would be challenging or effortless? What were your emotions upon completion?
6. What aspects of this specific piece of writing do you appreciate? What more modifications would you like to make?
7. Was there any specific revision strategy that was effective? By what means? Describe your strategy.
8. What is the target place and demographic of your project? By what means did you ascertain the target place and demographic? In what manner did the audience influence your paper's composition?
9. What went well in your writing experience? explain?
10. What aspects do you believe require improvement? Explain?

Descriptive Analysis

Our mission as instructors of these course extends beyond teaching the aesthetics of literature, aiming to equip students with a comprehensive education that encompasses environmental appreciation, knowledge acquisition, and a profound comprehension of their responsibility in safeguarding the environment for future generations. In both contexts, the syllabus used in each of these courses outlines how students can engage in real-world environmental issues and develop different critical thinking skills. For example, one of the writing assignments required visiting a local place on campus or in the local community and writing a place-based report. By engaging in practical exercises, interactive discussions, and in-depth research assignments, students will acquire skills to effectively utilize their knowledge to create a beneficial influence on the environment.

Both professors adhere to a mindset that highlights the significance of cultivating independent thinking and motivating students to question and evaluate materials. Fostering a classroom atmosphere that prioritizes critical thinking enables children to develop into capable individuals who can participate actively in addressing environmental issues. The instructors recognize that pedagogy is dynamic and constantly changes during critical teaching moments. Instructors also face challenges in addressing students' needs and adapting to the diverse academic learning styles and preferences of all learners. Both courses have extensively utilized a wide array of resources, including multimedia materials and interactive discussions, along with cutting-edge techniques to foster environmental consciousness and promote critical thinking among students.

The instructors selected English poems from distinct periods, including the Neoclassical, Romantic, and Modern eras, as their primary texts to explore varying perspectives on nature and the environment throughout literary history. These poems, including Pope's "An Essay on Man," Wordsworth's "The World is Too Much with Us," and Eliot's "The Waste Land," provide a basis for investigating the evolving perspectives on nature and the environment throughout history. In addition, the instructors showcased Neoclassical and Romantic paintings to exemplify the artistic portrayal of nature during distinct periods. Furthermore, both have encouraged subjective interpretation when engaging with environmentally focused memoirs, such as "Ecology of a Cracker Childhood" by Jane Ray and Terry Tempest Williams' "Refuge." In addition, the instructors included multimedia components that provided a broader understanding of environmental issues. The documentary films "David Attenborough: A Life on Our Planet" (2016) and "Before the Flood" (2016) are examples of documentaries. These films were chosen based on their ability to provide a comprehensive perspective on climate change and its effects. The pedagogical methods implemented in these courses promoted the cultivation of critical thinking abilities and reflective introspection.

Overall, the combination of film screenings and classroom discussions creates a well-rounded educational experience that fosters a deeper understanding of environmental issues. Classroom discussions are essential in both courses, as they allow students to articulate their understanding of the texts and relate environmental problems to their own lives and choices. These discussions also encourage students to engage with different perspectives and develop their communication skills. Classroom discussions are essential for students to actively participate in the course material, articulate their interpretations of the readings, and relate environmental issues to their personal experiences and decision-making. In a literature lesson centered on nature poetry, students can examine the works of the selected poets. They can then engage in discussions about how these poets effectively communicate ideas related to environmental responsibility and the bond between humans and the natural world.

Furthermore, the instructors agree that to foster a sense of responsibility and involvement in reading and wiring about the environment, encourage students to think critically, transfer knowledge, and create their own nature poetry or essays that reflect on their own relationships with the environment This is exemplified by the use of open-ended queries that generate debate, such as establishing connections between Wordsworth's depiction of nature and personal encounters with the natural environment. Additional question examples used by instructors and learners encompass deep inquires. Written assignments necessitate students to evaluate literature through an eco-critical lens and make connections to contemporary circumstances. Here are some example questions to illustrate this: What responsibilities do individuals possess for addressing environmental challenges? For example, a potential essay question could be: Examine the depiction of environmental decay in Eliot's 'The Waste Land' in contrast to a modern case of ecological disaster. What helpful insights may literature offer in understanding and addressing these issues?

According to Bell hooks, the act of educating is the exercise of freedom (hooks, 1994, p. 207). The primary texts selected for our courses mentioned previously serve as a basis for examining the different perspectives on nature and the environment throughout history. In addition to the main texts, both courses have also included a multimodal approach that incorporates multimedia assets to provide context for environmental issues. We showcase different documentary films to offer a current and comprehensive perspective on climate change and its consequences. By integrating environmental motifs into the instruction of English literature and composition writing, we can enable learners to make ecologically aware choices and engage in proactive measures within their personal lives and communities. This is also consistent with Freire's (1970) notion of "dialogic education", which highlights the importance of learners' knowledge and experiences in building transformative educational experiences.

Action Plan

Considering the instructor's teaching experience in these courses, we propose the following specific tasks to integrate environmental education into the English college curriculum: 1. Create a central database of lesson plans, assignments, and materials focused on teaching environmental protection through literature. 2. Encourage the integration of environmental literature into English courses at both departmental and institutional levels, emphasizing the importance of these topics to students' lives and future careers. 3. Conduct research on the results and most effective methods of teaching environmental protection based on literature, with the aim of improving teaching strategies and highlighting the importance of this interdisciplinary field.

CONCLUSION

Given the rapid degradation of the environment, it is important to integrate environmental awareness into higher education in various fields. Therefore, as English teachers, we have discovered that we cannot ignore the role of literature in teaching about the environment. A great approach to involve students in considering their relationship to environment and its part in producing conscious students is literature, both as a recurrent topic and as a tool for critical analysis and writing. By means of our individual experiences, we believe this strategy has been successful in enabling instructors and students to grow in knowledge and appreciation of environmental challenges. A key component of our group's research approach was definitely using reflective writing approaches. Reflective writing has given us insightful analysis of our professional growth and instructional strategies, therefore helping us to better know our experiences as teachers. Through the establishment of collaborative practice and engaging in frequent reflective writing, we effectively tackled our ideas and emotions pertaining to our teaching encounters, challenges, and achievements. Through reflections, we learned from each other and grew as educators. This has ultimately led to a more coherent and effective approach to addressing environmental issues in our educational practice. Lecturers have developed a reflective habit through our commitment to consistent reflective writing, which has enabled us to engage in transformative learning and understand the interdependence of social practice and education in addressing environmental issues.

Recommendation

Our experiences of including teaching nature-themed English literature texts into the first and second years of higher education institutions in Algeria and the United States lead us to propose the following suggestions for teachers who wish to include environmental themes into their courses: 1-Add environmental critical analysis as a fundamental component of literary studies and inspire students to investigate in a range of texts how nature, environmental issues, and ties to human nature are portrayed (Homaid, 2019). 2-Select books that advance a strong feeling of place. Select books that deftly capture certain locations and draw attention to the relationships of the characters to those places. Books covering personal and experiential levels can enable students to relate environmental problems to their own life (Christensen, 2000). 3- Incorporate readings from writers with different backgrounds, cultures, and experiences to offer a wide spectrum of ideas on environmental concerns, therefore strengthening the curriculum. Including many points of view will help students to critically evaluate several points of view and improve classroom debates. 4- Help students not just explore particular places and communities, but also acknowledge more general, worldwide environmental concerns, therefore stressing both local and global elements. 5-Guide them in seeing how the topics of the readings could fit their own life. 6-Create writing projects and assignments that inspire students to not only question what they are reading but also to consider their own experiences and actively participate in environmental concerns outside of the classroom. Personal narratives, site-specific research, and problem-solving essays among other assignments can inspire this kind of introspection and active engagement.7-Encourage honest and polite discussions on environmental issues by establishing a coherent and trustworthy classroom environment.8- Engaging in events that let students share their own opinions and experiences will help to create a supportive environment. 9-Work across institutions and disciplines. Promoting cooperation between several disciplines and educational institutions can help students to have a more complete awareness of environmental problems and solutions. Following these suggestions will help college teachers design courses that not only raise students' environmental consciousness but also equip them to be environmentally concerned writers and citizens. By means of literature analysis and writing practice, we can acquire the information, abilities, and values required to handle the urgent environmental issues of our day.

REFERENCES

Adichie, C. N. (2009, July). The Danger of a single Story [Video]. TED Conferences. https://www.ted.com/talks/chimamanda_ngozi_adichie_the_danger_of_a_single-story

Ahuja, A. (2016). *Essentials of Environmental Education.* Koros Press Limited.

Attenborough, D. (Director). (2020). David Attenborough: A Life on Our Planet [Documentary]. Netflix.

Bolton, G. (2010). *Reflective practice: Writing and professional development* (3rd ed.). SAGE Publications.

Braun, V., & Clarke, V. (2006). Using thematic analysis in psychology. *Qualitative Research in Psychology*, 3(2), 77–101. 10.1191/1478088706qp063oa

Brooke, R. (Ed.). (2003). *Rural voices: Place-conscious education and the teaching of writing*. Teachers College Press.

Christensen, L. (2008). Teaching for joy and justice: Re-imagining the language arts classroom. *Rethinking Schools.*

Clifton, L. (1993). The earth is a living thing. *The American Poetry Review*, 22(1), 20–20.

de Albuquerque Júnior, A. X., Andrade, E. P., de Morais, M. A. C., & Batista, W. M. N. (2021). Environmental Education and English Language: In search of a sustainable and liberating school. *Research. Social Development*, 10(13), e552101321465–e552101321465. 10.33448/rsd-v10i13.21465

Eliot, T. S. (1964). *The Waste Land 1922*. Grolier Club.

Farrell, T. S. C. (2012). Reflecting on Reflective Practice: (Re)visiting Dewey and Schön. *TESOL Journal*, 3(1), 7–16. 10.1002/tesj.10

Ferreira, J. A., Ryan, L., & Tilbury, D. (2007). Mainstreaming education for sustainable development in initial teacher education in Australia: A review of existing professional development models. *Journal of Education for Teaching*, 33(2), 225–239. 10.1080/02607470701259515

Finney, C. (2010). This land is Your Land, This Land is My Land: People and public lands redux. *The George Wright Forum*, 27(3), 247–254.

Garrard, G. (2007). Ecocriticism and education for sustainability. *Pedagogy*, 7(3), 359–383. 10.1215/15314200-2007-005

Goleman, D., Bennett, L., & Barlow, Z. (2012). *Ecoliterate: How educators are cultivating emotional, social, and ecological intelligence*. Jossey-Bass.

Gruenewald, D. A. (2003). The best of both worlds: A critical pedagogy of place. *Educational Researcher*, 32(4), 3–12. 10.3102/0013189X032004003

Holmes, A. J. (2021). Public Writing for Social Change. Writing Spaces. https://writingspaces.org/wp-content/uploads/2021/09/holmes.pdf

hooks, b. (2009). Belonging: A culture of place. Routledge.

Humes, B. (2008). Moving toward a liberatory pedagogy for all species: Mapping the need for dialogue between humane and anti-oppressive education. Green Theory & Praxis. *The Journal of Ecopedagogy*, 4(1), 65–85. 10.3903/gtp.2008.1.4

Jacobi, P. (2003). Educação Ambiental, Cidadania e Sustentabilidade. Cad. Pesquisa, n. 118. https://www.scielo.br/scielo.php?pid=S0100-

Jane, Castillo, & Mallillin. (2022). Environmental Education of Students Pursuing Higher Education: Probing on Climate Change Awareness. Journal of Climate Change. 10.3233/JCC220020

Kaza, S. (2008). *Mindfully green: A personal and spiritual guide to whole earth thinking*. Shambhala Publications.

Lopez, B. (1996). A literature of place. *US Society and Values*, 1(10), 10–12.

Macy, J., & Brown, M. Y. (1998). *Coming back to life: Practices to reconnect our lives, our world*. New Society Publishers.

Moon, J. A. (2006). *Learning journals: A handbook for reflective practice and professional development* (2nd ed.). Routledge. 10.4324/9780203969212

Moore, J. (2005). Is higher education ready for transformative learning? A question explored in the study of sustainability. *Journal of Transformative Education*, 3(1), 76–91. 10.1177/1541344604270862

Orr, D. W. (2004). *Earth in mind: On education, environment, and the human prospect*. Island Press.

Pope, A. (1798). *An Essay on Man*.

Powers, A. L. (2004). Teacher preparation for environmental education: Faculty perspectives on the infusion of environmental education into preservice methods courses. *The Journal of Environmental Education*, 35(3), 3–11.

Ray, J. (2023). *Ecology of a cracker childhood*. Milkweed Editions.

Shakespeare, W. (2006). *The sonnets* (Vol. 70). Cambridge University Press. 10.1017/9781139836623

Shiva, V. (2010). Earth democracy: Beyond dead democracy and killing economies. *Capitalism, Nature, Socialism*, 21(1), 83–95. 10.1080/10455751003655922

Shiva, V. (2013). Everything I Need to Know I Learned in the Forest. *NAMTA Journal*, 38(1), 273–276.

Silko, L. M. (1997). *Yellow woman and a beauty of the spirit: Essays on Native American life today*. Simon & Schuster.

Silva, C. H.Junior, Pessôa, A. C., Carvalho, N. S., Reis, J. B., Anderson, L. O., & Aragão, L. E. (2021). The Brazilian Amazon deforestation rate in 2020 is the greatest of the decade. *Nature Ecology & Evolution*, 5(2), 144–145. 10.1038/s41559-020-01368-x33349655

Smith, G. A. (2002). Place-based education: Learning to be where we are. *Phi Delta Kappan*, 83(8), 584–594. 10.1177/003172170208300806

Sobel, D. (2004). *Place-based education: Connecting classrooms and communities*. Orion Society.

Williams, S. (2019). The appearance of voice: EAP and academic literacies approaches to teaching reflective writing.

Wordsworth, W. (1979). *The world is too much with us*. Dreadnaught.

Wordsworth, W. (1980). Expostulation and reply. *Journal for the Education of the Gifted*, 4(1), 49–49. 10.1177/016235328000400108

Wordsworth, W., & Hayden, J. O. (1981). *The poems* (Vol. 1). Yale University Press.

Yeats, W. B., Hagstrom, F., & Gillespie, O. J. (1989). The second coming.

Zwozdiak-Myers, P. N. (2023). *The Teacher's Reflective Practice Handbook: Becoming an Extended Professional Through Enacting Evidence-informed Practice*. Taylor & Francis. 10.4324/9781003307143

APPENDIX

1) Questions and Answers

1-What are the challenges faced throughout the case of teaching environmentalism to literature students?

Developing environmental awareness among students reading different texts of literature could provide some difficulties. Among these are: strict syllabus and curriculum guidelines in certain countries, students' lack of interest in environmental problems, insufficient teaching materials and having interdisciplinary barriers.

2- What specific challenges do rigid curriculum guidelines present when trying to incorporate environmental consciousness into literature courses in Algerian universities?

Several challenges may be noted in relation to the attempt to develop environmental awareness to students of literature: A strict syllabus guideline limits the teaching of environmental issues to students. In the case of the Algerian University, curriculum changes must take place as instructors have to share proposals with the Ministry of Higher Education for any eventual change. This operation takes a period of a year.

3-How does the lack of student interest impact the effectiveness of teaching environmental issues in undergraduate literature courses?

Another challenge met throughout the study is the lack of interest in environmental issues by some of the students. This lack of interest can impede the effectiveness of the teaching process, making it difficult to engage students and foster a meaningful understanding of environmental issues.

4- What roles do insufficient teaching materials in certain universities play in hindering the development of environmental awareness among literature students, and what types of resources are most needed?

Insufficient availability of up-to-date and pertinent teaching materials or resources can impede the delivery of effective training on environmental subjects in certain locations. This includes educational materials such as printed books, interactive media, and digital resources that have the potential to enhance the process of acquiring knowledge.

5-How do conventional evaluation methods limit the assessment of students' understanding and engagement with environmental issues in literature courses?

Conventional evaluation methods frequently prioritize memorization and literary analysis, which may not accurately gauge students' comprehension or involvement with environmental issues. Creating alternative assessments that effectively measure critical thinking and the practical application of environmental principles might be challenging within current frameworks.

6-What interdisciplinary barriers exist when integrating environmental topics into literature courses, and how can educators overcome these challenges to create a more comprehensive curriculum?

- **Interdisciplinary Barriers**: The incorporation of environmental subjects into literature courses necessitates an interdisciplinary methodology that combines literature with science, ethics, and sociology. Nevertheless, opposition to this kind of integration among departments might restrict the extent of instruction, as literature programs may typically concentrate on canonical works and literary theory.
- **Instructor Readiness**: Educators may have insufficient preparation or resources to teach environmentalism effectively in the context of literature. They can require professional development opportunities to enhance their expertise and self-assurance in incorporating environmental themes into their curriculum.
- **Cultural Perceptions**: Diverse cultural views towards environmental issues can influence the willingness of students to accept and engage in these matters. Environmental problems may not be considered urgent in certain situations, which might affect students' involvement and the overall atmosphere in the classroom.

(2) An Epilogue and a List of Lessons Learned

If students are given the opportunity to explore life experiences through the concept of "place" and literature associated with the environment, they can develop a deeper understanding of how both physical and cultural surroundings influence individual and collective identities. Students can develop a more thorough sense of environmental consciousness and social responsibility through using the multimodal method and by employing a reflective methodology, which can improve their writing skills in a variety of genres. Providing students with opportunities to analyze and reflect on their surroundings through diverse life experiences and by addressing real-life problems in various projects and assignments can help prepare them to actively participate in society. Adopting place-based and ecologically sensitive English literature content in the first- or second-year college might have a significant and long-term impact on college students' academic and personal development.

(3) Lessons Learned

1. Interdisciplinary Integration Enhances Learning:

Explanation: Combining fictional literature, poetry, films, and personal narratives helps students build connections and makes their learning more relevant and comprehensive. This approach encourages critical thinking and the application of knowledge to real-world issues.

2. Place-Based Learning Deepens Engagement

Explanation: Focusing on "place" allows students to draw on personal experiences and local knowledge, making their writing more meaningful and engaging. This method also helps them understand the importance of context in shaping perspectives and narratives.

3. Diverse Writing Projects Foster Comprehensive Skills

Explanation: By engaging in various writing genres—from narrative essays to research reports and digital posters—students develop a broad range of skills. This versatility prepares them for diverse writing tasks in their academic and professional futures.

4. Reflective Practice Enhances Self-Awareness and Growth

Explanation: Reflective assignments encourage students to assess their progress, strengths, and areas for improvement. This practice promotes self-awareness and continuous personal and academic growth, making them more effective learners and writers.

5. Community Building through Shared Experiences

Explanation: Projects like the Hometown Presentation help build a sense of community among students by allowing them to share and learn from each other's diverse background and build a collective and enriching educational experience about the environment they are familiar

(3) A List of Resources for Further Study

Bonnett, M. (2023). Environmental consciousness, nature and the philosophy of education: matters arising. *Environmental Education Research*, *29*(10), 1377-1385.

Albrecht, D., & McInnis, N. F. (Eds.). (1975). *What Makes Education Environmental?* Data Courier.

Palmer, J. (2002). *Environmental education in the 21st century: Theory, practice, progress and promise*. Routledge.

Palmer, Joy, and Philip Neal. (2003). *The Handbook of Environmental Education*. New York: Routledge,

Uhl, christopher. (2004). *Developing Ecological Consciousness: Path to a Sustainable World*. Oxford: Rowman and Littlefield Publishers,

___ (2013). Developing Ecological Consciousness: The End of Separation. 2nd Ed. Rowman and Littlefield Publishers.

Wiley, J., ed. (1976). *Handbook of Environmental Education, With International Case Studies*. Madison: Winston University,

Chapter 7
Using Educational Technologies and Digital Media to Facilitate Environmental Education in Educational Institutions, Curricula, and Practice:
The Applications, Benefits, and Challenges

Shorif Mollah
https://orcid.org/0000-0001-7702-7154
University of Rome "Tor Vergata", Italy

EXECUTIVE SUMMARY

Today, the digitally driven world is continuously undergoing substantial environmental deterioration, climate change, melting icebergs in the polar zones (North and South poles), and risks of losing control over the health of the environment, climate, greenhouse system, and earth-surface, et cetera. Environmental issues such as climate change, pollution, resource depletion, and the conservation of nature and biodiversity are being viewed through technology-assisted approaches and lenses with possible novel solutions operated by the rapid advancement of digital innovations, tools, and strategies. Apart from the technologies, environmental education (EE)

DOI: 10.4018/979-8-3693-1710-5.ch007

plays a crucial role in raising awareness about pressing environmental concerns and inspires individuals to take action for the betterment of our planet. However, the advancements in technologies have notably been instrumental in integrating environmental education into various forms, platforms, and media, making information more accessible and engaging for teachers, researchers, practitioners, and learners.

INTRODUCTION (STATE OF THE ART)

Technology has become one of the determining forces in recent decades in how we interact with nature and our surrounding environment (Cho & park, 2023). The growing use of technology in educational practice has become a transformative force that has reshaped traditional teaching activities, styles, and learning methods across the globe (Morgan, 2022). Substantial advancements in technologies have broken the traditional boundaries of classrooms, expanding their scope beyond physical spaces and into the digital realm (Al-Ansi et al., 2023). This transition to the virtual space fosters a more interactive and dynamic educational experience in a larger spectrum than the traditional classroom setting. From smartboards and online learning platforms to virtual reality (VR) and artificial intelligence (AI), technology has enabled educators to tailor their content, knowledge, and training to meet the needs of students in a personalized manner (Al-Ansi et al., 2023). The integration of technology has introduced innovative pedagogical approaches, techniques, and tools, such as gamified learning experiences, personalized analytics, et cetera, empowering both educators and learners (Al-Ansi et al., 2023). Along with the mentioned areas in the education sector, the environmental education (EE) and research sector have significantly been influenced by the advancements of technologies and increased applications of technologies to practice education and reduce environmental pollution (e.g., carbon emissions) in the environment in direct ways.

For decades, the interconnections between humans and nature have always been a focal point in the pursuit of sustainable development and balance in the environment, ecology, and ecosystem (Morgan, 2022). Therefore, the education involved in the environmental phenomena strives to raise a population equipped with a comprehensive understanding of the biophysical environment and the problems involved, possessing the skills to contribute to problem-solving efforts, and being driven to actively engage in seeking environmental solutions (Stapp et al., 1969; Morgan, 2022). Conventional approaches to environmental education (EE) have been criticized due to their limited capability to alter students' attitudes and behavior toward environmental problems, concerns, and sustainability (Cao & Jian, 2024). In this regard, educational technology has emerged as a catalyst that has shaped public perspectives, nurturing a proactive stance toward environmental issues and protection (Cao & Jian, 2024). The use of digital technologies in EE extends to the area of plant identification, virtual hiking programs, citizen science initiatives, digital cartography, exploring various ecosystems, and so on (Lowan-Trudeau, 2023). However, EE heavily depends on direct experiences of natural phenomena outside the classroom (Ruchter et al., 2010). In situations where direct exposure to nature is limited, technology provides a virtual experience to those people with restricted access to nature (Willis et al., 2014).

Recent exploration into EE reveals that certain technological tools and emerging technologies are gaining prominence in facilitating environmental education (EE) through teaching-learning activities, in educational institutions and curricula. For example, existing studies show the increasing use of GPS for mapping, surveying, and navigation in various settings, AI and VR for virtual field trips and real-world simulation, and social media and digital cartography technology for mobilizing people for climate justice (Willis et al., 2014; BlockBirds, 2023; Lowan-Trudeau, 2023). GPS and virtual trips using AR technology allow the students to explore various ecosystems and understand their unique characteristics without leaving the classroom (BlockBirds, 2023). This is especially useful when they learn about inaccessible, remote locations (Jerowsky & Borda, 2022; Buchanan et al., 2018). Moreover, simulations using VR technology accurately replicate real-world scenarios, such as natural disasters or the impacts of climate change, enabling students to observe the repercussions firsthand (BlockBirds, 2023). Similarly, Augmented Reality (AR) and Virtual Reality (VR) expose students to immersive digital encounters that cannot be duplicated via conventional instructional approaches (Al-Ansi et al., 2023). Thus, the integration of existing and emerging technologies into EE is playing a transformative role by fostering awareness, cultivating positive attitudes and behaviors, and promoting environmental activism (Brečka et al., 2022; Cao & Jian, 2024).

Even though the appeal of technology is high in EE, the number of scholarly interventions is still limited in this area. Moreover, the majority of the empirical studies have investigated specific cases (micro-events) of technology integration into the EE curricula, lacking a comprehensive understanding and message of the current and emerging issues involved. In addition, existing studies have extensively highlighted the potential and advantages of integrating technology in EE but have largely neglected the challenges and limitations associated with the integration of technology in facilitating EE (Lowan-Trudeau, 2023). To address these gaps, the current study explores the advantages of integrating educational technologies in learning about the environment and the challenges associated with the integration.

Research Objectives

1. To explore and identify the advantages and benefits of integrating educational technologies in teaching and learning activities about environmental education (EE)
2. To point out the challenges and limitations associated with the integration of educational technologies in facilitating EE

ADVANTAGES AND PROSPECTS OF EDUCATIONAL TECHNOLOGIES IN FACILITATING ENVIRONMENTAL EDUCATION AND FIGHTING AGAINST CLIMATE

Educational technologies have made significant contributions to the field of EE by providing innovative tools, techniques, and platforms to enhance students' learning experiences (Cho & Park, 2023). These technologies are making education more interactive, accessible, and responsive to the evolving nature of environmental issues (Al-Ansi et al., 2023). For example, these educational technologies in environmental education deliver up-to-date information, real-time data, and multimedia resources, cultivating a dynamic understanding of environmental issues (Jardin, 2023; AL-Ansi et al., 2023). These technologies extend the scope of environmental education and encourage students to take an active role in tackling environmental issues with a sense of accountability towards the natural world. The exploration of the current study has found that the educators around the world are increasingly relying on technologies for delivering EE and solutions to students. A thorough analysis of the existing studies has revealed numerous benefits derived from the integration of technology to facilitate environmental education. The following section discusses the key advantages, prospects, and contributions of educational technologies in the teaching and learning about the environment through different curricula.

Increased Accessibility to Knowledge and Resources

Technological tools play a pivotal role in breaking down geographical barriers, and these tools make environmental education and knowledge accessible on a global scale (He, 2023; Morgan, 2022). For instance, online platforms, educational apps, and digital resources transcend borders and provide a platform for collaboration and knowledge-sharing among people with a shared interest in environmental issues (Kumpulainen et al., 2022; Willis et al., 2014). Online education allows educational institutions to extend their reach to a broader audience, expanding access to educational resources and opportunities (Morgan, 2022). Specifically, virtual classrooms, webinars, and online courses enable learners to participate in EE programs without being constrained by geographical location (Computeam, 2023; Ambe et al., 2024). By offering this borderless accessibility, technological tools act as a bridge to a more sustainable and interconnected world. The existing studies, thus, acknowledged the Internet as a valuable tool for making environment-related knowledge more accessible (Fernández et al., 2019). Adding to this, blogs and other online platforms are predominantly being used to demonstrate a positive relationship between technology and environmental education (Fernández et al., 2019; Ambe et al., 2024). Moreover, other technological tools, i.e., GPS, IoT, and mobile computers, are also found to be effective tools for increasing the accessibility to the wide collection of knowledge and resources of EE (Tabuenca et al., 2023; Ruchter et al., 2010; Willis et al., 2014).

VR and AR technologies are gaining popularity due to their ability for virtual transportation of people to different places (AIS, 2024). By integrating VR technology, individuals can virtually immerse themselves in a variety of locations, and gain valuable insights into the unique environmental challenges faced by various communities (He, 2023). Students have limited accessibility to some geographical areas due to distance, safety concerns, economic barriers, or ability (Jerowsky & Borda, 2022; He, 2023; Morgan, 2022). For instance, ecologically sensitive areas such as coral reefs or wetlands require preservation and thus remain inaccessible to the students (Jerowsky & Borda, 2022). Immersive technologies such as virtual reality (VR) and augmented reality (AR) offer an alternative means of experiencing these locations instead of being present physically. VR technologies transport individuals to distant environments such as the depths of the ocean or polar ice caps, to explore and interact with virtual ecosystems (AIS, 2024). It also helps to avoid physical geographical constraints such as seasonal variations and adverse weather conditions (Buchanan et al., 2018). Additionally, Morgan (2022) emphasized that internet-based environmental learning tools enable individuals with disabilities to experience natural environments accessible. Adding to this, Buchanan, Pressick-Kilborn, & Maher (2018) argued that VR tools are crucial for educating and connecting urban dwellers to physical nature and surroundings who lack the necessary access and scope for interaction with the natural environment in direct contact. This immersive encounter allows the users to connect with the environmental phenomena and issues and effectively encourages support for conservation efforts (Jerowsky & Borda, 2022; AIS, 2024).

Utilization of Diversified Resources and Technological Tools

Another significant role of technology in EE is optimizing the utilization of diversified web-based resources in environmental education in practice. Web-based learning resources integrate a diverse array of media types, including text, images, audio, animation, and video, seamlessly amalgamating them into comprehensive instructional materials (Morgan, 2022; Ambe et al., 2024). This helps to grow learners' interest and willingness to engage with the learning content. Existing studies have indicated that teachers have an array of technological tools at their disposal that provide opportunities for students to engage with the environment virtually (Willis et al., 2014). These tools typically include- webcams, digital cameras, video equipment, scanners, applications, GPS, geocaching, probe ware, web blogs, podcasts, mobile devices, and scanners (Willis et al., 2014; Ambe et al., 2024). These tools enable interactive, visual, and collaborative learning experiences and real-time observation and offer affordable and culturally diverse options (Morgan, 2022; Willis et al., 2014). When learners are presented with visual and verbal explanations, they are more prone to absorb the content compared to when the explanations are verbally delivered in lengthy segments (Morgan, 2022).

Bednarz (2004) explored the use of Geographic Information System (GIS) tools and concluded that GIS tools help visualize and analyze spatial data and enhance students' understanding of complex environmental patterns. Likewise, several other studies have explored the use of social media in EE. Social media provides platforms for sharing and discussing different environmental viewpoints and encourages active engagement in environmental protection (Lowan-Trudeau, 2023; Willis et al., 2014). Emerging technologies such as Augmented Reality (AR), Virtual Reality (VR), and Artificial Intelligence (AI) are found to be effective tools for combining different multimedia resources and information sources about environmental processes (Jardin, 2023; Lu & Liu, 2015; The Goan, 2022). Additionally, mobile technologies have also been proven useful in EE as they can accompany learners into the field, empower them to sense and record aspects of the local environment, take electronic notes, and facilitate the utilization of a diverse range of digital resources and representations (Rogers et al., 2009).

Enhanced Collaboration and Engagement

Digital platforms provide wider access to educational resources, enabling improved collaboration and feedback (Buchanan et al., 2018). In digital platforms, learners can interact, collaborate, and exchange ideas with experts and peers globally (Buchanan et al., 2018; BlockBirds, 2023). Online platforms, collaborative software, and virtual communication tools provide students with opportunities to connect, share ideas, and work collectively on environmental projects and group-work (BlockBirds, 2023). In addition, AR and VR enhance student-teacher collaboration by enabling group work in a virtual environment, fostering collaborative learning experiences, and enabling educators to gamify lessons (Al-Ansi et al., 2023). Moreover, technologies such as AR and VR are widely being utilized by educators to provide interactive content for EE due to their ability to broaden outreach and engage with new audiences (AIS, 2024).

A significant number of the existing studies have explored the role of AR and VR in environmental education. For example, Lu & Liu (2015) implemented an AR marine education program for elementary school goers. The results of this study indicate that AR activities not only enhance engagement and collaboration but also increase student participation, ensure the efficacy of learning, and enrich the overall learning experience. A similar finding was also reported by Jardin (2023), as he has found that augmented reality (AR) is a promising tool for enhancing engagement and experiential learning in environmental contexts. In contrast, the traditional teaching methods that rely heavily on passive learning, AR encourages hands-on exploration and problem-solving, paving the way for a higher level of engagement and motivation among the students (BlockBirds, 2023). Smith et al. (2018) proved this claim as they also found that students utilizing AR technology to study ecosystems exhibited greater levels of engagement and retention of learning in comparison to those utilizing conventional teaching methods.

Enhanced Learning Experience and Outcome

Technological tools enhance the learning experience in environmental education by providing immersive simulations of real-world scenarios, offering the students a firsthand understanding of environmental challenges and solutions (Buchanan et al., 2018). Integration of technological tools in the traditional classroom enhances interactivity in the learning process, which contributes to an improved learning experience (Baker, 2022). In addition, better educational outcomes are achieved through innovative technologies since they can depict and illustrate the impacts of certain environmental issues and policies beforehand (AIS, 2024). Technological tools achieve educational goals easily by motivating children to engage in environmental education, and activities, as well as increasing their environmental knowledge (Morgan (2022). Moreover, educational technologies can enhance the assessment process by offering the students personalized and interactive tests with challenges for better learning outcomes that are tailored to lead them to achieve unique learning levels (Al Ansi et al., 2023).

AR technologies enable students to replicate real-world scenarios (BlockBirds, 2023). The immersive capabilities of AR enable students to visualize complex concepts, engage with virtual elements, and thereby enrich their comprehension of environmental knowledge, problems, and solutions (Smith et al., 2018; Al-Ansi et al., 2023). Another way technologies improve the learning experience is by providing visual aids and scope for virtual tours alongside traditional text-based materials. It generates virtual tours of protected areas such as national parks and wildlife refuges, promoting conservation and raising awareness about threatened species and habitats while also vividly showcasing the impactful effects of environmental degradation, such as deforestation, and oil spills (He, 2023). ICT tools and portable mobile technologies further enhance the EE learning experience by empowering students themselves to generate scientific data and conduct environmental investigations directly in the field, developing skills in identifying and solving environmental problems (Fauville et al., 2014).

The primary goal of an effective environmental education program is to cultivate eco-citizenship by nurturing self-regulated strategies for autonomous engagement with environmental phenomena more interactively and dynamically (Khalifé et al., 2022). Scholars who delved into the utilization of technologies in EE argued that technology integration can be a catalyst in achieving educational goals. For instance, Aguayo and Eames (2023) investigated the efficacy of a mixed-reality immersive learning approach in marine education, utilizing a variety of digital technologies. Their findings suggest that employing XR-mediated marine education leads to enhanced learning outcomes. A similar finding was reported by Al-Ansi et al. (2023), who found that students who used AR and VR technologies demonstrated increased levels of motivation and engagement throughout their learning period and process. The utilization of AR and VR has improved the academic performance of the students compared to their peers who do not utilize such technologies.

Increased Awareness Among Stakeholders (Learners, Teachers, Researchers, Policymakers)

Existing studies have consistently affirmed and reaffirmed that the use of technology in EE plays a critical role in building environmental awareness. For instance, Fauville et al. (2014) explored the use of ICT tools in EE and concluded that ICT tools have the potential to raise awareness and help acquire an attitude of concern for the environmental issues. Similar is suggested by Li and Chen (2022) as they have also suggested that there is a strong link between the integration of technology in EE and positive environmental actions, beliefs, and meaningful environmental engagement. Additionally, Tabuenca et al. (2023) highlighted that IoT technology has the potential to enhance education and promote environmental awareness. The role of AI and VR in building environmental awareness has gained much exposure in several studies. Cao & Jian (2024) contended that AI and VR are effective in increasing ecological awareness and promoting sustainable practices.

The incorporation of artificial intelligence (AI) and virtual reality (VR) technologies in education has been recognized as a crucial factor in enhancing students' interest and motivation in understanding environmental concerns and planning for solutions (Bartolini & Bonatti, 2008). AI and VR technologies hold great promise for improving environmental education and public engagement because their combined use allows for the creation of realistic simulations that address the limitations of traditional teaching methods (Cao & Jian, 2024). AI and VR motivate students to engage in environmental issues by facilitating immersive experiences (Lee et al., 2022). Thus, the integration of AI and VR in education can increase students' awareness of environmental issues and inspire them to seek solutions proactively (Cao & Jian, 2024).

Adding to AI and VR, other emerging technologies, such as augmented reality (AR) and the Internet of Things (IoT) are utilized by several educators. For example, Tabuenca et al. (2023) explored the role of IoT technology in generating environmental awareness. The findings of Tabuenca et al. (2023) highlighted that the IoT system they used in their study utilized energy-saving strategies with solar panels and optimized water consumption based on humidity measurements, contributing to renewable energies and the conservation of resources. In addition, Kumpulainen et al. (2022) explored learners' experience with environmentally-themed digital storying in Finland with the aid of digital application. The findings of the study suggest that engaging in digital pedagogical practice can encourage students to explore their surrounding environments from new perspectives, rather than isolating them from them. They further argued that the use of digital storytelling with AR has the pedagogical potential to enhance learners' sensory experiences and aesthetic awareness of the environment.

Efficiency of Resources

Using technological tools represents a resource-efficient approach to delivering environmental education. One of the most notable benefits of educational technologies and tools lies in their ability to offer cost-effective and environment-friendly opportunities for learning (Buchanan et al., 2018). The affordability of educational technology is one of the primary factors contributing to the increasing adoption of technology in environmental education (Ambe et al., 2024). Existing studies have suggested that technological integration can significantly diminish the reliance on physical textbooks and printed materials, which are resource-intensive and expensive due to production and transportation costs (Computeam, 2023).

Moreover, the adoption of virtual classrooms and web conferencing tools reduces the need for physical travel, saving time and valuable resources and decreasing carbon emissions linked to daily commutes (Computeam, 2023; Al-Ansi et al., 2023). A similar finding has been suggested by He (2023) as he contends that emerging technologies, such as VR technologies, applied to simulate fieldwork scenarios in a controlled environment, reduce the cost of fieldwork. Similarly, IoT devices optimize energy use, and efficient data collection tools minimize the wastage of resources (DoForms, 2023). The cost of modern instruction technology is gradually decreasing as a result of constant technological innovation (Al-Ansi et al., 2023). Given their cost-effectiveness, low-cost immersive technologies (i.e., Oculus Rift, HTC VIVE, Microsoft HoloLens) have gained increasing popularity in recent times (Cho & Park, 2023).

Facilitate Positive Change in Behavior and Attitude

Educational technology plays a crucial role in fostering positive behavioral change in a learner toward eco-friendly attitudes and reducing adverse effects on the natural environment (Jerowsky & Borda, 2022; Bhang & Huh, 2023). Existing studies show that computer-mediated systems led to an increase in knowledge significantly and brought about behavioral changes positively among the participants (Ruchter et al., 2010). Several studies have experimented with the mentioned latest technologies and determined their role in forming environment-friendly attitudes and sustainable behavior among the subjects. Su et al. (2022) and Wang et al. (2023) utilized AI and machine learning algorithms in their smartphone application to offer personalized advice for reducing carbon emissions. The findings stressed that the App encouraged college students to adopt an environment-friendly lifestyle. Adding to this, Augmented Reality (AR) empowers students to employ their knowledge and abilities in the real world by replicating authentic situations (BlockBirds, 2023).

Additionally, VR technology enhances environmental education and awareness by offering diverse perspectives of species and ecosystems, promoting empathy and a comprehensive understanding of the environment (He, 2023). Moghaddam and Kunst (2023) explored the role of VR in education and argued that VR-based instructional programs improved both the environmental knowledge and attitudes of the students toward environmental phenomena and solutions. A similar conclusion has also been drawn by Cao & Jian (2024), who found that participating in VR simulations significantly increased college students' comprehension of environmental challenges, fostered the development of conservation values, and motivated individuals to participate actively in environmental advocacy.

They further showed that a VR simulation on climate change increased empathy and action inclination, while an AI chatbot in environmental education positively influenced the attitudes and behaviors of the subjects. Additionally, Boucenna, Benzouai, & Adli (2021) experimented with a VR simulation of plastic pollution and argued that the simulation heightened participants' commitment to reducing environmental impact. Li et al. (2022) conducted similar research as they developed a virtual reality simulation depicting an oil spill scenario to inspire individuals to proactively engage in environmental protection. The results indicate that exposure to the VR simulation has significantly increased students' preparedness to take concrete actions for environmental conservation. Thus, interactive simulations and games within VR educate individuals about environmental issues, provide immersive experiences of environmental challenges, and thereby encourage responsible behaviors among the stakeholders (He, 2023).

Ruchter et al. (2010), however, argued that technological tools are not necessarily effective in addressing the complex behavior and attitudes of humans. They examined the efficacy of a technology-rich medium in enhancing environmental literacy. Their study revealed that while the mobile computer system positively influenced environmental attitudes among adults, it did not have the same impact on children. This highlights the complexity of changing attitudes and suggests the need for long-term interventions that consider the emotional aspects of the children. The study proposes that the mobile computer system, being a relatively new technology, may not have adequately addressed the behavioral and attitudinal complexities unique to children.

TECHNICAL AND PEDAGOGICAL SHORTCOMINGS AND CHALLENGES OF EDUCATIONAL TECHNOLOGIES IN EE

Integration and adoption of educational technologies and tools in environmental education are not free from criticism and limitations. The number of research dealing with the potential and effectiveness of emerging technologies and tools in teaching EE is limited. However, the existing studies have shown several technical and pedagogical shortcomings and challenges in incorporating emerging technologies and tools in EE teaching and learning.

Difficulty in Maintaining Focus

Handling digital distractions effectively appears to be a significant hurdle in technology-integrated lessons. In Jardin (2023), both educators and students recognize that it is difficult to maintain focus in a technology-mediated learning environment where smartphones, laptops, and similarly dynamic and easily reachable gadgets play a pivotal role. Digital devices can be distracting because some students can browse the internet in the classroom or play video games instead of engaging with the lessons (Lightfoot-Wright, 2021). Thus, Lightfoot-Wright (2021) argued that the use of technology for educational purposes does not necessarily support academic improvement. A similar argument has also been established by Jakab, Zigová, & Pucherová (2019) as they found that using mobile phones in learning processes causes several hindrances in maintaining focus. In this regard, they specifically emphasized that the small-sized screens of smartphones cause difficulties in reading. In addition, mobile phones are not inherently designed for educational purposes, making them particularly challenging to use for completing classroom assignments (Jakab et al., 2019).

High Cost and Affordability Challenges

The digital divide and high cost are some of the key challenges of implementing technology-based environmental education in practice (Jardin, 2023). Virtual encounters offer inclusivity by providing opportunities for learners with limited mobility or resources (Buchanan et al., 2018). However, unequal access to technology can lead to disparities in students' learning experiences (Jardin, 2023). In addition, the high costs associated with building and maintaining digital infrastructures, such as AR and VR embedded platforms, are highly expensive and hinder the utilization of technologies in EE (Ambe et al., 2024; Al-Ansi et al., 2023). Khalifé, Chaker, and Gasparovic (2022) explored the integration of technologies and tools in the EE context in Lebanon. Their investigation has discovered that budgetary constraints limit the proper implementation of environmental awareness programs designed and run digitally in the country.

Institutions with more financial solvency show a higher inclination towards technology integration into the EE curricula. Al-Ansi et al. (2023) contended that the high costs associated with the purchasing of hardware and software and the maintenance of these technologies have been major barriers to implementing technology-based solutions in EE. They also acknowledged the inaccessibility and unaffordability of these emerging technologies by all students. Ambe et al. (2024) put forward similar findings as they identified resource constraints and unaffordability as major barriers to integrating advanced technologies in environmental education. Often, instructors have personal ownership of digital devices, but institutions do not purchase or subscribe to them. Because institutions are often slow to adopt rapidly advancing technologies due to the time required to train the instructors along with the financial and administrative barriers.

Lack of Quality Teaching Materials and Reluctance to Adopt Technologies

Existing studies have raised concerns about the quality of content and teaching materials for environmental education in the selected schools (Brečka et al., 2022). Ambe et al. (2024) explored the adoption of educational technologies in Nigerian higher education institutions. Their investigation revealed that the instructors' poor literacy and skills in computing resulted in low-quality and inadequate course materials for EE e-learning courses. The lack of interest of the instructors in emerging technologies makes it harder for them to adapt the technological innovations in the course and material design that is deployable to digital platforms. Likewise, some other studies find that teachers often hesitate to employ software in the classroom despite having sufficient equipment at their disposal, due to their reluctance and indirect resistance to the adoption of technologies (Brečka et al., 2022). The teachers' inclination towards traditional methods, rather than embracing technology-rich approaches, leads them to memorization-based strategies, thereby constraining students' comprehension of the broader environmental issues and impact (BlockBirds, 2023). In addition, teachers often lack the technical support or computer skills to address issues that arise when using tools and software in the classroom, which discourages them from adapting to new technologies (Fauville et al., 2014).

Risks of Content Privacy

The risk of privacy associated with the latest technology is another concern raised by many scholars. Kumpulainen et al. (2022) explored learners' experiences with digital tools in environmental education. They emphasized several challenges of integrating technologies and tools for educating the learners about natural phenomena. Kumpulainen et al. (2022) outlined three main categories of challenges: 'content risks' (exposure to harmful or age-inappropriate material), 'conduct risks' (cyberbullying), and 'contract risks' (data harvesting, commercial pressure, and encouragement to gamble). Thus, the learners are likely to be victims of both risks of inappropriate materials and content and privacy.

Limiting the Outdoor Experience

Field trips play a crucial role in EE by providing students with the opportunity to bridge theoretical textbook knowledge with real-world experiences (Cho & Park, 2023). VR technologies provide a useful means for conducting field trips to remote and inaccessible locations (Jerowsky & Borda, 2022; Buchanan et al., 2018). However, Ruchter et al. (2010) argue that VR technologies and virtual environments cannot replace the traditional field trips. In addition, the entertainment value inherent in these technologies often tempts both educators and learners to remain indoors, even when outdoor field trips are viable options (Fauville et al., 2014). As a result, emerging technologies and tools contribute to distancing students from the real world and surroundings.

CONCLUSION AND RECOMMENDATIONS

In conclusion, the integration of educational technologies in environmental education presents a myriad of benefits alongside notable challenges. The benefits include increased accessibility to the domain of knowledge, facilitation of positive change in behavior, resource efficiency, heightened environmental awareness, enhanced learning experiences and outcomes, improved collaboration and engagement, as well as the utilization of multiple technological tools and resources to deliver knowledge to facilitate environmental education and practice. However, these advantages are accompanied by challenges such as limiting outdoor experiences, privacy and security risks of stakeholders associated with digital content and resources and dependence, the scarcity of high-quality teaching materials, reluctance and resistance to adopt the technologies, affordability issues due to high costs, and the difficulty of staying focused on educational activities amid digital distractions. Despite the mentioned challenges, strategic planning and proper resource allocation can help environmental education and fight against the ever-deteriorating condition of the climate and environment as a whole.

The collective insights from various studies clearly show the significant role of integrating technologies in environmental education (EE) in institutions, curricula, and practice. There is a call for further research to explore metacognitive approaches, the impact of naive geography, and the alignment of GIS with cognitive processes to enhance spatial understanding among students (Bednarz, 2004). There is also a need for research on the professional development of environmental educators to develop digital competencies. The application of artificial intelligence, virtual reality, chatbots, and smartphone applications holds promise in raising students' environmental consciousness and engagement, albeit with challenges such as bias in AI algorithms, limited availability of VR technology, and ethical concerns (Cao & Jian, 2024). These issues related to emerging immersive technologies also demand further exploration. Moreover, future investigations should also delve into unexplored themes such as digital cartography and connections between environmental education, activism, and digital technologies. Finally, future research and environmental activism should focus on enhancing the usability, effectiveness, and inclusivity of technology-driven approaches to advance environmental literacy and action among diverse populations.

REFERENCES

Aguayo, C., & Eames, C. (2023). Using mixed reality (XR) immersive learning to enhance environmental education. *The Journal of Environmental Education*, 54(1), 58–71. 10.1080/00958964.2022.2152410

Al-Ansi, A. M., Jaboob, M., Garad, A., & Al-Ansi, A. (2023). Analyzing augmented reality (AR) and virtual reality (VR) recent developments in education. *Social Sciences & Humanities Open*, 8(1), 100532. 10.1016/j.ssaho.2023.100532

Ambe, B. A., Agbor, C. E., Amalu, M. N., Ngban, A. N., Bekomson, A. N., Etan, M. O., Ephraim, I. E., Asuquo, E. E., Eyo, O. E., & Ogunjimi, J. O. (2024). Electronic media learning technologies and environmental education pedagogy in tertiary institutions in Nigeria. *Social Sciences & Humanities Open*, 9, 100760. 10.1016/j.ssaho.2023.100760

Augmented Island Studios (AIS). (2024). Augmented Reality and Virtual Reality: A Game-Changer for Environmental Awareness. Retrieved from https://augmentedislandstudios.com/augmented-reality-and-virtual-reality-a-game-changer-for-environmental-awareness/

Baker, J. A. S. B. (2022). Impact Of Using Technology In Education Environment. *Journal of Pharmaceutical Negative Results*, 624–633.

Bartolini, S., & Bonatti, L. (2008). The role of social capital in enhancing factor productivity: Does its erosion depress per capita GDP? *Journal of Socio-Economics*, 37(4), 1539–1553. 10.1016/j.socec.2007.03.005

Bednarz, S. W. (2004). Geographic information systems: A tool to support geography and environmental education? *GeoJournal*, 60(2), 191–199. 10.1023/B:GEJO.0000033574.44345.c9

Bhang, K. J., & Huh, J. R. (2023). Effectiveness of Fine Dust Environmental Education on Students' Awareness and Attitudes in Korea and Australia Using AR Technology. *Sustainability (Basel)*, 15(22), 16039. 10.3390/su152216039

BlockBirds. (2023, November 19). Revolutionizing Environmental Education: Exploring the Power of AR Technology. Retrieved from https://blockbirds.de/en/uncategorized/ar-in-environmental-education/

Boucenna, M. R., Benzouai, M. C., & Adli, B. (2021). Effect of foreign direct investment on agriculture productivity: A PMG panel ARDL approach. *International Journal of Green Economics*, 15(1), 42. 10.1504/IJGE.2021.117676

Brečka, P., Valentová, M., & Tureková, I. (2022). Digital Technologies in Environmental Education. *TEM Journal*, 11(2), 726–730. 10.18421/TEM112-28

Buchanan, J., Pressick-Kilborn, K., & Maher, D. (2018). Promoting environmental education for primary school-aged students using digital technologies. *Eurasia Journal of Mathematics, Science and Technology Education*, 15(2), em1661.

Cao, F., & Jian, Y. (2024). The Role of integrating AI and VR in fostering environmental awareness and enhancing activism among college students. *The Science of the Total Environment*, 908, 168200. 10.1016/j.scitotenv.2023.16820037918744

Cho, Y., & Park, K. S. (2023). Designing Immersive Virtual Reality Simulation for Environmental Science Education. *Electronics (Basel)*, 12(2), 315. 10.3390/electronics12020315

Computeam. (2023, December 19). How Technology in Education Can Help Tackle Environmental Challenges. Retrieved from https://www.computeam.co.uk/videos-and-blog/article/how-technology-in-education-can-help-tackle-environmental-challenges

DoForms. (2023, September 13). 5 Technologies That Help the Environment & How Electronic Forms Preserve The Planet's Wellbeing. Retrieved from https://www.doforms.com/technologies-that-help-environment/

Fauville, G., Lantz-Andersson, A., & Säljö, R. (2014). ICT tools in environmental education: Reviewing two newcomers to schools. *Environmental Education Research*, 20(2), 248–283. 10.1080/13504622.2013.775220

Fernández, A. H., Camargo, C. D. B., & Nascimento, M. S. L. D. (2019). Technologies and Environmental Education: A Beneficial Relationship. *Research in Social Sciences and Technology*, 4(2), 13–30. 10.46303/ressat.04.02.2

He, M. (2023). VR in Environmental Science. Retrieved February 2, 2024 from https://www.vrwiki.cs.brown.edu/applications-of-vr/vr-in-environmental-science

Jakab, I., Zigová, M., & Pucherová, Z. (2019). Modernization of Environmental Education with the Use of Project-Based Learning, Outdoor Education, and Mobile Learning Supported by Information and Communication Technology. *Universities in the Networked Society: Cultural Diversity and Digital Competences in Learning Communities*, 223-248

Jardin, E. J. (2023). Exploring the Use of Technology in Environmental Education: Teachers' and Students' Perspectives. *Excellencia: International Multi-disciplinary Journal of Education (2994-9521), 1*(1), 118-126.

Jerowsky, M., & Borda, A. (2022, July 27). Virtual reality can support and enhance outdoor environmental education. *The University of British Columbia Magazine.* Retrieved from https://magazine.alumni.ubc.ca/2022/environment-technology/virtual-reality-can-support-and-enhance-outdoor-environmental-education

Khalifé, M., Chaker, R., & Gasparovic, S. (2022). Environmental education and digital solutions: An analysis of the Lebanese context's existing and possible digital actions. *Frontiers in Education*, 7, 958569. 10.3389/feduc.2022.958569

Kumpulainen, K., Kajamaa, A., Erstad, O., Mäkitalo, Å., Drotner, K., & Jakobsdóttir, S. (2022). *Nordic childhoods in the digital age: Insights into contemporary research on communication, learning and education*. Taylor & Francis. 10.4324/9781003145257

Lee, C., Chang, Y., & Wang, E. (2022). Crossing the rivers by feeling the stones: The effect of China's green credit policy on manufacturing firms' carbon emission intensity. *Energy Economics*, 116, 106413. 10.1016/j.eneco.2022.106413

Li, H., Chai, J., Qian, Z. F., & Chen, H. (2022). Cooperation strategies when leading firms compete with small and medium-sized enterprises in a potentially competitive market. *J Manag Sci Eng*, 7(3), 489–509. 10.1016/j.jmse.2022.02.003

Li, R., & Chen, Y. (2022). The influence of a green credit policy on the transformation and upgrading of heavily polluting enterprises: A diversification perspective. *Economic Analysis and Policy*, 74, 539–552. 10.1016/j.eap.2022.03.009

Lightfoot-Wright, A. D. (2021). Environmental Education, Digital Technologies, and Confronting New Realities in the Time of COVID-19. [MA dissertation, California State University]. Retrieved from https://scholarworks.calstate.edu/downloads/9k41zk91t

Lowan-Trudeau, G. (2023). Digital technologies and environmental education. *The Journal of Environmental Education*, 54(1), 1–7. 10.1080/00958964.2022.2152413

Lu, S. J., & Liu, Y. C. (2015). Integrating augmented reality technology to enhance children's learning in marine education. *Environmental Education Research*, 21(4), 525–541. 10.1080/13504622.2014.911247

Moghaddam, H., & Kunst, R. M. (2023). The Role of Natural Gas in Mitigating Greenhouse Gas Emissions: The Environmental Kuznets Curve Hypothesis for Major Gas-Producing Countries. *Sustainability (Basel)*, 2023(15), 4266. 10.3390/su15054266

Morgan, A. (2022, August 28). Information and Communication Technology impacts on Environmental Education Sustainable Developments of EU and USA. NEWS. Retrieved from https://www.uspa24.com/bericht-21509/information-and-communication-technology-impacts-on-environmental-education.html

Rogers, Y., Connelly, K., Hazlewood, W. R., & Tedesco, L. P. (2009). Enhancing learning: A study of how mobile devices can facilitate sensemaking. *Personal and Ubiquitous Computing*, 14(2), 111–124. 10.1007/s00779-009-0250-7

Ruchter, M., Klar, B., & Geiger, W. (2010). Comparing the effects of mobile computers and traditional approaches in environmental education. *Computers & Education*, 54(4), 1054–1067. 10.1016/j.compedu.2009.10.010

Smith, J., Johnson, A., & Thompson, D. (2018). The Impact of Augmented Reality on Student Engagement: A Case Study Analysis. *Journal of Educational Technology Systems*, 47(1), 122–139.

Stapp, W. B., Bennett, D., Bryan, W., Fulton, J., MacGregor, J., Nowak, P., & Havlick, S. (1969). The concept of environmental education. *The Journal of Environmental Education*, 1(1), 30–31. 10.1080/00139254.1969.10801479

Su, X., Pan, C., Zhou, S., & Zhong, X. (2022). Threshold effect of green credit on firms' green technology innovation: Is environmental information disclosure important? *Journal of Cleaner Production*, 380, 134945. Advance online publication. 10.1016/j.jclepro.2022.134945

Tabuenca, B., Moreno-Sancho, J. L., Arquero-Gallego, J., Greller, W., & Hernández-Leo, D. (2023). Generating an environmental awareness system for learning using IoT technology. *Internet of Things : Engineering Cyber Physical Human Systems*, 22, 100756. 10.1016/j.iot.2023.100756

The Goan. (2022, July 12). Virtual reality can support, enhance outdoor environmental education. Retrieved from https://www.thegoan.net/tg-life/%EF%BB%BFvirtual-reality-can-support-enhance-outdoor-environmental-education/86403.html

Wang, F., Du, L., & Tian, M. (2023). Does Agricultural Credit Input Promote Agricultural Green Total Factor Productivity? Evidence from Spatial Panel Data of 30 Provinces in China. *International Journal of Environmental Research and Public Health*, 20(1), 529. Advance online publication. 10.3390/ijerph2001052936612851

Willis, J., Weiser, B., & Kirkwood, D. (2014). Bridging the gap: Meeting the needs of early childhood students by integrating technology and environmental education. *International Journal of Early Childhood Environmental Education*, 2(1), 140–155.

Zheng, S., Zhang, X., & Wang, H. (2023). Green credit policy and the stock price synchronicity of heavily polluting enterprises. *Economic Analysis and Policy*, 77, 251–264. 10.1016/j.eap.2022.11.011

APPENDIX

Warm-Up Questions and Answers of the Chapter

1. What is the overall problem presented throughout the chapter (case)?

Answer: One of the biggest and most frequently faced problems identified in the current project was the lack of appropriate resources and framework to apply essential technologies involved in helping environmental education among the learners.

2. What are the factors affecting the problem(s) related to the current case?

Answer: The major factors creating the problem include insufficient exposure to essential educational technologies, lack of skills and training, and lack of institutional support. In addition, policies and practices have not yet adopted practicable initiatives in the classroom.

3. Discuss managerial, organizational, and technological issues and resources related to this case.

Answer: There are many gaps and weaknesses in managerial and organizational responses and initiatives to apply educational technologies to facilitate environmental education in a real classroom environment. The policymaking levels have not yet been successful in addressing these issues and forming a legitimate framework for applying educational technologies in the real environment.

4. What role do different players (decision-makers) play in the overall planning, implementation, and management of applying educational technologies in environmental education?

Answer: To date, leaders have been prevalent in policymaking, execution, monitoring, assessment, and reporting on the current situation of running environmental education through the use of educational technologies at the institutional, national, and global levels. They play the roles of forming the policies and pushing the required measures to introduce and adopt the policies into execution. Thus, they play the role of driving forces.

5. What are the alternatives that could be a possible solution to existing practices in a real classroom environment?

Answer: One of the pivotal alternatives is encouraging and developing project-based learning (PBL) courses and providing the required level of training for the stakeholders to harness the practical benefits of using technologies while teaching environmental education to the learners.

6. What are some of the emerging technologies that should be considered in solving the gaps and problems related to applying educational technologies while teaching EE?

Answer: Proper use of artificial intelligence (AI), augmented reality (AR), virtual reality (VR), Software, tools, and platforms can solve a wide range of problems and issues arising from challenges while teaching complex environmental issues with the features of simulations and demonstrations in the classroom settings.

7. What is the final solution that can be recommended to the management body of the organization addressed in this study?

Answer: Policies related to environmental education should be thoroughly revised and modified according to the needs of the time, model, and teaching-learning activities. If the policies cannot address the current problems, issues, and solutions, effective teaching and knowledge delivery cannot be assured, and the use of technologies and tools cannot also be practically aligned.

Epilogue and Lessons Learned

Technology has significantly impacted the dynamics of the educational landscape and practices. Similarly, educational technologies have also impacted immeasurably not only environmental education but also the whole educational sector worldwide. Environmental education has already experienced several uses of tools, platforms, software, and relevant technologies during teaching, resourcing materials, demonstrations, simulations, and many other activities. These technologies have brought a variety of opportunities and challenges when applied in real classroom environments. The current project has identified some significant strengths and shortcomings of applying these technologies to facilitate environmental education. However, the attention and measures from policymakers and major actors can solve the existing problems in the near future.

1st Lesson: Introduction to the use of educational technologies in environmental education

2nd Lesson: Sample class with demonstrations of using educational technologies in EE

3rd Lesson: Introduction of some essential platforms, databases, Apps, software, and tools used in EE for further use in the future

3. List of Additional Sources for Further Studies

The current study has covered some major issues involved with the current area of study and research. However, the readers, learners, teachers, and researchers can find additional information and guidelines through the following websites, platforms, and resources.

1. https://www.inaturalist.org/
2. https://ebird.org/home
3. https://naturalstart.org/
4. https://www.projectnoah.org/
5. https://merlin.allaboutbirds.org/
6. https://identify.plantnet.org/
7. https://www.ispotnature.org/
8. https://natureshare.org/
9. https://observation.org/
10. https://gobotany.nativeplanttrust.org/
11. https://www.vrwiki.cs.brown.edu/applications-of-vr/vr-in-environmental-science
12. https://blockbirds.de/en/uncategorized/ar-in-environmental-education/
13. https://augmentedislandstudios.com/augmented-reality-and-virtual-reality-a-game-changer-for-environmental-awareness/
14. file://www.thegoan.net/tg-life/virtual-reality-can-support-enhance-outdoor-environmental-education/86403.html
15. https://bio2lab.com.au/bringing-augmented-reality-to-environmental-education/
16. https://en.wikipedia.org/wiki/Educational_technology
17. https://en.wikipedia.org/wiki/Environmental_technology
18. https://www.canberra.edu.au/unit/11775/1/2024.html
19. https://magazine.alumni.ubc.ca/2022/environment-technology/virtual-reality-can-support-and-enhance-outdoor-environmental-education
20. https://pubmed.ncbi.nlm.nih.gov/37918744/
21. https://www.nsta.org/nstas-official-positions/environmental-education
22. https://www.moenv.gov.tw/en/7668B5A92350AB53
23. https://mgiep.unesco.org/article/building-a-technology-driven-environment-for-collaborative-learning-at-schools
24. https://www.mhskids.org/blog/technology-enhance-learning-environment/
25. https://www.doforms.com/technologies-that-help-environment/

26. https://www.uspa24.com/bericht-21509/information-and-communication -technology-impacts-on-environmental-education.html
27. https://nursing-informatics.com/blog/reducing-the-environmental-impacts-of -using-technologies/
28. https://epale.ec.europa.eu/en/blog/technology-planet-tools-and-applications -help-raise-environmental-awareness
29. https://edinburghsensors.com/news-and-events/environmental-technology-the -impact-of-technology-on-the-environment-and-environmental-technology/
30. https://medium.com/@isabelloucao18/the-role-of-technology-in-environmental -protection-e2f1519af074
31. https://eepro.naaee.org/research/eeresearch/reviewing-technology-use-classroom-environmental-education
32. https://www.computeam.co.uk/videos-and-blog/article/how-technology-in -education-can-help-tackle-environmental-challenges
33. https://naturalstart.org/feature-stories/bridging-gap-integrating-technology-and -environmental-education

Chapter 8
The Validity and Reliability of a Tool for Measuring Ecological Literacy

Fatma Alkan
https://orcid.org/0000-0003-2784-875X
Hacettepe University, Turkey

Esra Cantürk
Hacettepe University, Turkey

EXECUTIVE SUMMARY

Unconscious resource consumption and environmental problems pose a threat to humanity. Therefore, it is necessary for young individuals to gain ecological literacy knowledge and reflect it in their daily lives in order to know only the changing and developing world. The aim of this research is to adapt a scale to determine the ecological literacy levels of high school students. The sample of the study consists of approximately 530 volunteer high school students. In terms of construct validity, exploratory and confirmatory factor analyzes were performed on the data. Factor analysis of the scale was performed. Sixteen items with an item load of less than 0.45 and burdening more than one factor were removed from the scale, and the remaining 24 items were found to be grouped under five factors. A Cronbach's alpha (α=0.816) scale with high reliability, consisting of five factors and 24 items, including ecological behavioral literacy, information literacy, emotional literacy, ethical literacy, mindfulness literacy, was obtained. The reliability value of the scale is high.

DOI: 10.4018/979-8-3693-1710-5.ch008

INTRODUCTION

Scientific and technological developments facilitate human life on the one hand, and on the other hand increase the interaction between man and nature. This interaction, which has increased as a result of people making excessive use of the natural environment, has become a situation that harms nature (Eruz, 1989). Along with technological advances, people's unconscious production and consumption behaviors and careless and insensitive activities in industrialization have upset the ecological balance and caused concerns in people. Consequently, ecological issues have surfaced, encompassing pollution, depletion of natural resources, extinction of species, reduction of arable land, nuclear hazards, rapid population growth, drought, acid rain, global warming, climate change, as well as pollution of air, water, and soil, water scarcity, soil and forest degradation, toxic substances, radioactive pollution, species extinction, and hunger. These challenges stem from a combination of climatic, hydrological, anthropogenic, soil, and vegetation factors, altering the geographic landscape and posing concerns for the world and all living organisms (Eruz, 1989; Çabuk & Karacaoğlu, 2003). The aforementioned ecological problems, as a result of their occurrence anywhere in the world, negatively affect the entire ecosystem, including living and non-living elements, and disrupt the ecological balance. The deterioration of the ecological balance threatens the sustainability of humans and other living things. From any angle, the world has to deal with ecological problems (Akkuzu, 2016). The biggest culprit of this threat is humans. In order to solve ecological problems, the main reason of which is human behavior, individuals need to correct their behaviors and lifestyles, and create a sustainable society with an environmentalist approach. In this context, it is important to take responsibility for ecological problems and to change bad consumption habits that harm sustainable ecosystems. Otherwise, the person with the highest power of perceiving, understanding and thinking about events may face irreparable threats by destroying the place where they live (Kağar & Çakır Arıca, 2018; Özgün, 2018). Individuals try to understand themselves and nature, in other words, the world. He seeks answers by forming some questions in his own mind and tries to transfer the answers he finds to his life (Akdağ, 2014).

As ecological problems increase, more importance has been given to the science of ecology (Zeydanlı & Tuğ, 2008). Ecology is an important and developing science that examines the relations of human beings with the natural environment, of which they gain meaning in the natural environment and of which they form a part (Keleş & Hamamcı, 2002). With this importance, it is necessary to provide an effective environmental education in order to raise individuals with environmental awareness. With environmental education, individuals can both acquire ecological information and develop positive attitudes towards the environment and transform their attitudes into positive behaviors (Erten, 2005). In Turkey, environmental education is included in life studies, social studies and science courses in primary and secondary schools. When various curricula are examined within the framework of environmental education, it is seen that sufficient importance is given to this subject, but the curricula cannot reach their final goal due to deficiencies or inadequacies in the practices. In other words, it is seen that the education given about nature and environment is insufficient (Meydan, 2014). It is seen that individuals have cognitive and affective awareness with environmental education, but do not exhibit behavioral characteristics. Based on this, literate individuals who have a love of nature from a young age, respect all living things, love, protect and think about them, adopt nature-friendly approaches, and are aware of what their behaviors can lead to should be raised. For this reason, individuals should receive education in environments where they can recognize the natural existence of nature and the environment at a young age, develop themselves by living rather than theoretically, and get to know the environment in which they live. Environmental and human relations and ecological problems that arise as a result of the deterioration of these relations have become one of the most important issues of applied ecology. Thus, environmental science emerged. Environmental scientists examine how people and other living species are in a relationship with each other and their inanimate environment, and the problems and solutions that arise when these relations are broken (Yıldız et al., 2007).

Literacy encompasses the comprehension, interpretation, and meaningful attribution of individual experiences, objects, and societal interactions, alongside the engagement in reading and writing activities (Aşıcı, 2009). Ecological literacy, on the other hand, delves into the understanding of human interactions with both society and natural systems, promoting sustainable practices (Orr, 1992). Essentially, ecological literacy entails grasping the intricate natural processes vital for life on Earth and exhibiting a comprehensive set of knowledge, skills, attitudes, and values towards nature (TEMA Foundation, 2015). It can also be described as the capacity to apply ecological insights, thoughts, and behaviors in daily life, recreational pursuits, and environmental studies (Berkowitz et al., 2004). Mastery of the language of nature constitutes the proficient aspect of ecological literacy. The concern of entrusting the world and life to conscious ecologically literate generations is an issue that is gaining importance day by day. With ecological literacy, individuals should be provided with life skills in harmony with nature, and they should develop positive attitudes towards the environment and participate in the solution process of ecological problems. Thus, by expanding the domain of ecological literacy, individuals will be able to take more part in ecological dilemmas, develop their worldview, initiate changes in their personal habits regarding sustainability, and make decisions as ecologically literate (Balgopal & Wallace, 2009). Ecologically literate individuals can use their cognitive, affective and behavioral competencies effectively. Developing ecological awareness in individuals, teaching them the principles of ecology and establishing a positive relationship with nature will be possible with ecological literacy.

This study aims to introduce a scale for assessing the ecological literacy of secondary school students to the existing body of literature. Given the numerous benefits associated with scale adaptation studies over scale development endeavors (Hambleton & Patsula, 1999; Ha, Huang et al., 2022), the objective was to adapt the "Ecological Literacy" scale into Turkish and subsequently conduct a rigorous examination of its validity and reliability.

What Is the Overall Problem Presented in This Case?

Global environmental issues today have reached dimensions that threaten the future of all living beings. Guiding societies worldwide towards protecting and enhancing the environment, and finding common ground, are essential. Emphasis should be placed on innovative perspectives on environmental conservation and development, as well as the importance of education in environmental matters. Research is needed to explore what should be done to entrust the world and life to ecologically literate generations with awareness. Based on this, the theme of ecological literacy should be comprehensively addressed.

What Are the Factors Affecting the Problem Related to This Case?

The concept of ecological literacy, which forms a common foundation worldwide for the protection and enhancement of the environment, needs to be central. Ecological literacy refers to understanding the natural processes that sustain life in nature, or demonstrating knowledge, skills, attitudes, values, and understanding towards nature. It also encompasses the necessary knowledge, skills, and values for understanding nature and living sustainably. Ecological literacy can be defined as a competence that enables individuals to understand environmental issues, grasp the complexity of natural systems, and generate solutions based on sustainability principles. This concept encourages both individuals and societies to become environmentally conscious and take action.

Actions within the framework of ecological literacy include utilizing education to raise awareness about environmental issues. To achieve this, environmental education programs should be developed in educational institutions such as schools, public institutions where work continues, and other workplaces to enhance ecological literacy. When developing environmental education programs, national and international environmental policies should be reviewed, and current, practical content should be created accordingly. When emphasizing the importance of ecological literacy, eco-friendly technologies must be mentioned, and the positive contributions of innovative technologies to the environment should be highlighted. Another crucial aspect is individuals' consumption behaviors, which should be thoroughly addressed in environmental education programs, detailing what to consider when purchasing products.

Developing environmental education programs should outline responsibilities to increase ecological literacy levels among individuals in society, leave a livable world for future generations, include steps to reduce individuals' negative impacts on the environment, and ultimately contribute to solving global environmental issues.

What Role Do Different Decision-Makers Play in the Overall Planning, Implementation, and Management of the Information Technology Applications?

In today's information age, leveraging information technology has become essential to enhance and develop ecological literacy, which is crucial for addressing environmental issues. Information technology can illuminate the environmental impacts from the production to consumption of goods, as well as from raw material sourcing to waste management. It serves as a powerful tool to inform large audiences about environmental issues, raise awareness, and illustrate the consequences of negative trends. Educational materials, news, and videos developed through information technology are accessible to people worldwide, facilitating global outreach and engagement.

What Are Some of the Emerging Technologies That Should Be Considered in Solving the Problem(s) Related to The case?

Emerging technology's impact on environmental problems and ecological literacy is of great importance in terms of both its direct environmental effects and its role in environmental awareness and solution development processes. The potential of these technologies to support sustainability and environmental protection can play a critical role in achieving the goal of leaving a more livable world to future generations.

What Is the Final Solution That Can Be Recommended to the Management of the Organization Described in the Case?

There are many technological approaches to reduce environmental problems and improve ecological literacy. Some important technologies to consider in this context are renewable energy technologies, waste management and recycling technologies, agriculture and water resources management technologies. Expanding the use of renewable energy resources such as solar, wind, hydro, geothermal and biomass, which are renewable energy technologies, will reduce the need for fossil fuels and help reduce the carbon emission rate. In order to reduce the amount of waste and prevent waste of raw materials by recycling waste, waste management and recycling technologies should be taught at all levels of education and their importance should be drawn. Another important point is agriculture and water resources management technologies. In order to make agricultural production more efficient and environmentally friendly, attention should be drawn to the use of smart irrigation technologies during agricultural production, and emphasis should be placed on the management of water resources for sustainable agriculture. Effective and efficient use of information technology plays an important role in improving ecological literacy and ultimately reducing environmental problems.

RESEARCH METHOD

The study is a methodological examination focusing on the validity and reliability of a measurement instrument. Conducted in 2022 by Ha, Huang et al., this research aims to assess the ecological literacy levels among adults. Its objective is to adapt the measurement tool titled "Assessing ecological literacy and its application based on linguistic ecology: a case study of Guiyang City, China" to Turkish and apply it to high school students. The process of scale adaptation entails adjusting scales developed within distinct cultural contexts to fit different languages and cultures (Kılıçer & Odabaşı, 2010). Therefore, scale adaptation studies involve not only translating the scale into another language but also conducting necessary analyses to ensure validity, reliability, and cultural appropriateness.

Study Group

The study involves high school students enrolled in institutions under the Ministry of National Education for the academic year 2022-2023. Approximately 530 volunteer high school students from different grades in Ankara constitute the study sample. Literature suggests having 5-10 times more participants than the number of items in a scale for scale studies (Grimm & Yarnold, 1995; Worthington & Whittaker, 2003). With the ecological literacy scale comprising 40 items, the inclusion of 530 participants was considered sufficient for the scale adaptation process. Moreover, efforts were made to include high school students from diverse schools and classes to ensure the validity and reliability of the measurement tool across a broad spectrum of the target audience.

The study employed a suitable sampling approach in selecting its sample. Utilizing the convenient sampling method, the researcher opted for a sample closely aligned with the research problem and readily accessible (Fraenkel & Wallen, 2012; Yıldırım & Şimşek, 2016). This approach entails selecting groups that are easily reachable and manageable, considering potential constraints related to finances, time, and labor (Büyüköztürk et al., 2014). Table 1 provides the frequency and percentage distributions of the demographic characteristics within the sample group.

Table 1. Demographic Information of Sampling

Demographic Features		f	%
Gender	Female	286	56,2
	Male	223	43,8
School type	Anatolian high school	370	72,7
	Science high school	14	2,8
	Vocational high School	21	4,1
	Imam Hatip high school	24	4,7
	Private high school	80	15,7
Class	9	91	17,9
	10	146	28,7
	11	55	10,8
	12	217	42,6
Total		500	100

Data Collection Tools

Since this research is a scale adaptation study, information about the scale adapted from the original to Turkish is given in this section.

Original scale. The Ecological Literacy Scale, introduced by Ha, Huang, Zhang, and Dong in China in 2022, aimed to evaluate adults' competence across five dimensions: ecological information literacy, ecological awareness literacy, ecological ethical literacy, ecological emotional literacy, and ecological behavioral literacy.

During its development, the framework and dimensions of "ecological literacy" were established, encompassing the aforementioned dimensions. Questions were then devised under each dimension to compile the item pool. The creation of the item pool drew inspiration from the "Chinese Urban Public Environmental Awareness Survey" of 2005, developed by the State Environmental Protection Administration Public

Environmental Awareness Research Group and the Public Environmental Awareness Research Group of the Chinese Academy of Social Sciences. Additionally, insights were gleaned from the ecological literacy assessment scale and survey questions by Pitman, Daniels, and Sutton (2018) from the University of South Australia.

The initial draft of the scale underwent rigorous examination by experts, leading to the elimination of items deemed inappropriate for the sub-dimensions. Subsequently, a pilot study was conducted, during which the scale's content was meticulously analyzed and refined based on feedback and suggestions. Following the pilot study's findings, the final version of the scale was formulated.

The Ecological Literacy Scale employs a five-point Likert scale, ranging from "Never" to "Very familiar," corresponding to values 1 to 5, respectively. It consists of 40 items divided into five sub-dimensions: ecological information literacy, ecological awareness literacy, ecological ethical literacy, ecological emotional literacy, and ecological behavioral literacy, each comprising eight items. Validity and reliability evaluations were conducted with a sample size of 494 participants. The scale demonstrates strong internal consistency, indicated by a Cronbach's Alpha coefficient of 0.888 for the entire scale. Scores on the scale range from a minimum of 40 to a maximum of 200.

Scale Adaptation Process

Prior to commencing the adaptation process of the Ecological Literacy Scale into Turkish, the original authors, Ha, Huang et al., were contacted via email to seek their opinion and obtain permission for the adaptation study. Following their approval and guidance, the adaptation study adhered to the eight-step framework outlined by Şeker and Gençdoğan (2006). These steps involved the translation of the English version into Turkish by separate experts, followed by verification by a Turkish language teacher, researcher, and chemistry educator. Subsequently, the Turkish version underwent further editing and scrutiny by an expert proficient in English, along with review by a field education specialist. This meticulous process ensured the accuracy of translation and language validity by rectifying any items that were ambiguous or inaccurately expressed. Both the English and Turkish versions of the scale were administered to a group of high school students proficient in a foreign language, with a one-week interval between administrations. The correlation coefficients between the scores obtained from the English and Turkish versions were then compared. Subsequently, researchers implemented the scale in secondary schools, collecting pilot application data in line with the insights garnered from this process.

Analysis of Data

Following the administration of the translated ecological literacy scale to high school students, data collection commenced. The collected data were then input into a computerized environment, marking the initiation of the analysis process. Before determining the sub-factors and reliability levels of the measurement tool, known as the "Ecological Literacy Scale," utilized in the research, a thorough check was conducted to identify any missing or erroneous data within the dataset. Subsequently, the validity and reliability of the data acquired from the measurement tool were rigorously tested.

Validity. The construct validity of the scale underwent scrutiny through exploratory factor analysis(EFA). This analysis aimed to explore and identify the underlying factor structure within the data. Subsequently, the validity of the determined factor structure was assessed using confirmatory factor analysis(CFA), which aimed to ascertain whether the identified model adequately represented the data. The determined structure was then further verified through confirmatory factor analysis(CFA), ensuring its robustness and appropriateness for the given dataset.

Exploratory factor analysis. Exploratory factor analysis(EFA) was performed to determine the construct validity of the scale. Before starting the analysis, it was checked whether the data set met the exploratory factor analysis assumptions such as sample size, sample homogeneity, linear correlation between variables, normality, and outlier (Tabachnick & Fidell, 2007). As part of these assumptions, 15 data were excluded because they were missing. The remaining 500 data meet the assumptions.

The data's suitability for factor analysis was assessed through Bartlett's Test of Sphericity and the Kaiser-Meyer-Olkin(KMO) test. Bartlett's test evaluates whether the data derive from a multi-dimensional normal distribution, with significance indicating suitability for factor analysis. Meanwhile, the KMO test evaluates the adequacy of the sample size, with values above 0.60 considered acceptable (Field, 2013). The decision to proceed with factor analysis is typically based on a significant Bartlett's test and a KMO value exceeding 0.60 (Tabachnick & Fidell, 2007).

Principal Components Analysis (PCA) was utilized to conduct Exploratory Factor Analysis (EFA) and explore the factor structure of the scale. Varimax rotation was applied to the principal components to facilitate interpretation. Common variance values ideally fall within the range of 0.5, although lower common variances can be acceptable with sample sizes of 500 or more (Tabachnick & Fidell, 2013). Additionally, items with anti-image correlation values below 0.5 were scrutinized for potential exclusion from the analysis (Field, 2013). When determining the number of factors, those with eigenvalues exceeding 1 were retained, with reference to the Scree plot diagram for additional guidance. Furthermore, items were included in the scale if their factor loadings were 0.45 or higher, ensuring their contribution to the underlying factors (Field, 2013).

Confirmatory factor analysis(CFA) plays a critical role in assessing the construct validity of the scale. It verifies whether the factor structure identified through exploratory factor analysis (EFA) aligns with the scale, providing additional validation for the factor structure revealed by EFA. CFA is commonly employed in the scale development process to validate the underlying factor structure.

In the analysis conducted using the Amos 23 program, the Chi-square goodness-of-fit ($\chi 2$) statistic, a fit index calculated for the CFA model, was initially examined. Ideally, this value should not be significant; however, it's important to consider that the Chi-square test is sensitive to sample size. To address this sensitivity, the $\chi 2/sd$ ratio was evaluated, with values below 3 typically indicating a perfect fit for large samples.

Furthermore, several other goodness-of-fit indices were assessed, including the root mean square error of approximation(RMSEA), comparative fit index(CFI), goodness fit index(GFI), adjusted goodness fit index(AGFI), standardized root mean square error(SRMR), normed fit index(NFI), and incremental fit index(IFI). These indices were evaluated based on predefined fit intervals, providing comprehensive insights into the adequacy of the model fit.

Table 2. Standard Goodness-Of-Fit Indices

Fit Measure	Perfect Fit	Acceptable Fit	
χ^2/df	$0 \leq \chi^2/df \leq 3$	$3 \leq \chi^2/df \leq 5$	Kline, 2005; Sümer, 2000
RMSEA	$0 \leq$ RMSEA $\leq .05$	$.05 \leq$ RMSEA $\leq .08$	Hu and Bentler, 1999; Jöreskog and Sörbom, 1993; Hair, Black, Babin and Anderson, 2009
CFI	$.95 \leq$ CFI ≤ 1.00	$.90 \leq$ CFI $\leq .95$	Hu and Bentler, 1999; Sümer, 2000
GFI	$.95 \leq$ GFI ≤ 1.00	$.90 \leq$ GFI $\leq .95$	Baumgartner and Homburg, 1996; Bentler, 1980; Marsh, Hau, Artelt, Baumert and Peschar, 2006
AGFI	$.90 \leq$ AGFI ≤ 1.00	$.85 \leq$ AGFI $\leq .90$	Schermelleh-Engel and Moosbrugger, 2003
NFI	$.95 \leq$ NFI ≤ 1.00	$.90 \leq$ NFI $\leq .95$	Hu and Bentler, 1999; Tabachnick and Fidell, 2013;
IFI	$.95 \leq$ IFI ≤ 1.00	$.90 \leq$ IFI $\leq .95$	Hu and Bentler, 1999

Reliability. In assessing the reliability of the scale, Cronbach's α coefficient was computed. This reliability coefficient evaluates the level of internal consistency within the scale. Typically, a Cronbach's α value of 0.70 or higher is considered indicative of satisfactory reliability (Horn, 1965).

RESEARCH RESULTS

Findings Regarding Validity

Findings Related to Exploratory Factor Analysis

The data's suitability for factor analysis was thoroughly assessed using Bartlett's Test of Sphericity and the Kaiser-Meyer-Olkin(KMO) measure. The KMO value obtained was 0.871, indicating a high level of adequacy for factor analysis (Field, 2013). Additionally, Bartlett's test produced a significant result ($\chi2 = 2775.479$, p $< .001$), suggesting that the data did not originate from a spherically distributed population, thus confirming the suitability of the data for factor analysis (Tabachnick and Fidell, 2007). These findings are summarized in Table 3.

Table 3. KMO and Bartlett Test Results

Tests	χ^2	Sd	p
Barlett	2775.479	276	.000
Kaiser-Meyer-Olkin	.871		

Upon examination of the table, it was evident that the data exhibited a multivariate normal distribution, affirming its suitability for factor analysis. To explore the factor structure of the scale, exploratory factor analysis(EFA) was conducted using Principal Components Analysis. Varimax rotation was applied to facilitate interpretation based on the principal components.

The common variance value fell within the acceptable range, indicating conformity with the default values for factor analysis. Items with anti-image correlation values below 0.5 were carefully considered, ensuring adherence to analysis standards. When determining the number of factors, those with eigenvalue statistics greater than 1 were retained, guided by the Scree plot diagram. Additionally, items with factor loadings below 0.45 were excluded from further analysis.

Following rigorous assessment, 6 items (4, 9, 21, 25, 31, 33) with factor loadings below 0.50 and 10 items (8, 10, 14, 15, 8, 10, 14, 15) exhibiting non-conjunctive loadings under two factors, differing by more than 0.1, were excluded from the analysis one at a time. This iterative process of item refinement was repeated multiple times until the final form of the items was determined, ensuring the scale's optimal distribution.

Ultimately, the scale was deemed to have an ideal distribution in its final form, confirming its five-dimensional structure as acceptable. The developed scale, based on the results of the exploratory factor analysis, accounted for 49.92% of the total variability. Consequently, a structure comprising 5 factors and 24 items was established. The items categorized under each factor, along with their respective factor loadings, are presented in Table 4.

Table 4. Items and Factor Loads in the Scale

	Factor-1	Factor-2	Factor-3	Factor-4	Factor-5
Item 36	**.746**				
Item 39	.733				
Item 38	.703				
Item 35	.650				
Item 37	.646				
Item 40	.637				
Item 5		.675			
Item 7		.655			
Item 6		.609			
Item 2		.590			
Item 3		.556			
Item 1		.409			
Item 27			.705		
Item 26			.665		
Item 28			.657		
Item 29			.530		
Item 17				.659	
Item 20				.575	
Item 24				.561	
Item 18				.560	
Item 22				.546	
Item 13					.745
Item 11					.553
Item 12					.512
Eigenvalue	5.002	2.976	1.531	1.316	1.155
Total variance explained (Total=% 49.92)	13.15	11.29	8.83	8.8	7.85

Upon examining the table, it becomes apparent which items constitute each factor. These factors were named in alignment with the original scale: "Ecological Behavioral Literacy" (Factor 1), "Ecological Information Literacy" (Factor 2), "Ecological Emotional Literacy" (Factor 3), "Ecological Ethical Literacy" (Factor 4), and "Ecological Awareness Literacy" (Factor 5).

Factor 1 comprises six items, namely Item 36-39-38-35-37 and Item 40, with factor loadings ranging from 0.637 to 0.746. The eigenvalue of Factor 1 in the overall scale is 5.002, contributing 13.15% to the total variance.

Factor 2 consists of six items, including Items 5- 7- 6 – 2- 3, and Item 1, with factor loadings ranging from 0.409 to 0.675. The eigenvalue of Factor 2 in the overall scale is 2.967, contributing 11.29% to the total variance.

Factor 3 comprises four items, namely Item 27- 26- 28, and Item 29, with factor loadings ranging from 0.530 to 0.705. The eigenvalue of Factor 3 in the overall scale is 1.531, contributing 8.83% to the total variance.

Factor 4 consists of five items, including Item 17- 20- 24-18, and Item 22, with factor loadings ranging from 0.546 to 0.659. The eigenvalue of Factor 4 in the overall scale is 1.316, contributing 8.8% to the total variance.

Factor 5 comprises three items, namely Item 13-11- and Item 12, with factor loadings ranging from 0.512 to 0.745. The eigenvalue of Factor 5 in the overall scale is 1.155, contributing 7.85% to the total variance.

Upon examining the total variance of the items and factors included in the scale, it was determined that these items and factors accounted for 49.92% of the total variance.

When the scree plot given in Figure 1 is examined, it is seen that 24 items have a 5-factor structure.

Figure 1. Scree Plot Graph for the Number of Factors in the Scale

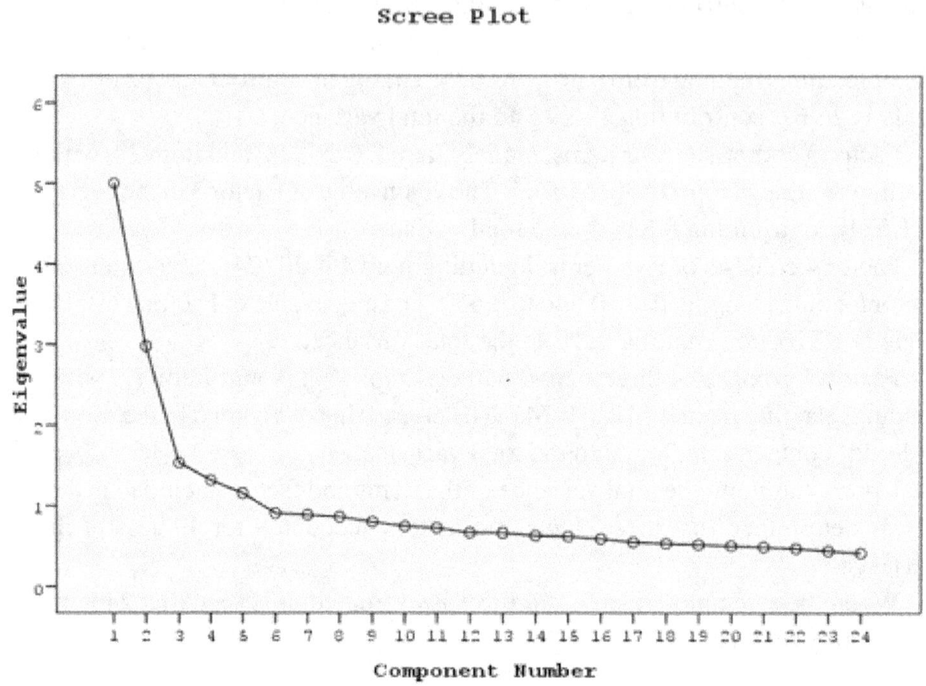

Findings on Confirmatory Factor Analysis

Confirmatory factor analysis (CFA) was conducted to elucidate the relationships between the items within each factor of the scale, which was determined to comprise five factors through exploratory factor analysis (Yılmaz & Çelik, 2009; Raykov & Marcoulides, 2006). The maximum likelihood technique was employed in this CFA, and multiple fit values were reported, in accordance with recommendations in the literature (Kaiser, 1958; O'Rourke, Psych, & Hatcher, 2013). Consequently, the obtained fit values were interpreted and reported based on Table 5 (Kline, 2005).

Table 5. Fit Values

Fit measure	Perfect fit	Acceptable fit	Fit value obtained from the scale	Fit degree
χ^2/df	$0 \leq \chi^2/df \leq 3$	$3 \leq \chi^2/df \leq 5$	2.377	Perfect Fit

continued on following page

Table 5. Continued

Fit measure	Perfect fit	Acceptable fit	Fit value obtained from the scale	Fit degree
RMSEA	$0 \leq$ RMSEA $\leq .05$	$.05 \leq$ RMSEA $\leq .80$	0.053	Perfect Fit
CFI	$.95 \leq$ CFI ≤ 1.00	$.90 \leq$ CFI $\leq .95$	0.867	Acceptable Fit
GFI	$.95 \leq$ GFI ≤ 1.00	$.90 \leq$ GFI $\leq .95$	0.891	Acceptable Fit
AGFI	$.90 \leq$ AGFI ≤ 1.00	$.85 \leq$ AGFI $\leq .90$	0.866	Acceptable Fit
NFI	$.95 \leq$ NFI ≤ 1.00	$.90 \leq$ NFI $\leq .95$	0.793	Acceptable Fit
IFI	$.95 \leq$ IFI ≤ 1.00	$.90 \leq$ IFI $\leq .95$	0.868	Acceptable Fit

When Table 6 is examined, as a result of the confirmatory factor analysis performed without any limitations using the maxibmum probability technique, the values of goodness of fit $\chi 2(sd=265)= 629.875$, $p<.000$, RMSEA= 0.053, RMR= 0.086, S-RMR= 0.042, GFI= 0.891, AGFI= 0.866, CFI= 0.867, NFI= 0.793 and IFI= 0.868. According to these values, it can be said that $\chi 2/df$ and RMSEA goodness-of-fit values show perfect fit, and all other goodness values show acceptable fit. The factors of this model obtained in line with this information are verified by the data. The factorial model of the scale and the t values of the factor-item relationship are given in Figure 2.

Figure 2. Confirmatory Factor Analysis Diagram

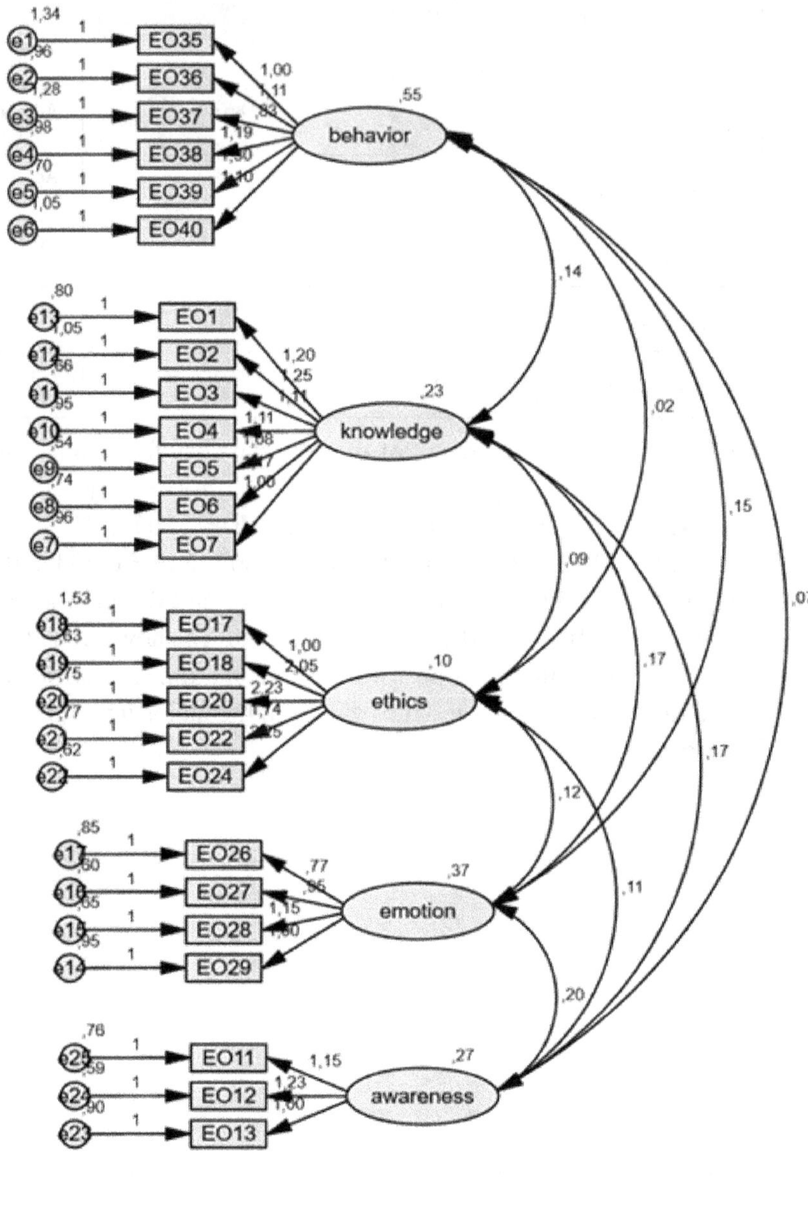

Findings Regarding the Reliability of the Scale

The reliability of the scale was assessed through internal consistency and stability analyses conducted on the obtained data. Below, I present the procedures and findings of these analyses.

Table 6. Internal Consistency Levels

Factors	Number of items		Cronbach's Alpha
Ecological Behavioral Literacy (Factor 1)	6		0.790
Ecological Information Literacy (Factor 2)	6		0.699
Ecological Emotional Literacy (Factor 3)	4		0.639
Ecological Ethical Literacy (Factor 4)	5		0.666
Ecological Awareness Literacy (Factor 5)	3		0.579
Total	24		0.816

Upon examination of Table 6, it is evident that the Cronbach's Alpha internal consistency coefficient for the entire scale, consisting of five factors and a total of 24 items, is $\alpha = 0.816$, indicating a high level of reliability for the scale as a whole. Moreover, the Cronbach's Alpha values for the individual factors range between 0.579 and 0.790.

Specifically, the reliability coefficient for the first factor is $\alpha = 0.790$, for the second factor is $\alpha = 0.699$, for the third factor is $\alpha = 0.639$, for the fourth factor is $\alpha = 0.666$, and for the fifth factor is $\alpha = 0.579$. These values demonstrate that while there may be slight variations in reliability across the factors, the overall reliability of the scale remains high. These internal consistency coefficients ensure that the scale can consistently measure the constructs it intends to assess.

DISCUSSION

As a result of this study, which was carried out with the aim of adapting a measurement tool for the assessment of ecological literacy, the result of this study was ecological behavioral literacy, ecological information literacy, ecological emotional literacy, ecological ethical literacy, ecological awareness literacy, consisting of five factors and 24 items, a Cronbach's Alpha (α=0.816) scale with high reliability was obtained.

To evaluate the construct validity of the scale, both exploratory and confirmatory factor analyses were conducted. Initially, factor analysis was employed to uncover the scale's underlying factor structure. Principal Component Analysis, alongside the Varimax vertical rotation technique, facilitated this process. Consequently, 16 items with a loading below 0.45 and exhibiting cross-loadings were eliminated from the scale. Subsequently, the remaining 24 items were categorized into five factors. Factor loadings ranged from 0.409 to 0.746, elucidating 49.92% of the total variance. This percentage, surpassing the 40% threshold, provides sufficient evidence for the scale's usage (Büyüköztürk, 2011; Eroğlu, 2008; O'Rourke, Psych & Hatcher, 2013). Confirmatory factor analysis was then conducted to validate the obtained five-factor structure. The resulting fit index values indicated both perfect and acceptable fit levels.

When assessing the reliability of the scale, Cronbach's Alpha internal consistency coefficient was computed for both the individual factors and the entire scale. The internal consistency coefficients for the factors ranged from 0.579 to 0.790. These values indicate a high level of reliability for the scale as a whole.

When delving into ecological literacy and its fundamental constituents, five key factors emerge: ecological information literacy, ecological awareness literacy, ecological ethical literacy, ecological emotional literacy, and ecological behavioral literacy. Ecological literacy essentially serves as a compass for fostering sustainable ecological behaviors, acquiring and disseminating ecological knowledge, promoting ecological conservation awareness, and ultimately attaining a heightened level of ecological literacy (Ha, Huang et al., 2022). Put differently, these five dimensions of ecological literacy form a cohesive whole, with each holding significant theoretical importance, interplaying interactively. Among them, ecological information literacy serves as the bedrock, ecological awareness literacy points towards actionable directions, ecological ethical literacy underscores moral standards, ecological emotional literacy acts as an internal driving force, and ecological behavioral literacy represents the ultimate objective. Ecological literacy is indispensable for comprehending the intricacies of the contemporary world, understanding the nature of ecological challenges, and crafting sustainable lifestyles (Boehnert, 2013). It aims to comprehensively grasp how individuals and societies can become agents of destruction (Orr, 1992). Ecologically literate individuals are adept at leveraging their cognitive, affective, and behavioral competencies effectively. Fostering ecological awareness, imparting ecological principles, and fostering a harmonious bond with nature are all attainable through ecological literacy initiatives.

When the literature is examined, environmental attitude (Lee, 2008), environmental concern (Mostafa, 2009), environmentally friendly consumption knowledge (Mohr et al., 1998), environmentally friendly consumption behavior (Lee, 2008), environmental literacy at primary, secondary and university levels. There are many scales such as (Öcal & Önsüz, 2020; Atabek-Yiğit et al., 2014). Before starting a course in high school, students' perceptions of the subject should be determined and course contents should be arranged according to these perceptions (Altundağ et al., 2022). Anxiety towards a lesson or subject will decrease as the level of knowledge on that subject increases (Koçak Altundağ & Yücel, 2022). Revealing the relationship between courses and daily life in the high school curriculum will increase the knowledge of teachers and students on subjects such as chemistry and the environment, while also improving their information literacy (Mustafaoğlu & Yücel, 2022a, b). The change in literacy will lead to greater focus on environmental and ecology issues. This scale adaptation study can be used as a resource that can be used while carrying out studies such as the ecological literacy scale at the high school level, environment, sustainability, and ecological intelligence. The research data is limited to 500 high school students participating in the research, it can be said that the application of the scale to different sample groups and the evaluation of the obtained data will contribute to the literature.

CONCLUSIONS AND RECOMMENDATIONS

Students studying in secondary education have been deficient in ecology knowledge due to problems such as ecological problems, environmental pollution, and pandemics they have witnessed during the current period, they have the perspective required by the age and have problems in accessing resources. The main problem to be addressed in the study is to examine in detail the issue of ecological literacy, which is important for a sustainable future. When the literature is examined, there is no survey on ecological literacy. Ecological literacy scale was adapted to Turkish in order to eliminate the deficiency in this subject.

The research is limited to the data collected from the Ecological Literacy Scale and high school students in the sample group of the research. The validity and reliability study of this scale for middle school, high school and university students can be examined. The relationship between ecological literacy and different variables should be investigated.

EPILOGUE

The greatest threat humanity faces today is environmental issues. These problems manifest in our lives as climate change, biodiversity loss, and pollution of water, air, and soil. Dealing with environmental problems requires going back to basics. It should start with educating humans, who are the primary culprits, to tackle these issues. Therefore, research should focus on ecological literacy. Ecological literacy involves acquiring and spreading ecological knowledge, increasing awareness of ecological conservation, and ultimately guiding ecological behavior towards sustainable development.

Ecological literacy evolves individuals to understand the world, respond to social and environmental challenges, and develop new skills by transferring ecological knowledge, attitudes, and behaviors. Individuals should be educated as ecological literates to evaluate various environmental and social challenges faced by modern society and to generate sustainable solutions. Ecological literacy also enables individuals to understand environmental issues, learn sustainable use of natural resources, and adopt environmentally friendly behaviors.

Technological advancements and scientific discoveries can play a critical role in this process, but the greatest change will come from increasing individuals' and communities' sensitivity and responsibility towards the environment. To reduce environmental problems and build a sustainable future, everyone must contribute and take conscious steps.

REFERENCES

Akdağ, H. (2014). Sosyal Bilgilerin Tanımı, Amacı, Önemi ve Türkiye'deki Yeri. Sosyal Bilgiler Öğretiminde Yeni Yaklaşımlar I (2. Basım). (s.2-21). Ankara: Pegem Akademi.

Akkuzu, N. (2016). Towards a profound ecological understanding: Statistical attempts to measure our ecological intelligence. *International Journal of Social Sciences and Education*, 6(2), 198–216.

Altundağ, C. K., Mustafaoğlu, F. T., & Yücel, A. S. (2022). Developing a scale: Perceptions of high school students about the relationship between chemistry and daily life. *Jurnal Ilmiah Peuradeun*, 10(3), 721–746. 10.26811/peuradeun.v10i3.718

Aşıcı, M. (2009). Kişisel ve sosyal bir değer olarak okuryazarlik. *Değerler Eğitimi Dergisi*, 7(17), 9–26.

Atabek-Yiğit, E., Köklükaya, N., Yavuz, M., & Demirhan, E. (2014). Development and validation of environmental literacy scale for adults (ELSA). *Journal of Baltic Science Education*, 13(3), 425–435. 10.33225/jbse/14.13.425

Balgopal, M. M., & Wallace, A. M. (2009). Decisions and dilemmas: Using writing to learn activities to increase ecological literacy. *The Journal of Environmental Education*, 40(3), 13–26. 10.3200/JOEE.40.3.13-26

Baumgartner, H., & Homburg, C. (1996). Applications of structural equation modeling in marketing and consumer research: A review. *International Journal of Research in Marketing*, 13(2), 139–161. 10.1016/0167-8116(95)00038-0

Bentler, P. M. (1980). Multivariate analysis with latent variables: Causal modeling. *Annual Review of Psychology*, 31(1), 419–456. 10.1146/annurev.ps.31.020180.002223

Berkowitz, A. R., Ford, M. E., & Brewer, C. A. (2005). A Framework for Integrating Ecological Literacy, Civics Literacy and Environmental Citizenship in Environmental Education. In Johnson, E. A., & Mappin, M. J. (Eds.), *Environmental Education or Advocacy: Perspectives of Ecology and Education in Environmental Education* (pp. 227–265). Cambridge University Press.

Boehnert, J. (2013). *The visual communication of ecological literacy: Designing, learning and emergent ecological perception*. PhD, University of Brighton

Büyüköztürk, Ş. (2011). *Manual of data analysis for social sciences* [Sosyal bilimler için veri analizi el kitabı]. Ankara, Pegem Publishing.

Büyüköztürk, Ş., Kılıç Çakmak, E., Akgün, Ö. E., Karadeniz, Ş., & Demirel, F. (2014). *Scientific research methods* [Bilimsel araştırma yöntemleri]. Pegem Publishing.

Çabuk, B., & Karacaoğlu, C. (2003). Üniversite öğrencilerinin çevre duyarlılıklarının incelenmesi. Ankara Üniversitesi, Eğitim Bilimleri Fakültesi Dergisi, 36(1-2).

Çakır Arıca, Ş., & Kağar, C. (2018). The key of leaving a life to the future generations: Ecological literacy. *Doğanın Sesi*, 2, 31–42.

Erten, S. (2005). Okul öncesi öğretmen adaylarinda çevre dostu davranişlarin araştirilmasi. *Hacettepe Üniversitesi Eğitim Fakültesi Dergisi*, 28, 91–100.

Eruz, E. (1989). Kentleşmede ekolojik yaklaşım. *İstanbul Üniversitesi Deniz Bilimleri ve Coğrafya Enstitüsü Bülteni,* 79-83

Field, A. (2013). *Discovering Statistics Using IBM SPSS Statistics* (4th ed.). SAGE.

Fraenkel, J. R., Wallen, N. E., & Hyun, H. H. (2012). *How to design and evaluate research in education* (8th ed.). McGraw Hill.

Grimm, L. G., & Yarnold, P. R. (1995). *Reading and understanding multivariate statistics*. American Psychological Association.

Ha, C., Huang, G., Zhang, J., & Dong, S. (2022). Assessing ecological literacy and its application based on linguistic ecology: A case study of Guiyang City, China. *Environmental Science and Pollution Research International*, 29(13), 18741–18754. 10.1007/s11356-021-16753-734704227

Hair, J. F.Jr, Black, W. C., Babin, B. J., & Anderson, R. E. (2009). *Multivariate Data Analysis* (7th ed.). Prentice Hall.

Hambleton, R. K., & Patsula, L. (1999). Increasing the validity of adapted tests: Myths to be avoided and guidelines for improving test adaptation practices. *Journal of Applied Testing Technology.*, 1(1), 1–30.

Horn, J. L. (1965). A rationale and test for the number of factors in factor analysis. *Psychometrika*, 30(2), 179–185. 10.1007/BF0228944714306381

Hu, L. T., & Bentler, P. M. (1999). Cutoff criteria for fit indexes in covariance structure analysis: Conventional criteria versus new alternatives. *Structural Equation Modeling*, 6(1), 1–55. 10.1080/10705519909540118

Jöreskog, K. G., & Sörbom, D. (1993). *LISREL 8: Structural equation modeling with the SIMPLIS command language. Scientific Software International.* Lawrence Erlbaum Associates, Inc.

Kaiser, H. F. (1958). The varimax criterion for analytic rotation in factor analysis. *Psychometrika*, 23(3), 187–200. 10.1007/BF02289233

Kılıçer, K., & Odabaşı, H. F. (2010). Bireysel yenilikçilik ölçeği (BYÖ): Türkçeye uyarlama, geçerlik ve güvenirlik çalışması. *Hacettepe Üniversitesi Eğitim Fakültesi Dergisi, 38*(38).

Kline, R. B. (2005). *Principles and practice of structural equation modeling* (2nd ed.). Guilford Press.

Koçak Altundağ, C., & Yücel, A. S. (2022). Development of an anxiety scale for chemistry. *Journal for Educators. Teachers and Trainers*, 13(1), 1–8. 10.47750/jett.2022.13.01.001

Lee, E. B. (2008). Environmental attitudes and information sources among African American college students. *The Journal of Environmental Education*, 40(1), 29–42. 10.3200/JOEE.40.1.29-42

Marsh, H. W., Hau, K. T., Artelt, C., Baumert, J., & Peschar, J. L. (2006). OECD's brief self-report measure of educational psychology's most useful affective constructs: Cross-cultural, psychometric comparisons across 25 countries. *International Journal of Testing*, 6(4), 311–360. 10.1207/s15327574ijt0604_1

Meydan, A. (2014). Sosyal Bilgiler Öğretimde Gezi-Gözlem ve Doğa Eğitimi. Sosyal Bilgiler Öğretiminde Yeni Yaklaşımlar I (s. 242-260). Ankara: Pegem Akademi.

Mohr, L. A., Ero lu, D., & Ellen, P. S. (1998). The development and testing of a measure of skepticism toward environmental claims in marketers' communications. *The Journal of Consumer Affairs*, 32(1), 30–55. 10.1111/j.1745-6606.1998.tb00399.x

Mostafa, M. M. (2009). Shades of green: A psychographic segmentation of the green consumer in Kuwait using self-organizing maps. *Expert Systems with Applications*, 36(8), 11030–11038. 10.1016/j.eswa.2009.02.088

Mustafaoğlu, F.M., & Yücel, A.S. (2022a). Developing context-based teaching competencies of chemistry teachers: designing and implementing context-based activities. *Jurnal Penelitian dan Pembelajaran IPA, 8*(2), 126-152. 10.30870/jppi.v8i2.16491

Mustafaoğlu, F. M., & Yücel, A. S. (2022b). Context-based teaching experiences of chemistry teachers: Expectations, gains and applicability conditions. *Journal of Turkish Science Education*, 19(3), 958–978. 10.36681/tused.2022.158

O'Rourke, N., Psych, R., & Hatcher, L. (2013). *A step-by-step approach to using SAS for factor analysis and structural equation modeling*. Sas Institute.

Öcal, E. E., & Önsüz, M. F. (2020). Reliability and validity of ecological literacy scale for primary school students. *European Journal of Public Health*, 30(Supplement_5), 5. 10.1093/eurpub/ckaa166.103

Orr, D. (1992). *Ecological literacy*. State of New York Press.

Özgün, B. B. (2018). *Öğretmen Adaylarının Eko-Okuryazarlık Düzeylerinin İncelenmesi. Yayımlanmamış doktora tezi*. İnönü Üniversitesi Eğitim Bilimleri Enstitüsü.

Pitman, S. D., Daniels, C. B., & Sutton, P. C. (2018). Ecological literacy and psychographics: Lifestyle contributors to ecological knowledge and understanding. *International Journal of Sustainable Development and World Ecology*, 25(2), 117–130. 10.1080/13504509.2017.1333047

Pituch, K. A., & Stevens, J. P. (2016). *Applied multivariate statistics for the social sciences: Analyses with SAS and IBM's SPSS* (6th ed.). Routledge.

Raykov, T., & Marcoulides, G. A. (2006). *A first course structural equation modeling*. Lawrence Erlbaum Assocation Inc. Publishers.

Schermelleh-Engel, K., Moosbrugger, H., & Müller, H. (2003). Evaluating the Fit of Structural Equation Models: Tests of Significance and Descriptive Goodness-of-Fit Measures. *Methods of Psychological Research Online*, 8(2), 23–74.

Şeker, H., & Gençdoğan, B. (2006). *Measurement Tool Development in Psychology and Education* [Psikolojide ve Eğitimde Ölçme Aracı Geliştirme]. Nobel Press.

Sümer, N. (2000). Yapısal Eşitlik Modelleri: Temel Kavramlar ve Örnek Uygulamalar. *Türk Psikoloji Yazilari*, 3, 49–73.

Tabachnick, B. G., & Fidell, L. S. (2007). *Using Multivariate Statistics* (5th ed.). Allyn and Bacon.

Tabachnick, B. G., & Fidell, L. S. (2013). *Using multivariate statistics* (6th ed.). Allyn and Bacon.

TEMA (2011-2015) Türkiye Erozyonla Mücadele Vakfı Genel Müdürlüğü Arşivi, İstanbul.

Ullman, J. B. (2013). Structural Equation Modeling. In Tabachnick, B., & Fidell, L. (Eds.), *Using Multivariate Statistics* (pp. 681–785). Pearson Education Inc.

Worthington, R. L., & Navarro, R. L. (2003). Pathways to the future: Analyzing the contents of a content analysis. *The Counseling Psychologist*, 31(1), 85–92. 10.1177/0011000002239402

Yıldırım, A., & Şimşek, H. (2013). *Qualitative research methods in the social sciences* (9th ed.). Seçkin Press.

Yıldız, K., Sipahioğlu, Ş., & Yılmaz, M. (2007). *Çevre Bilimi* (Genişletilmiş 2. Basım). Ankara: Gündüz Yayıncılık.

Yılmaz, V., & Çelik, E. H. (2009). *Structural equation modeling with Lisrel: basic concepts, applications, programming* [Lisrel ile yapısal eşitlik modellemesi: temel kavramlar, uygulamalar, programlama]. Pegem Publishing.

Zeydanlı, U., & Tuğ, S. (2008). Biyolojik Çeşitlilik ve Orman Ekosistemlerimdeki Önemi. Orman ve Biyolojik Çeşitlilik (s. 15-37). Ankara: Doğa Koruma Merkezi.

ADDITIONAL READING

Chambers, J. M. (2007). Ecological Literacy Materials for Use in Elementary Schools: A Critical Analysis. Unpublished doctoral thesis. Alberta University, Edmonton, Kanada.

Chin, L. S., & Wahid, R. (2020). Digital game-based learning in enhancing recycling awareness. *International Journal of Academic Research in Progressive Education and Development*, 9(2), 269–275. 10.6007/IJARPED/v9-i2/7300

Cook, E. L., & Ormsby, A. A. (2020). Measuring and developing ecological literacy to conserve the critically endangered Mariana Crow. *Applied Environmental Education and Communication*, 20(3), 288–300. 10.1080/1533015X.2020.1859418

Norazlan, S. A., & Said, I. (2022). Understanding children's preferences: Developing ecological literacy in the village landscape. *Journal of Adventure Education and Outdoor Learning*, 1–2.

Noviana, E., Kurnianam, O., Huda, M. N., Ramadhan, Z. H., Erlinda, S., Hidayati, R., & Rusmiati, T. (2018). How to foster students' ecoliteracy in primary school? *International Conference on Elementary Education Universitas Pendidikan Indonesia*, 120-126.

Putri, S. S., Japar, M., & Bagaskorowati, R. (2019). Increasing ecoliteracy and student creativity in waste utilization. *International Journal of Evaluation and Research in Education*, 8(2), 255–264.

Salimi, M., Dardiri, D., & Sujarwo, S. (2021). The profile of students' eco-literacy at nature primary school. *Cypriot Journal of Educational Sciences*, 16(4), 1450–1470. 10.18844/cjes.v16i4.5999

KEY TERMS AND DEFINITIONS

Ecological Literacy: It is the understanding of how people and societies interact reciprocally and with natural systems and how they can do this sustainably.

Ecology: Environmental science is a group of branches of science that examines the causal relationships and interactions between living beings and their biotic and abiotic environments, the circulation of matter and energy that ensures the continuity of life, within the framework of self-renewing functional space units.

Literacy: Along with reading and writing activities, it is the individual's ability to perceive, understand and interpret the life he lives and the objects and events in this life, and to give meaning to his relationships in his social life.

Chapter 9
The Place of Ecological Literacy in High School Biology Curriculum:
The Example of Turkey

Dilek Sultan Acarlı
https://orcid.org/0000-0003-1090-8912
Hacettepe University, Turkey

EXECUTIVE SUMMARY

For a sustainable future, it is crucial to cultivate individuals with ecological literacy skills. As in every field, the process that begins within the family to increase the number of ecologically literate individuals must be supported and sustained through education in schools. Consequently, curricula and courses should be structured to promote ecological literacy. This study examines the presence of ecological literacy in the high school biology curriculum using the document analysis method. The research findings indicate that topics and learning outcomes related to ecological literacy are limited in the overall program and are not included at all grade levels. Moreover, it has been determined that the outcomes related to the ethical, emotional, and behavioral dimensions of ecological literacy are insufficient. Recommendations have been made to ensure that the curriculum is structured to support ecological literacy.

DOI: 10.4018/979-8-3693-1710-5.ch009

INTODUCTION

In the 21st century, the world expects individuals to have literacy skills in various areas. So, what does it mean to be literate? The concept of literacy, which first emerged in the late 19th century, initially referred to the the ability to read and write in its simplest form (Stibbe, 2009). Today, however, the concept of literacy encompasses the process of meaning-making, including identity formation, agency, and using and learning literacy skills (Häggströmi & Schmidt, 2020). Therefore, this concept has evolved beyond the ability of just reading and has gained a deeper meaning that includes learning and understanding by using these skills. One of the literacy skills expected of individuals in today's world is environmental literacy. The concept of environmental literacy was first used by Charles Roth (McBride et al., 2013; Roth, 1968; Roth, 1992). Roth (1968) defined environmental literacy as an individual's level of environmental knowledge and awareness. In a broader sense, environmental literacy encompasses the combination of individuals' knowledge about the environment and their attitudes and behaviors towards the environment (Hsu & Roth, 1998; Pe'er et al., 2007; Roth, 1992). The concept of ecological literacy, which is addressed in this study, is more narrowly focused and has evolved from environmental literacy, and these two concepts are inseparable. Ecology is the scientific discipline that examines the relationships of living organisms with their environment and other living beings (Brewer, 1994). Ecological literacy, which is a combination of the words ecology and literacy, generally means understanding the natural processes that make life on earth possible (Orr, 1992). However, researchers differ in defining the meaning and scope of the concept of ecological literacy, which is an abstract concept. Throughout the historical development of the concept, many researchers have discussed ecological literacy in various ways (Braus & Wood, 1993; Bruyere, 2008; Capra, 1999; Huang & Ha, 2021; Huang & Zhao, 2019; Orr, 1992; Pitman & Daniels, 2016). According to Orr (1992), ecological literacy consists of the components of knowing, caring, and applying. Braus and Wood (1993) define ecological literacy as a complex concept involving skills and attitudes related to environmental issues and participation in environmental matters. Capra (1999) describes ecologically literate individuals as those who understand the fundamental principles of ecology and act in accordance with these principles in their daily lives. Berkowitz et al. (2005; p228) define ecological literacy as "*the ability to use ecological understanding, thinking, and habits of mind for living in, enjoying, and/or studying the environment.*" In recent years, there has been an observed increase in the number of studies that examine the concept of ecological literacy by categorizing it into different components. For example, Bruyere (2008) categorizes ecological literacy into ecological knowledge, ecological attitudes, and ecological behaviors. Huang and Zhao (2019) address ecological literacy in four categories: ecological knowledge literacy, ecological ethics literacy, ecological emotional literacy, and ecological behavioral literacy. One of the most recent studies on this subject belongs to Ha et al. (2022). Ha et al. (2022), by adding the "ecological awareness literacy" category to the 4 categories identified by Huang and Zhao (2019), discussed ecological literacy in five dimensions, each considered theoretically almost equally important: (1) ecological knowledge literacy; (2) ecological awareness literacy; (3) ecological ethics literacy; (4) ecological emotional literacy; and (5) ecological behavioral literacy. These dimensions influence each other interactively. Of these, ecological knowledge literacy is foundational, ecological awareness literacy indicates the direction of action, ecological ethics literacy emphasizes moral standards, ecological emotional literacy is the internal driving force, and ecological behavioral literacy is the ultimate goal (Ha et al., 2022).

In these rapidly changing days of our planet, there is a need for conscientious citizens who can understand the inherent complexity of ecological systems to tackle the environmental challenges emerging. At this point, ecological literacy can play an important role in restoring the disrupted world balance and preventing climate change. As a matter of fact, ecologically literate individuals are those who are aware of the principles and limits of nature, adopting a lifestyle in harmony with nature. Therefore, it is expected that they act with an understanding of natural balances and the importance of ecosystems, causing less harm to the environment. Orr (2005) stated that ecological literacy requires knowledge, caring, and practical competence. Ecological literacy implies a broad understanding of how humans are sustainably connected with each other and natural systems, requiring an awareness of interconnectedness of life and an understanding of how the world functions as a physical system. In fact, ecological literacy involves understanding our place in the evolutionary history. An ecologically literate individual becomes aware of the destructive impact of human activities, understanding the dimensions of the crisis in species extinction, soil loss, deforestation, climate change, desertification, ozone depletion, population growth, resource depletion, air and water pollution, toxic and radioactive contamination, and the use of resources and energy (Orr, 2005).

LITERATURE REVIEW

Studies emphasizing the importance of ecological literacy suggest that ecological literacy education should be provided from an early age. Research indicates that ecological literacy education given at a young age increases children's knowledge and awareness of nature, fosters positive attitudes toward the environment, and instills a willingness to participate in environmental activities (Barberán et al., 2016; Cutter-Mackenzie & Smith, 2003; Irianto et al., 2020; Johnson & Manoli, 2010;

Lewis et al., 2010). Hardiansyah et al. (2023) have considered ecological literacy as a teaching tool for science learning in primary schools in their research. They reported that outside the classroom, in a learning environment interacting with the environment, students' learning motivation increased. In this regard, it is observed that ecological literacy education also benefits students in terms of learning motivation. Environmental education, which forms the basis of ecological literacy education, is an educational process that begins in the family, continues during school, and aims to create awareness about the environment after leaving school (Akınoğlu & Sarı, 2009). For environmental education to be effective, the process starting in the family needs to be effectively supported in schools. In other words, as in every field, the role of education in increasing the number of ecologically literate individuals in society is significant. Especially curricula and primary education occupy an important place in this regard. For this reason, curricula and courses need to be organized in a way that supports environmental education. When reviewing the literature, it is noticeable that there is a considerable amount of research on developing ecological literacy in preschool and primary school children, while there is a limited number of studies at the secondary education level. However, existing research indicates that the knowledge of secondary education students in ecology is not at the desired level (Kutay & Güngör Cabbar, 2022; Niklanović et al., 2014; Prastiwi et al., 2019). Furthermore, when research results are examined, it can be seen that attitudes towards the environment and environmental problems change negatively as age increases. For example, Oğuz et al. (2010) concluded that university students had lower awareness of environmental problems and responsible behavior towards the environment than expected in their research. Tuncer et al. (2009) discussed the role of pre-service teachers in raising environmentally literate students and ultimately becoming an environmentally literate society, and reported that the majority of pre-service teachers participating in their research did not have sufficient knowledge to be classified as having an acceptable level of environmental knowledge. Similarly, Güven and Aydoğdu (2012) reported, in their research with pre-service teachers, that their behavior towards environmental problems was moderate. In their study examining the environmental literacy and self-efficacy levels of elementary school pre-service teachers, Saribas et al. (2014) stated that although the participants had relatively high attitudes, anxiety, and perceptions towards the environment, they did not have sufficient environmental knowledge or self-efficacy beliefs related to environmental education. Çabuk and Haktanır (2017) revealed in their research that the levels of knowledge among pre-service preschool teachers regarding ecological literacy were significantly below the expected level.

There are also studies conducted in secondary education reporting that students do not have sufficient environmental and ecology knowledge and attitudes at the desired level. For example, Kutay and Güngör Cabbar (2022) examined the attitudes of high school students towards environmental problems. According to the findings obtained from their research, it was determined that the attitudes of high school students towards the environment were generally moderate and low. When the average scores of students' attitudes towards environmental problems were evaluated according to the class levels, it was observed that the average scores increased as the class level increased. This finding was interpreted as indicating that the school has a role in positively influencing attitudes towards environmental problems. In their study to determine the environmental literacy of high school students, Can et al. (2016) found that there was a statistically significant difference in favor of female students in terms of attitudes, interest, and use towards the environment, whereas there was a statistically significant difference in favor of male students in terms of environmental knowledge. They also found that as the class level increased, students' environmental knowledge, interest levels, and positive behaviors towards environment increased, but emphasized that students' environmental literacy skills need to be developed. Holden (2007) conducted a study with primary and secondary school students and stated that primary school students believed that environmental problems needed to be solved through environmental behavior, while the majority of secondary school students stated that there was no need for any environmental behavior towards environmental problems. However, studies conducted at the primary and secondary school level indicate that attitudes towards the environment and environmental problems are higher than at the high school and university level (Atasoy & Ertürk, 2008; Aydın & Çepni, 2012; Nalçacı & Beldağ, 2012; Peker & Ceylan, 2020; Teyfur, 2008; Soğukpınar & Korucu, 2020). Nevertheless, graduates of secondary education are now expected to be ready to take their place in society as adults. Therefore, the ecological literacy skill that is expected to be acquired in primary education should be supported in the secondary education process in order to maintain its continuity. In this regard, it is crucial to have subjects, learning outcomes, and guidelines in secondary education programs that aim to educate and support students as ecologically literate individuals.

Aim of the Study

This study aims to analyze to what extent ecological literacy is addressed in the high school biology curriculum in Turkey (Ministry of National Education [MofNE], 2018) and to discuss the results and develop recommendations accordingly. The research questions to be answered for this purpose are as follows:

- What are the learning outcomes related to ecological literacy in the biology curriculum?
    - In which units of which class are these outcomes included?

- o To which dimension of ecological literacy do these outcomes relate?

METHOD

Document analysis, one of the qualitative research methods, was used in the research. Document analysis is a series of processes that occur during the examination and evaluation of both printed and electronic materials (Bowen, 2009). Through document analysis, various forms of written text, which are the primary source of research data, are collected, reviewed, questioned, and analyzed (O'Leary, 2017). In this process, a lot of information can be accessed in the field without the need for observation and interviews. In this respect, it provides time-saving for the researcher (Yıldırım & Şimşek, 2013). In document analysis, the process begins with selecting the document and accessing it. Depending on the subject of the research, the selected document is limited and all or the parts to be examined are determined (Kıral, 2020). The document used as a data source in the research is the high school biology curriculum that was accepted and implemented in Turkey in 2018. Content analysis method (Mayring, 2002) was used in the analysis of the curriculum. The strongest feature of content analysis is the controlled and systematic analysis of data. The data material is examined according to category systems developed on the material based on theory. Categories can be created in two ways: by creating categories based on theory and by obtaining categories from the data material. In this study, a deductive structured content analysis was used, which employed pre-existing categories created based on theory. Within the framework of structured content analysis, the biology curriculum was analyzed in the context of ecological literacy according to the theoretical framework proposed by Ha et al. (2022). To do this, first, the definition, dimensions, indicators, and criteria that distinguish the dimensions of ecological literacy from each other were reviewed. Then, the learning outcomes related to ecological literacy in the curriculum were categorized under the relevant dimensions of ecological literacy.

A database was created in the MS Excel Program to code the learning outcomes within the theoretical framework, track and report the coding. Firstly, class, unit name, subject, and outcome statements were entered into the MS Excel program. Then, the outcome statements were coded under the relevant dimension of ecological literacy. At this stage, the coding made by two field experts was compared, and any differences in coding were discussed and agreed upon.

Below, the general structure of the biology course curriculum, which is the document analyzed, and place of environmental issues in the program are explained. In addition, the theoretical structure used as a criterion in the content analysis in examining how ecological literacy is addressed in the program is briefly introduced:

High School Biology Curriculum

The High School Biology Curriculum (MofNE, 2018), which is the document of the research, is based on a unit-based approach. The program includes three units at the 9th grade level, three units at the 10th grade level, two units at the 11th grade level, and four units at the 12th grade level. Table 1 provides information on the units by grade level, the number of learning outcomes, and the predicted course hours for each unit, and their percentages at the grade level.

Table 1. Units, Number of Learning Outcomes, and Duration of High School Biology Curriculum by Grade Level (MofNE, 2018)

Grade	Unit	Number of learning outcomes	Course time (hour)	Rate (%)
9	9.1. Life Science Biology	3	26	36.1
	9.2. Cell	3	22	30.6
	9.3. World of Living Things	5	24	33.3
	Total	11	72	100
10	10.1. Cell Division	5	18	25
	10.2. General Principles of Inheritance	2	30	41.7
	10.3. Ecosystem Ecology and Current Environmental Issues	10	24	33.3
	Total	17	72	100
11	11.1. Human Physiology	29	116	80.6
	11.2. Community and Population Ecology	5	28	19.4
	Total	34	144	100
12	12.1. From Gene to Protein	8	56	38.9
	12.2. Energy Transformations in Living Organisms	8	32	22.2
	12.3. Plant Biology	11	44	30.6
	12.4. Living Things and Environment	2	12	8.3
	Total	29	144	100

In terms of the environment, in the curriculum, the 9th grade includes the World of Living Things, the 10th grade includes Ecosystem Ecology and Current Environmental Issues, and the 11th grade includes Community and Population Ecology units. World of Living Things unit in the 9th grade consists of two subjects: "Biodiversity and Classification of Living Things" and "Kingdoms of Living Things and Their Characteristics". Ecosystem Ecology and Current Environmental Issues unit in the 10th grade includes three subjects: "Ecosystem Ecology", "Current Environmental Issues and Humans" and "Conservation of Natural Resources and Biodiversity". Community and Population Ecology unit in the 11th grade consists of two subjects: "Community Ecology" and "Population Ecology". Additionally, the 12th grade includes a unit titled "Living Things and Environment", which focuses on genetic changes in living things due to environmental conditions. Therefore, the curriculum includes three units related to ecological literacy, with a total of seven subjects. When looking at the distribution of these units and subjects across the grades, it is observed that environmental and ecological subjects are predominantly addressed in the 10th and 11th classes, which are intermediate grades. Furthermore, under the heading "Issues to Consider in the Implementation of the Curriculum", it is emphasized that students should be directed to research, observe, and examine the living things around them and their interactions with the environment they live in, as well as to participate in activities such as field trips, botanical garden visits, natural history museum visits, nature reserve visits, and factory visits. However, it is mentioned that the successful biology projects in national and international project competitions should be evaluated in terms of their contributions to science, society, technology, environment and economy.

The curriculum presents learning outcomes using a 4-digit numbering system that includes grade level, unit number, subject title number, and outcome number. Furthermore, some outcomes are accompanied by explanatory sentences (as a., b., etc.) regarding the content limitations of the outcome, points to consider during instruction, and methods and applications that can be used during teaching and learning process.

Theoretical Framework

In this study, while examining the unit and learning outcomes related to ecological literacy in the biology curriculum, the five dimensions in the definition of ecological literacy by Ha et al. (2022) have been taken as a basis. Ha et al. (2022) has suggested that the concept of ecological literacy can be examined under five dimensions. These dimensions, which they defined as the first-level indicators of ecological literacy, are expressed as ecological knowledge literacy, ecological awareness literacy, ecological ethics literacy, ecological emotional literacy, and ecological behavioral literacy. In order to facilitate the investigation and measurement of these indicators in real situations, a hierarchical process was followed, and second-level indicators were identified under the five first-level indicators. For ecological knowledge literacy, the second-level indicators are "Ecosystem knowledge", "Knowledge of damage to the ecological environment", "Knowledge of the relationship between humans and nature", and "Ecological and environmental protection knowledge". For ecological awareness literacy, the second-level indicators are "Ecological environmental protection behavior subject consciousness", "Ecological environmental protection value awareness", "Awareness of the severity of current ecological and environmental problems", and "Making judgments on the ecological environment damage encountered". For ecological ethics literacy, the second-level indicators are "Correctly recognizing the relationship between humans and nature", "The ethics and morality of protecting the ecological environment", "Affirming the role of nature", and "Respecting and cherishing all living things". For ecological emotional literacy, the second-level indicators are "Awe of the natural environment", "Love for the natural environment", "Sensitivity to natural environment protection", and "Ability to take responsibility for ecological and environmental issues". For ecological behavioral literacy, the second-level indicators are "Daily practice of environmental protection", "Participation in environmental education activities", "Scientific environmental protection skills and methods", and "Positive influence on the environmental protection behavior of others". Additionally, in the hierarchical process they followed in their study, Ha et al. (2022) suggested 8 questions for measuring each of these 5 dimensions, with answer options of 5-point Likert type. Under ecological knowledge literacy, questions related to ecosystems, the importance of living organisms in ecosystem balance, waste classification, and low-carbon living are included. Under ecological awareness literacy, there are questions related to awareness of the importance of ecological environmental protection and consciousness on this issue. Under the title of ecological ethics literacy, questions related to the role of nature, perception of the human-nature relationship, respect and value for other living things are included. In the questions under the title of ecological emotional literacy, love, admiration, and protective approach towards the natural environment are addressed. Under the title of ecological behavioral literacy, questions related to direct or indirect activities related to environmental protection are included.

In this study, the High School Biology Curriculum in Turkey (MoNE, 2018) was examined within the theoretical framework of ecological literacy as described above. The study identified which learning outcomes of the curriculum are related to ecological literacy and which dimension of ecological literacy they are related to, by considering the relevant questions for determining the second-level indicators and dimensions (Ha et al., 2022).

FINDINGS

As a result of the document analysis, the relevant learning outcomes categorized under the 5 dimensions of ecological literacy are presented in Table 2. Here, on the right side of the table, the first and second-level indicators constituting the theoretical framework of ecological literacy (Ha et al., 2022) are given, and on the left side, the learning outcomes determined as the counterparts of these indicators in the curriculum analysis are provided. The numbering system used for the outcomes is as given in the curriculum (The first digit represents the grade, the second digit represents the unit number, the third digit represents the subject number, the fourth digit represents the outcome number, and if available, an explanation related to the outcome is provided (e.g., a, b, etc.)). The outcomes are discussed along with their explanations. In other words, some outcomes correspond to multiple categories of ecological literacy due to the explanations under them. In such cases, considering the explanations and the content of the subject, some outcomes were classified in more than one category.

Table 2. Learning Outcomes Related to Ecological Literacy in the Biology Curriculum and Their Corresponding Theoretical Dimensions

Theoretical Framework of Ecological Literacy (Ha et al., 2022)		**High School Biology Curriculum (MofNE, 2018)**
First-level indicators	**Second-level indicators**	**Learning outcomes related to ecological literacy ***
Ecological knowledge literacy	1- Ecosystem knowledge 2- Knowledge of damage to the ecological environment 3- Knowledge of the relationship between humans and nature 4- Ecological and environmental protection knowledge	10.3.1.1**. Describes the relationship between the living and non-living components of the ecosystem. *a. The relationship between population, community, and ecosystem is explained with examples.*
		10.3.1.2. Explains the nutrition modes in living organisms with examples.
		10.3.1.3. Analyzes the flow of matter and energy in the ecosystem. *a. The role of producers, consumers, and decomposers in the flow of matter and energy is examined. b. The flow of matter and energy in ecosystems is exemplified by relating it to food chains, food webs, and trophic pyramids.*
		10.3.1.4. Establishes the relationship between matter cycles and the sustainability of life.
		10.3.3.2. Questions the importance of biological diversity for life. *a. The factors that make Turkey rich in terms of biodiversity are discussed.*

continued on following page

Table 2. Continued

Theoretical Framework of Ecological Literacy (Ha et al., 2022)		**High School Biology Curriculum (MofNE, 2018)**
First-level indicators	**Second-level indicators**	**Learning outcomes related to ecological literacy ***
		11.2.1.1. Explains the factors affecting the structure of the community. *It is emphasized that the biological diversity contained in communities is related to latitude in terrestrial ecosystems, and to the depth and pollution of water in aquatic ecosystems.*
		11.2.1.2. Explains intraspecific and interspecific competition in the community with examples.
		11.2.1.3. Explains the symbiotic relationships between species in the community with examples.
		11.2.1.4. Explains the succession in communities with examples.
Ecological awareness literacy	1- Ecological environmental protection behavior subject consciousness 2- Ecological environmental protection value awareness 3- Awareness of the severity of current ecological and environmental problems 4- Making judgments on the ecological environment damage encountered	9.3.2.2. Explains the contributions of living organisms to biological processes, economy, and technology with examples.
		10.3.1.1. Explains the relationship between the living and non-living components of the ecosystem. *b. Emphasizes the possible consequences of any changes that may occur in the ecosystem.*
		10.3.1.3. Analyzes the flow of matter and energy in the ecosystem. *c. Investigates and discusses the negative effects of biological accumulation on human health and other organisms.*
		10.3.2.1. Evaluates the causes and possible consequences of current environmental issues. *a. Current environmental issues (decrease in biodiversity, air pollution, water pollution, soil pollution, radioactive pollution, noise pollution, acid rain, global climate change, erosion, destruction of natural habitats and forest fires) are summarized and the negative effects of these problems on living organisms are stated.* *b. Emphasizes the diseases that occur due to environmental problems.*
		10.3.2.2. Questions his role as an individual in the emergence of environmental issues. *a. Applications related to ecological footprint, water footprint, and carbon footprint are made.*
		10.3.2.3. Proposes solutions for the prevention of environmental pollution in local and global contexts. *a. Examples of studies carried out to prevent environmental pollution in local and global contexts are given.* *b. Human activities that harm the environment on a local and global scale are discussed.* *c. Provides examples of how biology is related to other disciplines in the prevention of environmental pollution.*

continued on following page

Table 2. Continued

Theoretical Framework of Ecological Literacy (Ha et al., 2022)		High School Biology Curriculum (MofNE, 2018)
First-level indicators	**Second-level indicators**	**Learning outcomes related to ecological literacy ***
		10.3.3.1. Explains the importance of sustainability of natural resources. *a. The importance of environmental awareness is emphasized by exemplifying successful practices for the sustainability of natural resources in Turkey.*
		10.3.3.2. Questions the importance of biodiversity for life. *b. The value and importance of endemic species in terms of our country's biodiversity are emphasized and examples of their contributions to health and economy are given.*
		10.3.3.3. Proposes solutions for the conservation of biodiversity. *a. Studies carried out to protect endangered species and endemic species in Turkey are exemplified.* *c. The necessity of gene banks is stated.*
		11.2.2.1. Analyzes the factors affecting population dynamics. *a. Human age pyramids are emphasized.* *b. Different growth curves (S and J) related to population growth are drawn.* *c. Population change in the world and in our country is analyzed through graphics and possible consequences are discussed.*
Ecological ethics literacy	1- Correctly recognizing the relationship between humans and nature 2- The ethics and morality of protecting the ecological environment 3- Affirming the role of nature 4- Respecting and cherishing all living things	10.3.3.1. Explains the importance of the sustainability of natural resources. *b. The necessity of not wasting natural resources in order to leave a livable and healthy world for future generations is emphasized.*
		10.3.3.2. Questions the importance of biodiversity for life. *c. It is emphasized that biological diversity and endemic species are a global and national heritage.* *ç. The importance of respecting biological diversity and the natural functioning of the ecosystem and the need to avoid interference is emphasized by highlighting that every living organism plays important roles in nature.* *d. It is emphasized that extinct species are an irreplaceable loss in terms of biodiversity.*
Ecological emotional literacy	1- Awe of the natural environment 2- Love for the natural environment 3- Sensitivity to natural environment protection 4- Ablility to take responsibility for ecological and environmental issues	10.3.2.2. Questions his role as an individual in the emergence of environmental issues. *b. The development of solution suggestions to reduce ecological, water, and carbon footprints is encouraged.*

continued on following page

Table 2. Continued

Theoretical Framework of Ecological Literacy (Ha et al., 2022)		**High School Biology Curriculum (MofNE, 2018)**
First-level indicators	**Second-level indicators**	**Learning outcomes related to ecological literacy ***
Ecological behavioral literacy	1- Daily practice of environmental protection 2- Participation in environmental education activities 3- Scientific environmental protection skills and methods 4- Positive influence on the environmental protection behavior of others	10.3.1.1. Explains the relationship between the living and non-living components of the ecosystem. *c. Students are encouraged to prepare a presentation introducing an ecosystem of their choice.*
		10.3.1.3. Analyzes the flow of matter and energy in the ecosystem. *ç. Students are encouraged to construct a food web that illustrates the feeding relationships among organisms.*
		10.3.2.3. Proposes solutions for the prevention of environmental pollution in local and global contexts.
		10.3.3.3. Proposes solutions for the conservation of biodiversity. *b. Solutions for the conservation of biodiversity and the prevention of biopiracy are discussed.*

* Considering the explanations and the content of the subject, some learning outcomes were classified in more than one category.

**The numbering system of the learning outcomes is as given in the curriculum. The first number is the class, the second number is the unit number, the third number is the subject number and the fourth number is the outcome number.

When Table 2 is examined, it is seen that there are a total of 26 learning outcomes related to ecological literacy in the program. Table 3 provides the distribution of these outcomes according to the grade level and dimensions of ecological literacy.

Table 3. Distribution of Learning Outcomes by Grade Level and Ecological Literacy Dimensions

Dimensions of ecological literacy	**Frequency of learning outcomes according to grade levels ***				**Total**
	9th grade	**10th grade**	**11th grade**	**12th grade**	
Ecological knowledge literacy	-	5	4	-	9 (34.6%)
Ecological awareness literacy	1	8	1	-	10 (38.5%)
Ecological ethics literacy	-	2	-	-	2 (7.7%)
Ecological emotional literacy	-	1	-	-	1 (3.8%)
Ecological behavioral literacy	-	4	-	-	4 (15.4%)
	1 (3.9%)	20 (76.9%)	5 (19.2%)	-	26 (100%)

* Since learning outcomes are considered with their explanations, some of them were coded multiple times in different dimensions. The numbers in the table represent the extent to which ecological literacy is addressed.

When Tables 2 and 3 are considered together, the findings can be summarized as follows:

Ecological knowledge literacy: There are 9 learning outcomes in the curriculum in the context of ecological knowledge literacy. 4 of them are related to the subject of Ecosystem Ecology in the unit "Ecosystem Ecology and Current Environmental Issues", which is the 3rd and last unit of the 10th grade. These outcomes are more oriented towards ecosystem knowledge such as the relationship between living and non-living elements in the ecosystem, the nutrition of living things and material cycles. In the same unit, one learning outcome related to ecological knowledge literacy is about understanding the factors affecting biodiversity within the subject of the Conservation of Natural Resources and Biodiversity. The other 4 outcomes are on the subject of Community Ecology in the "Community and Population Ecology" unit, which is the second and last unit of Grade 11. Upon examining the content of these outcomes, it is observed that they are more related to ecosystem knowledge and information about environmental damage, such as the structure of the community, relationships between organisms, and negative impacts on biodiversity. There are no outcomes related to ecological knowledge literacy in the 9th and 11th grade.

Ecological awareness literacy: There are 10 learning outcomes related to ecological awareness literacy in the program. One of them is under the subject of the Kingdoms of Living Things and Their Characteristics of the "World of Living Things" unit, which is the 3rd and last unit of the 9th grade, and is related to the awareness of the value of ecological environmental protection. Here, there is an awareness of the contribution of ecological environmental protection to the economy and technology as well as biological processes. In the 3rd and final unit of the 10th grade, titled "Ecosystem Ecology and Current Environmental Issues", there are 2 outcomes on the subject of Ecosystem Ecology, 3 outcomes on the subject of Current Environmental Issues and Humans, and 3 outcomes on the subject of Conservation of Natural Resources and Biodiversity. Upon examining these outcomes, it is evident that they are aimed at creating awareness of the seriousness of existing/potential ecological and environmental problems, ecological environmental conservation value awareness, and awareness of ecological environmental conservation behavior. There is also one outcome related to ecological awareness literacy in the 2nd and final unit of the 11th grade, "Community and Population Ecology" unit, under the subject of Population Ecology. This outcome mainly focuses on population change and aims to raise awareness about the importance of population control. There is no learning outcome related to ecological awareness literacy in 12th grade.

Ecological ethics literacy: There are only 3 learning outcomes related to ecological ethics literacy in the curriculum. All these three outcomes are on the subject of Conservation of Natural Resources and Biodiversity at the 10th grade level. These outcomes reflect the indicators of ecological ethics literacy such as ethics and morality of protecting the ecological environment, respecting and valuing all living things.

Ecological emotional literacy: In relation to ecological emotional literacy, there is only one learning outcome in the curriculum at the 10th grade level on the subject of Current Environmental Issues and Humans. In this learning outcome, it is aimed for the student to gain the ability to take responsibility for the environment, to develop suggestions and be willing to reduce the ecological footprint, water footprint and carbon footprint as an individual.

Ecological behavioral literacy: In the curriculum, there are only 4 learning outcomes related to ecological behavioral literacy at the 10th grade level. Two of them are on the subject of Ecosystem Ecology, one on the subject of Current Environmental Issuses and Humans, and one on the subject of Conservation of Natural Resources and Biodiversity. The learning outcomes on Ecosystem Ecology are mostly oriented towards students' activities of introducing ecological concepts to others. The learning outcome on Current Environmental Issues and Humans aims to develop solutions to environmental problems. Another outcome on Conservation of Natural Resources and Biodiversity is about the direct or indirect effects of students on the conservation of biodiversity and prevention of biotrafficking.

DISCUSSION AND CONCLUSION

In today's world, where humanity has developed a consumption habit beyond its needs, the fundamental things consumed are natural resources. Natural resources are rapidly depleting, and there is a struggle to have access to clean water, clean soil, and clean air. One of the concepts emphasized for the fair sharing of these resources and for a cleaner future is ecological literacy. The need for individuals who think ecologically in all aspects of life is increasing day by day. In the functioning of nature, especially for the solution of problems caused by humans, it is crucial for every individual to be ecologically literate and to ensure continuity in this regard. In addition to the education provided in the family, schools have significant responsibilities in raising individuals with ecological thinking for a sustainable future. Within the framework of the provided qualified formal education, it is important to facilitate the process of ecological literacy education and ensure the continuity of this education. At this point, the importance of the curriculum emerges. It is crucial that the curricula developed in line with education policies are instructive, supportive, and encouraging in this regard. Therefore, when preparing curricula, the high-level relationships of the subjects with ecological literacy should be ensured, and individuals should be supported to think and live ecologically.

In this study, the subjects and learning outcomes of the High School Biology Curriculum (MoNE, 2018) were examined in the context of the theoretical framework of ecological literacy. As a result of the research, it was observed that there are 26 learning outcomes related to ecological literacy in the curriculum and the majority of them (76.9%) are in the 10th grade. The lack of subjects and outcomes related to ecological literacy in the 9th grade, which represents the transition from primary education to a new level of education, or in the 12th grade, which represents the end of secondary education and the beginning of adulthood, is noteworthy. This situation is also incompatible with the spiral principle of the curriculum. Furthermore, when the findings are examined, it is seen that the outcomes related to ecological literacy are mostly in the dimensions of ecological knowledge literacy (34.6%) and ecological awareness literacy (38.5%), while there are significantly fewer outcomes in the dimensions of ecological ethics literacy (7.7%), ecological emotional literacy (3.8%), and ecological behavioral literacy (15.4%). This finding overlaps with the results of the study by Güngör Cabbar et al. (2020) in certain aspects. Güngör Cabbar et al. (2020) analyzed the environmental learning outcomes in the 2018 biology curriculum according to the revised Bloom's taxonomy. As a result of the analysis, they indicated that there are 5 outcomes in the conceptual knowledge dimension (3 outcomes at the level of understanding, 1 outcome at the level of analysis and 1 outcome at the level of evaluation), 1 outcome in the operational knowledge dimension at the level of analysis, and 3 outcomes in the metacognitive knowledge dimension (1 outcome at the level of analysis and 2 outcomes at the level of creation). Therefore, parallel to the findings of this study, they revealed that there are not enough outcomes related to environmental subjects in the curriculum in terms of operational and metacognitive knowledge dimensions, as well as application, evaluation, and creation levels. The researchers reported that there are 12 outcomes related to the environment in the curriculum, with 10 of them in the 10th grade and 2 in the 12th grade. They considered the outcomes only superficially and did not touch upon the outcome explanations or the content of the subjects. For this reason, they did not evaluate the outcomes in the units of Community and Population Ecology in the 11th grade or World of Living Things in the 9th grade within the scope of their research. Additionally, the two outcomes they considered in the Living Things and Environment unit in 12th grade related to the environment are more about the impact of the environment on the genetic structure of organisms, in other words, outcomes related to evolutionary processes. Therefore, these outcomes in the 12th grade were not evaluated in the context of ecological literacy in this study. this point, there are differences with the results of this study.

The research suggests that some dimensions of ecological literacy are not adequately addressed in the curriculum. While the outcomes in ecological knowledge and awareness literacy generally meet the theoretical framework and are considered sufficient, there is a noticeable lack of content on ecological knowledge literacy, such as waste classification and low-carbon living. Furthermore, there are fewer outcomes related to ecological ethics and emotional literacy. However, researchs on environmental issues (Aydın & Selvi, 2020; Karakuş & Çimen, 2020; Soğukpınar & Korucu, 2020) indicates that knowledge about environmental problems is not adequately reflected in behavior and attitudes. Kutay and Güngör Cabbar (2022) reported that high school students have a high level of awareness of environmental issues, but they are at moderate level in the dimensions of valuing, organizing and personalizing. They interpreted this findig as a lack of sufficient transformation from knowledge and awareness of environmental issues to behavior and attitude. The literature indicates the necessity of prolonged exposure for the development and change of behavior and attitude (Erbasan & Erkol, 2020; Peker & Ceylan, 2020; Soğukpınar & Korucu, 2020). Additionally, the curriculum should be designed to support changes in students' attitudes and behaviors. As revealed by this research, this biology curriculum is not sufficient to develop students' ecological literacy, especially in the ethical, emotional, and behavioral contexts. Therefore, the program needs to be revised to develop and support students' ecological ethical, emotional, and behavioral literacy.

In order to develop ecological ethics- and emotional literacy, more content is needed to encourage students to empathize with nature, understand the human-nature relationship correctly, and raise awareness about loving, respecting, admiring, and appreciating the natural beauty. Students should acquire knowledge about the plants, animals, soil, climate, and life cycles in their environment and learn how their lives are affected when they value other forms of life and nature. Students should be encouraged to observe and 'read' nature, while also seeing their communities as an ecosystem and understanding the relationships and processes necessary for the health of the ecosystem, as well as the role of humans in these processes. Additionally, exploring and observing the culture of local communities can be beneficial. For this purpose, research assignments, web-based applications, and social networks can be utilized.

The use of web-based applications and social networking sites that enable the monitoring of ecological issues in different regions of the world, sharing information about the source of the problem and potential solutions, can be integrated into the curriculum. This will make many ecological issues more visible to students from environmental, health, equality, and social justice perspectives. In this process, students can acquire knowledge about the environment, follow different perspectives, and develop a sense of responsibility towards others. Students can also be encouraged to develop foresight and long-term thinking. As the process continues, they can be asked to develop solutions to real problems based on what they have learned. If students are made aware of their own roles in solving environmental problems and experience a sense of participation, belief and desire to make a difference will increase.

It is known that students comprehend the seriousness of ecological problems better through activities in nature (Ha et al., 2022), and that it is beneficial to include content and field studies that include learning and teaching together in ecological literacy education (Berkowitz et al., 2005). Okyay et al. (2022) have found that the ecological literacy education they provided through various activities significantly increased children's awareness and attitudes towards the environment. Therefore, to support ecological literacy, it is beneficial to organize trips to natural areas, zoos, botanical gardens, animal shelters, and animal rescue centers, where students can have close contact with plants and animals, and to involve them in field projects related to environmental protection, habitat restoration, and recycling, to make their involvement in activities that raise environmental awareness effective, and to give them responsibilities in these areas as much as possible and to integrate what they have learned into classroom lessons. At this point, the importance of the outdoor environment as a teaching resource should be emphasized, the curriculum should provide guidance in this regard, and include materials that will guide the teacher.

Integrating ecological literacy into the high school graduation requirements and adding it as one of the essential skills to be acquired by students in the biology curriculum would be beneficial. At this point, easily applicable and evaluable measurement tools for determining levels of ecological literacy can be integrated into the curriculum. Also, it would be beneficial to review the learning outcomes related to ecological literacy within a theoretical framework and to increase number of outcomes, particularly in the context of ethics, emotions, and behaviors. In this process, if attention is paid to the distribution of outcomes according to grades and content intensity, possible difficulties that may be experienced due to the intensive curriculum will be prevented. At this point, the successful integration of the subject into guide materials such as textbooks is crucial for teaching the subject in the desired way. Additionally, developing and presenting various guide teaching materials for teachers that include ecological literacy subjects and also meet academic standards would be effective.

Certainly, just improving the program will not be sufficient for effective ecological literacy education. Niklanović et al. (2014) indicated in their research that their interdisciplinary teaching model designed for ecology education was more efficient than the traditional approach. As in many other fields, in the teaching of ecological literacy, presenting knowledge holistically and meaningfully, along with daily life and other disciplines, will be beneficial. Therefore, it is recommended that the subject be approached in a related manner in the curricula of other relevant disciplines such as geography, chemistry, and physics as well as the biology curriculum, and that the curricula be reviewed and organized holistically in the context of ecological literacy. For instance, Firdausi and Wulandari (2021) have concluded that their developed web-based science learning module is effective in enhancing ecological literacy understanding. Irianto et al. (2020) provided learning experiences focusing on scientific activities such as observation, analysis, research, problem-solving, and project creation using multiliteracy learning models for elementary school students. (The multiliteracy learning model is a learning approach that enables students to gain new knowledge, understanding, and skills based on their learning activities (Allison & Goldston, 2018; Silvers et al., 2010)). Researchers have reported that scientific thinking activities enriched with multiliteracy increased students' awareness and strengthened their knowledge about the environment. They also noted that strong environmental knowledge influences students' responsible behaviors towards the environment and enhances their ecological literacy abilities in terms of knowledge, attitudes, and skills development. As indicated by such studies in the literature, various teaching practices and activities integrated with nature enhance student motivation and are effective in developing ecological literacy. It's essential not to overlook the importance of teacher education in this regard. Teachers must be willing and knowledgeable about using different methods to provide effective instruction in ecological literacy and increase student motivation during the ecological literacy process.

REFERENCES

Akınoğlu, O., & Sarı, A. (2009). Environmental education in primary school curriculum. *Atatürk Education Faculty Journal of Educational Sciences*, 1(30), 5–29.

Allison, E., & Goldston, M. J. (2018). Modern scientific literacy: A case study of multiliteracies and scientific practices in a fifth grade classroom. *Journal of Science Education and Technology*, 27(3), 270–283. 10.1007/s10956-017-9723-z

Atasoy, E., & Ertürk, H. (2008). A field study about environmental knowledge and attidudes of elementary school students, *Erzincan University Journal of Education Faculty, 10*(1), 105-122.

Aydın, E., & Selvi, M. (2020). Development of ecosystem, biodiversity and environmental problems achievement test for secondary school students. *Journal of Research in Education and Society*, 7(2), 661–682.

Aydın, F., & Çepni, O. (2012). Investigation of primary education second grade students' attitudes towards environment in terms of various variables (Karabük City Case). *Dicle University Journal of Ziya Gokalp Education Faculty*, 18, 189–207.

Barberán, A., Hammer, T. J., Madden, A. A., & Fierer, N. (2016). Microbes should be central to ecological education and outreach. *Journal of Microbiology & Biology Education*, 17(1), 23–28. 10.1128/jmbe.v17i1.98427047584

Berkowitz, A. R., Ford, M. E., & Brewer, C. A. (2005). A framework for integrating ecological literacy, civicsliteracy, and environmental citizenship in environmental education. In Johnson, E. A., & Mappin, M. J. (Eds.), *Environmental education and advocacy: Changing perspectives of ecology and education* (pp. 227–266). Cambridge University Press.

Bowen, G. A. (2009). Document analysis as a qualitative research method. *Qualitative Research Journal*, 9(2), 27–40. 10.3316/QRJ0902027

Braus, J., & Wood, D. (1993). *Environmental education in the schools: Creating a program that works!* (Vol. 2). North American Association for Environmental Education.

Brewer, R. (1994). *The science of ecology* (2nd ed.). Saunders College Publishing.

Bruyere, B. L. (2008). The effect of environmental education on the ecological literacy of first-year college students. *Journal of Natural Resources and Life Sciences Education*, 37(1), 20–26. 10.2134/jnrlse2008.37120x

Çabuk, B., & Haktanır, G. (2017). *Okul öncesi öğretmenliği öğretmen adaylarının ekolojik okuryazarlıkla ilgili bilgi düzeylerinin incelenmesi* [Examining the knowledge levels of pre-school teacher candidates regarding ecological literacy]. Ankara Üniversitesi Çocuk Kültürü Araştırma Uygulama Merkezi.

Can, D., Üner, S., & Akkuş, H. (2016). Determination of the secondary school students' environmental literacy. *Ondokuz Mayis University Journal of Education Faculty*, 35(2), 23–35.

Capra, F. (1999). *Ecoliteracy. The challenge for education in the next century.* Liverpool Schumacher Lectures.

Cutter-Mackenzie, A., & Smith, R. (2003). Ecological literacy: The 'missing paradigm' in environmental education (part one). *Environmental Education Research*, 9(4), 497–524. 10.1080/1350462032000126131

Erbasan, Ö., & Erkol, M. (2020). Investigation of environmental knowledge, attitudes and behaviors of primary school teachers. *OPUS International Journal of Society Researches*, 15(24), 2443–2471. 10.26466/opus.619973

Firdausi, A., & Wulandari, F. E. (2021). Development of web-based science learning module in improving the students' understanding of eco-literacy. *Jurnal Penelitian Dan Pengkajian Ilmu Pendidikan: E-Saintika*, 5(3), 241–252. 10.36312/esaintika.v5i3.426

Güngör Cabbar, B., Gültekin, S., Güneş, E., Aytaç, E., & Daşgın, F. (2020). Analysis of environmental achievements in 2018 science and biology courses curriculum according to the revised bloom taxonomy. *Necatibey Faculty of Education Electronic Journal of Science and Mathematics Education*, 14(1), 504–527. 10.17522/balikesirnef.702537

Güven, E., & Aydoğdu, M. (2012). Development of environmental problems behaviour scale and determination of teacher candidates' behaviour levels. *Journal of Uludag University Faculty of Education*, 25(2), 573–589.

Ha, C., Huang, G., Zhang, J., & Dong, S. (2022). Assessing ecological literacy and its application based on linguistic ecology: A case study of Guiyang City, China. *Environmental Science and Pollution Research International*, 29(13), 18741–18754. 10.1007/s11356-021-16753-734704227

Häggströmi, M., & Schmidt, C. (2020). Enhancing children's literacy and ecological literacy through critical placebased pedagogy. *Environmental Education Research*, 26(12), 1729–1745. 10.1080/13504622.2020.1812537

Hardiansyah, F., Ar, M. M., Hidayatillah, Y., & Astutik, C. (2023). Utilization of eco-literacy in science learning as a teaching media in elementary schools. *Pendas: Jurnal Ilmiah Pendidikan Dasar*, 8(3), 1384–1396.

Holden, C. (2007). Young people's concerns. In Hicks, D., & Holden, C. (Eds.), *Teaching the global dimension: Key principles and effective practice* (pp. 49–60). Routledge.

Hsu, S. J., & Roth, R. E. (1998). An assessment of environmental literacy and analysis of predictors of responsible environmental behavior held by secondary teachers in the Hualien Area of Taiwan. *Environmental Education Research*, 4(3), 229–249. 10.1080/1350462980040301

Huang, G. W., & Ha, C. C. (2021). The relationship between ecological literacy and ecolinguistics. *Foreign Language Education*, 1, 15–19.

Huang, G. W., & Zhao, R. H. (2019). *What is ecolinguistics*. Shanghai Foreign Language Education Press.

Irianto, D. M., Yunansah, H., Mulyati, T., Herlambang, Y. T., & Setiawan, D. (2020). Multiliteracy: Alternative learning models to improve ecological literacy of primary school students. *PalArch's Journal of Archaeology of Egypt/Egyptology, 17*(9), 614-632.

Johnson, B., & Manoli, C. C. (2010). The 2-MEV scale in the United States: A measure of children's environmental attitudes based on the theory of ecological attitude. *The Journal of Environmental Education*, 42(2), 84–97. 10.1080/00958964.2010.503716

Karakuş, G., & Çimen, O. (2020). The examination of ethical approaches of candidate teachers towards environmental problems. *Gazi University Journal of Gazi Educational Faculty*, 40(2), 411–435.

Kıral, B. (2020). Document analysis as a qualitative data analysis method. *Siirt University Journal of Social Sciences Institute, 8*(15), 170-189.

Kutay, Y., & Güngör Cabbar, B. G. (2022). Investigation of high school students' attitudes of environmental problems. *Journal of Balıkesir University Institute of Science and Technology*, 24(1), 29–46. 10.25092/baunfbed.912537

Lewis, E., Mansfield, C., & Baudains, C. (2010). Going on a turtle egg hunt and other adventures: Education for sustainability in early childhood. *Australasian Journal of Early Childhood*, 35(4), 95–100. 10.1177/183693911003500412

Mayring, P. (2002). *Einführung in die qualitative sozialforschung* (5th ed.). Beltz Verlag.

McBride, B. B., Brewer, C. A., Berkowitz, A. R., & Borrie, W. T. (2013). Environmental literacy, ecological literacy, ecoliteracy: What do we mean and how did we get here? *Ecosphere*, 4(5), 1–20. 10.1890/ES13-00075.1

Mof, N. E. (2018). *Ortaöğretim Biyoloji Dersi (9, 10, 11 ve 12. Sınıflar)*. Öğretim Programı. https://mufredat.meb.gov.tr/Dosyalar/20182215535566-Biyoloji%20d%C3%B6p.pdf

Nalçacı, A., & Beldağ, A. (2012). Identifying the environmental attitudes of 7th and 8th grade elementary students (Erzurum sample). *Eastern Geographical Review*, 17(28), 141–154.

Niklanović, M., Miljanović, T., & Pribićević, T. (2014). A model of interdisciplinary teaching of ecology in the high school. *Archives of Biological Sciences*, 66(3), 1291–1297. 10.2298/ABS1403291N

O'Leary, Z. (2017). *The essential guide to doing your research project.* SAGE Publications Inc.

Oğuz, D., Çakci, I., & Kavas, S. (2010). Environmental awareness of university students in Ankara, Turkey. *African Journal of Agricultural Research*, 5(10), 2629–2636.

Okyay, Ö., Sayın, A., Demir, Z. G., & Özdemir, K. (2022). The impact of ecological literacy training on environmental awareness and attitudes of pre-school childrens. *Journal of National Education*, 51(236), 2851–2870.

Orr, D. (2005). Ecological literacy. In Pretty, J. N. (Ed.), *The Earthscan reader in sustainable agriculture* (pp. 21–29). Earthscan.

Orr, D. W. (1992). *Ecological literacy: Education and the transition to a postmodern world.* SUNY Press.

Pe'er, S., Goldman, D., & Yavetz, B. (2007). Environmental literacy in teacher training: Attitudes, knowledge, and environmental behavior of beginning students. *The Journal of Environmental Education*, 39(1), 45–59. 10.3200/JOEE.39.1.45-59

Peker, R., & Ceylan, S., (2020). The investigation of primary school fourth grade student's attitudes and behaviors towards the environment, Bilecik Şeyh Edebali University Journal of Social Sciences, *5*(TBMM 100. Year Special Issue), 60-90. 10.33905/bseusbed.747762

Pitman, S. D., & Daniels, C. B. (2016). Quantifying ecological literacy in an adult western community: The development and application of a new assessment tool and community standard. *PLoS One 11*(3), e0150648, 1-18. 10.1371/journal.pone.0150648

Prastiwi, L., Sigit, D. V., & Ristanto, R. H. (2019). Ecological literacy, environmental awareness, academic ability and environmental problem-solving skill at Adiwiyata school. *Indonesian Journal of Science and Education*, 3(2), 82–92. 10.31002/ijose.v3i2.1114

Roth, C. E. (1968). Curriculum Owerwiev for Developing Environmentally Literate Citizens. (ERIC Reproduction Service No. ED 032982).

Roth, C. E. (1992). Environmental Literacy: Its roots, evolution and directions in the 1990s. (ERIC Reproduction Service No. ED 348235).

Saribas, D., Teksoz, G., & Ertepinar, H. (2014). The relationship between environmental literacy and self-efficacy beliefs toward environmental education. *Procedia: Social and Behavioral Sciences*, 116, 3664–3668. 10.1016/j.sbspro.2014.01.820

Silvers, P., Shorey, M., & Crafton, L. (2010). Critical literacy in a primary multiliteracies classroom: The hurricane group. *Journal of Early Childhood Literacy*, 10(4), 379–409. 10.1177/1468798410382354

Soğukpınar, R., & Korucu, D. K. (2020). Investigation of elementary school students' attitudes towards environment. *Mehmet Akif Ersoy University Journal of Education Faculty*, 53, 583–606.

Stibbe, A. (2009). *The handbook of sustainability literacy: Skill for a changing world*. Green Books, Dartington.

Teyfur, E. (2008). The effect of academic achievement and enviroment club studies on primary school students towards the enviroment attitudes (Pattern of İzmir). *Ege Journal of Education*, 9(1), 131–149.

Tuncer, G., Tekkaya, C., Sungur, S., Cakiroglu, J., Ertepinar, H., & Kaplowitz, M. (2009). Assessing pre-service teachers' environmental literacy in Turkey as a mean to develop teacher education programs. *International Journal of Educational Development*, 29(4), 426–436. 10.1016/j.ijedudev.2008.10.003

Yıldırım, A., & Şimşek, H. (2013). *Sosyal bilimlerde nitel araştırma yöntemleri* [Qualitative research methods in the social sciences]. Seckin Publishing Company.

ADDITIONAL READING

Abdullah, K. H. (2023). Eco-literacy and social media: A bibliometric review. *Journal of Scientometric Research*, 12(3), 631–640. 10.5530/jscires.12.3.061

Hammarsten, M., Askerlund, P., Almers, E., Avery, H., & Samuelsson, T. (2019). Developing ecological literacy in a forest garden: Children's perspectives. *Journal of Adventure Education and Outdoor Learning*, 19(3), 227–241. 10.1080/14729679.2018.1517371

Koyama, D., & Watanabe, T. (2023). Why a dispositional view of ecological literacy is needed. *Teaching in Higher Education*, 28(5), 1108–1117. 10.1080/13562517.2023.2198637

Martin, P. (2008). Teacher qualification guidelines, ecological literacy and outdoor education. *Journal of Outdoor and Environmental Education*, 12(2), 32–38. 10.1007/BF03400868

Nadiroh, N., Zulfa, V., & Yuliani, S. (2021). Learning transformation of the 21st century curriculum for prospective teacher in term of eco-literacy. In *IOP Conference Series: Earth and Environmental Science*. IOP Publishing. 10.1088/1755-1315/802/1/012009

Pitman, S. D., Daniels, C. B., & Sutton, P. C. (2018). Ecological literacy and psychographics: Lifestyle contributors to ecological knowledge and understanding. *International Journal of Sustainable Development and World Ecology*, 25(2), 117–130. 10.1080/13504509.2017.1333047

Soetens, N. (2022). *Measures and guidelines for increasing the ecological literacy in students aged 12-18.* [Master's thesis], Utrecht University.

KEY TERMS AND DEFINITIONS

Content Analysis: Content analysis is a research tool used to determine the presence of certain words, themes, or concepts within some given qualitative data.

Curriculum: Curriculum is what is taught in a given course or subject. Curriculum refers to an interactive system of instruction and learning with specific goals, contents, strategies, measurement, and resources.

Document Analysis: Document analysis is the process of reviewing or evaluating documents both printed and electronic in a methodical manner.

Ecological Literacy: Ecological literacy is the ability to understand the organization of natural systems and the processes that maintain the healthy functioning of living systems and the life on Earth.

Ecology: The branch of biology that deals with the relations of organisms to one another and to their physical surroundings.

Ethics: Moral principles that govern a person's behaviour or the conducting of an activity.

Learning Outcomes: Learning outcomes describe what learners should know, be able to do, and value as a result of integrating knowledge, skills, and attitudes learned throughout the course.

APPENDIX

Questions and Answers

In this study, it was examined to what extent ecological literacy is addressed in the high school biology curriculum in Turkey (Ministry of National Education [MofNE], 2018). The questions and answers addressed in this context can be summarized as follows:

1. What are the learning outcomes related to ecological literacy in the high school biology curriculum?
   a) In which units of which class are these outcomes included?

As a result of the research, it was observed that there are 26 learning outcomes related to ecological literacy in the curriculum and the majority of them (20 outcomes; 76.9%) are in the 10th grade. The 10th grade is followed by the 11th grade (5 outcomes; 19.2%) and the 9th grade (1 outcomes; 3.9%), respectively. It is noteworthy that there are no subjects or learning outcomes related to ecological literacy in the 12th grade, which represents the end of secondary education and the beginning of adulthood.

b) To which dimension of ecological literacy do these outcomes relate?

When the findings are examined, it is seen that the outcomes related to ecological literacy are mostly in the dimensions of ecological knowledge literacy (34.6%) and ecological awareness literacy (38.5%), while there are significantly fewer outcomes in the dimensions of ecological ethics literacy (7.7%), ecological emotional literacy (3.8%), and ecological behavioral literacy (15.4%).

Epilogue and Lessons Learned

It is expected that the biology curriculum will support changes in students' attitudes and behaviors regarding ecological literacy. However, as revealed by this research, biology curriculum examined is not sufficient to develop students' ecological literacy, especially in the ethical, emotional, and behavioral contexts. Therefore, the program needs to be revised to develop and support students' ecological ethical, emotional, and behavioral literacy.

Chapter 10

The Introduction, Current State, and Policies for the Future of Environmental Education in Curricula and Educational Institutions in Bangladesh:

Perspectives From Policies and Practice

A. N. Bushra
University of Rajshahi, Bangladesh

Shorif Mollah
https://orcid.org/0000-0001-7702-7154
University of Rome "Tor Vergata", Italy

Farzana Akter Munni
University of Rajshahi, Bangladesh

EXECUTIVE SUMMARY

The climate over the Earth's surface has remarkably been affected by various inconsiderate usages of natural resources and human-made disasters worldwide, such as substantial growth of fossil fuel consumption and coal-energy-dependent

DOI: 10.4018/979-8-3693-1710-5.ch010

infrastructures, abundant use of synthetic materials, and the absence of environment-friendly technologies and solutions. In addition, in recent years, people seem to be more indulged in hyper-consumerism, quickly changeable products and fashion, and unsustainable urban design and living habits, which generates more waste and gives birth to numerous complex environmental problems. For the last few decades, the delicate balance of the Earth's ecosystems has continued to face constant threats, ranging from the profound impacts of climate change to the depletion of biodiversity and the severe consequences of pollution.

INTRODUCTION

The climate over the Earth's surface has remarkably been affected by various inconsiderate usages of natural resources and human-made disasters worldwide, such as substantial growth of fossil fuel consumption and coal-energy-dependent infrastructures, abundant use of synthetic materials, and the absence of environment-friendly technologies and solutions. In addition, in recent years, people seem to be more indulged in hyper-consumerism, quickly changeable products and fashion, and unsustainable urban design and living habits, which generates more waste and gives birth to numerous complex environmental problems (Bocken & Short, 2021; Sangwan & Vyas, 2020). For the last few decades, the delicate balance of the earth's ecosystems has continued to face constant threats, ranging from the profound impacts of climate change to the depletion of biodiversity and the severe consequences of pollution (Baste & Watson, 2022). As a result, there has been a notable global focus on environmental protection, with growing concerns about the current state and conservation of nature, environment, and diverse ecosystems for the future taking the central spot in the global development agendas (Kremen et al., 1994; Arthington et al., 2018; Baste & Watson, 2022).

Achieving a sustainable future relies heavily on a reciprocal relationship between environmental stewardship and economic development strategy (Salequzzaman & Stocker, 2001). Environmental education (EE) plays a crucial role in fostering the needed balance by enhancing public awareness, knowledge, and positive attitudes toward environmental protection and sustainability (Sultana et al., 2017). Thus, EE is gradually becoming a new phenomenon in development thinking in response to the contemporary environmental challenges (Chowdhury, 2014). Hence, governments and organizations across the globe are now progressively allocating resources to environmental protection and awareness-building initiatives to build a population that is knowledgeable, responsible, environmentally aware, and capable of tackling the challenges of climate change, biodiversity loss, and pollution (Rovňák et al., 2021; Clark & Whitford, 2011).

EE involves organized initiatives to teach the learners about the functions of the natural environment, with a focus on how humans can adopt sustainable practices to manage their behavior and ecosystems (Chowdhury, 2014). It aims to raise social and scientific awareness, pinpoint symptoms and root causes of environmental degradation, promote environmental ethics, understand the ecological interdependence of economic, social, and political factors, and instill consciousness (Chowdhury, 2014). The Environmental Protection Agency (EPA) (2016) outlines key components of environmental education (EE), which include- fostering awareness, acquiring knowledge, cultivating positive attitudes, motivating enhancement of environmental education, developing problem-solving skills, and encouraging active participation in addressing environmental issues. Increased environmental knowledge fosters attitudinal and behavioral changes that empower and mobilize communities, thereby improving environment-friendly lifestyles (Chowdhury, 2014). Thus, going beyond knowledge transfer, EE is a dynamic learning process that promotes environmental awareness, cooperation, and a balanced exploration of perspectives to address ecological challenges and instill the concept of political action to resolve the issues (Chowdhury, 2014; Islam et al., 2016).

United Nations Educational, Scientific and Cultural Organization (UNESCO) and the United Nations Environment Programme (UNEP) took the pioneering step to incorporate EE into the curricula. In 1975, they launched the International Environmental Education Programme (IEEP) to integrate EE at all levels of education globally (MEF, 2016). Thereafter, IEEP emphasized continuously on incorporating environmental education into primary and secondary curricula, teacher education, university general education, technical and vocational education, and non-formal education (MEF, 2016). Since then, multiple efforts by the UN agencies and other global actors have facilitated the inclusion of environmental education in the national curricula of Bangladesh (MEF, 2016). After 1987, EE became an integral part of the national educational policy of Bangladesh in alignment with the global imperative for 'sustainable development' (Chowdhury, 2014).

However, scholarly investigations into the evolution and current situation of EE in Bangladesh indicate that the state of EE in Bangladesh has historically been inadequate (Salequzzaman & Stocker, 2001). Criticism is directed at the school-level EE curricula, characterized as brief, random, and fragmented, lacking continuity, logical progression, and a clear goal for learners (Chowdhury, 2014). In addition, the current syllabus for environmental education falls short of adequately addressing contemporary environmental challenges, i.e., global warming, and climate change (Chowdhury, 2014; Islam, 2020). Moreover, at the university level, the disciplines of ecology and environmental science have traditionally been integrated as components within broader fields of study such as botany, zoology, geology, and civil engineering rather than being offered as a comprehensive and independent field of study (Salequzzaman & Stocker, 2001).

Bangladesh is vulnerable to climate change, which stems from a combination of geographical, environmental, and socio-economic factors, making it one of the most affected countries globally (Jahan & Quddusi, 2013; Kagawa, 2022). The country faces recurrent natural disasters such as cyclones, flooding, and riverbank erosion, exacerbated by rising sea levels and changing weather patterns attributed to climate change (Mojid, 2020). In such a situation, environmental education could play a pivotal role in addressing the involved issues and tackling the challenges. A detailed study and analysis of the current state, trends, and future direction of EE in Bangladesh demand further exploration due to the vulnerability of climate change and the limitations of EE in the country. In addition, there is a pressing need for new studies and perspectives, warranting further research on EE due to the continuously evolving environmental challenges and issues. Based on this backdrop, the present study is designed to explore the historical context, current state, and future directions of environmental education in Bangladesh. By keeping focus on these objectives, the current study set out to provide a holistic understanding of the evolving trends of EE in the country, expand the scope, and offer valuable insights for policymakers, professionals, environmentalists, experts, and educators.

Research Objectives

1. To explore the historical context of introducing environmental education (EE) in Bangladesh
2. To find out and sum up the current state of environmental education in Bangladesh
3. To find out the plans and future directions of environmental education in Bangladesh

HISTORICAL BACKGROUND OF ENVIRONMENTAL EDUCATION IN BANGLADESH

The origins of environmental education can be attributed to early naturalists such as Jean-Jacques Rousseau and Louis Agassiz, who emphasized the importance of learning directly from nature (McCrea, 2006). However, the modern environmental education movement truly flourished in the 20th century (McCrea, 2006). The 1960s and 70s witnessed a surge in environmental awareness sparked by groundbreaking publications, such as Rachel Carson's "Silent Spring" and the global phenomenon of Earth Day (Anguish, 2015). This momentum culminated in the landmark 1972

United Nations Conference on the Human Environment in Stockholm, Sweden. During this event, environmental education was acknowledged as a vital instrument in confronting worldwide environmental issues (Dorn, 2020; UN, n.d.). Tracing a single starting point for EE in Bangladesh is difficult as it has progressed gradually from traditional practices and early initiatives to formalized programs and national strategies to the current situation. Traditional practices in rural Bangladesh have often embedded environmental understanding through daily life, such as respecting specific trees or water bodies (Chadwick et al., 1998). Since the independence of Bangladesh in 1971, this country has undertaken substantial policy efforts to promote EE in different ways. The following section discusses the major policies, measures, and developments to introduce and establish EE in Bangladesh.

Global Efforts to Introduce Environmental Education in Formal Spaces

Global actors have long felt the need to transform traditional educational systems with a systematic approach to address emerging environmental problems (MEF, 2016). The first organized effort to integrate EE into all levels of education was initiated by UNESCO and UNEP in 1975. In 1975, the International Environmental Education Program (IEEP) was launched globally to integrate EE into the curricula. In the same year, the IEEP organized a global workshop leading to the creation of the 'Belgrade Charter,' which focused on developing frameworks for EE (Gough, 2016). As a result, in 1977, the Tbilisi Inter-Governmental Conference on Environmental Education was organized that served as the foundation for EE strategies in various developed countries such as New Zealand, the United Kingdom, Australia, and the United States, et cetera (Biswas et al., 2009; Sarkar, 2011). In 1992, the United Nations Conference on Environment and Development adopted a chapter advocating for education, public awareness, and training, endorsing environmental education as a key mechanism for attaining sustainability. Following that, in late 1997, UNESCO organized and held a conference in Thessaloniki, Greece, on the International Conference on Environment and Society: Education and Public Awareness for Sustainability. The mandates arising from the conferences have shaped the narratives and development of literature on environmental education locally and globally (Biswas et al., 2009).

UN Decade of Education and Sustainable Development

To facilitate the implementation of the agendas set in these conferences, the UN declared the UN Decade of Education for Sustainable Development (DESD), spanning between 2005 and 2014 in 2002. The DESD resulted in a convergence of sustainable development goals and education into the existing curricula around the globe. According to the DESD final report, during this period, many member states have actively initiated the process of aligning education policies, curricula, and plans with the principles of environmental protection and sustainable development (UNESCO, 2014). As a participant of the DESD, Bangladesh has undertaken substantial steps to promote sustainable development and environmental protection through education (MFE, 2016). For instance, the National Adaptation Program of Action (NAPA) identified environmental awareness and education as crucial for adapting to climate change in 2005, in alignment with the goals of the DESD.

Introduction of EE in the Primary and Secondary School Levels

The concept of EE was first introduced in Bangladesh to a limited extent within environmental and education policies during the mid-80s (MFE, 2016). In the beginning, the National Curriculum and Textbook Board (NCTB) took the initiative to reform school curricula, developed teaching materials and aids, and trained in-service teachers to teach EE, with assistance from UNDP, UNICEF, and SIDA (Biswas et al., 2009). The Environment Policy of 1992 by the Government of Bangladesh (GoB) outlined specific directives on environmental education and public awareness, which included fostering mass awareness about environmental protection and resource utilization in a sustainable way, advocating for sustainable resource use, integrating environmental knowledge through media and education systems, and encouraging public engagement in management activities.

This policy advocated for the integration of environmental knowledge in formal and non-formal education systems and the media, as well as active public participation in environmental activities (Salequzzaman & Stocker, 2001). Adding to this, in the Fourth Five-Year Plan of 1996, the education policy of the GoB articulated a commitment to provide environmental education to teachers and students across all levels of education. In addition, the National Adaptation Programme of Action - NAPA (2005) and the Bangladesh Climate Change Strategy and Action Plan (BCCSAP)-2009 stressed the importance of knowledge generation on climate change issues and the inclusion of climate change-related issues in the curriculum at secondary and tertiary educational institutions (MFE, 2005; MEF, 2009).

Towards the first introduction of the environment-related curriculum, Salequzzaman & Stocker (2001) stressed that environmental issues were first incorporated into the primary and junior-level curricula in Bangladesh in 1978 through a subject named "Environmental Studies" (Salequzzaman & Stocker, 2001). However, Sarkar (2011) showed a different view as he proposed that EE was primarily introduced in grade three, where students encountered two units in environmental education: "Introduction to Environment: Science" and "Introduction to Environment: Social

Science" (NCTB, 2003; Sarkar, 2011). During this introduction phase, the teaching and learning content covered in the primary level of the EE syllabus included some rudimentary and recurrent issues such as the causes and effects of environmental degradation, the necessity and approaches for environmental conservation, and efforts to reduce waste, prevent pollution, and conserve water resources (Salequzzaman & Stocker, 2001). In secondary-level textbooks, chapters related to geography and environmental studies were first incorporated in the 1980s (Rithu et al., 2016). Later on, in 1995, the NCTB brought about a new textbook called 'Geography and Environment,' which was integrated as an elective subject for students studying arts and humanities at the secondary and higher secondary levels (Rithu et al., 2016).

At the secondary level, environmental education was taught through different disciplines including Language, Social Science, General Science, and Biology, encompassing multiple environment-related themes (Sarkar, 2011). The secondary level General Science curriculum (in 2006), which was designed to be studied by the students studying Humanities and Business Studies, included five chapters in the textbook that cover some significant topics such as Population and Environment, Energy, Fuel, Ecology, and Disaster Management and Bangladesh (Haque et al., 2007; Sarkar, 2011). Similarly, the Social Science textbook designed for students studying Science explores environmental themes, covering topics such as the earth's surface, changes in the earth's crust, atmosphere, isolation, temperature, cyclones, atmospheric pressure, humidity, precipitation, and the impact of population growth on economic development (Begum, Shikdar, & Das, 2007).

At the same time, the biology curriculum addressed environmental issues through chapters on organisms and the environment, pollution, conservation of the environment, and natural diversities in Bangladesh (Sarkar, 2011). Salequzzaman & Stocker (2001) argued that in the beginning, there was substantial coverage of environmental education at the primary stage within the formal school system, but the treatment at the lower secondary stage (grades 6-8) remained somewhat superficial. Adding to this, Chowdhury (2014) pointed out that school-level textbooks in Bangladesh have neglected long-standing environmental issues such as global warming, despite their growing significance over the past three decades. Additionally, an activity-oriented method of teaching has been absent in the school-level EE curriculum (Salequzzaman & Stocker, 2001).

Environmental Education at Tertiary Level in Bangladesh

At the university level, environmental education is offered in four sub-disciplines, namely- Environmental Science, Environmental Engineering, Environmental Studies, and Environmental Management to the date (Biswas et al., 2009). Until 1995, the coverage of environmental studies in higher education was fragmented. As one of the first tertiary institutions, in 1996, Khulna University established the environmental science discipline, which started to offer undergraduate courses the immediate next year (MEF, 2016; Salequzzaman & Stocker, 2001). The syllabus of this discipline covered key areas such as ecology, ecosystem management, environmental issues, technologies, sustainability, social science, economics, environmental ethics, water resources, land use management, environmental management with mathematics and computer applications, environmental engineering, sustainability of energy technologies, environmental impact assessment, legislation and planning, and environmental auditing (Salequzzaman & Stocker, 2001). This curriculum aimed to establish a robust foundation in the principles and techniques of environmentally sustainable development, highlighting relevant physical, chemical, biological, social, economic, and ethical processes (Salequzzaman & Stocker, 2001; Salequzzaman & Davis, 2003; MEF, 2016).

During the same period, some other universities in Bangladesh have incorporated disciplines related to the environment. In 1995, Shahjalal University of Science and Technology introduced an undergraduate-level course on Civil and Environmental Engineering, and the Department of Environmental Studies at the North-South University of Bangladesh started its journey. In the following year (1996), the Independent University of Bangladesh (IUB) introduced the Department of Environmental Science and Management. Following the trend, both the University of Dhaka and Jahangirnagar University, too, established undergraduate programs in environmental education (Salequzzaman & Stocker, 2001). However, universities that pioneered environmental studies at the tertiary level in the 90s, have faced challenges with insufficient teaching staff and limited internet access, hampering students' access to updated environmental information (Salequzzaman & Stocker, 2001; Salequzzaman & Davis, 2003).

While many of the tertiary institutions have maintained discipline and focus on environmental studies as an independent field of study, several other higher education institutions in Bangladesh have incorporated environmental studies and programs within various other disciplines. At the undergraduate level, the University of Dhaka, University of Rajshahi, University of Chittagong, Jahangirnagar University, Khulna University, Shahjalal University of Science and Technology, the Open University, and National University, included the environmental studies and related programs under the departments of botany and zoology or similar departments (Salequzzaman & Stocker, 2001). In alignment, the postgraduate programs in botany and zoology included compulsory units covering the studies and knowledge of ecology, ecosystems, and the environment. Some universities offer master's programs in environmental management and earth science under the Department of Geography/Geology (Salequzzaman & Stocker, 2001). For example, Bangladesh Agricultural University (BAU), the Life Science School of Khulna University, and the Civil Engineering

Department of Bangladesh University of Engineering and Technology (BUET) have also offered comprehensive environment-related programs at both the undergraduate and postgraduate levels (Salequzzaman & Stocker, 2001).

THE CURRENT STATE OF POLICIES, ACTIVITIES, PRACTICES, AND APPLICATIONS

In Bangladesh, environmental education is increasingly getting recognized as a crucial component of national development efforts day by day. The government has taken steps to integrate environmental education into the national curriculum, with courses and programs, such as environmental science and studies being included at various educational levels (Salequzzaman & Stocker, 2001; Biswas et al., 2009). Non-governmental organizations (NGOs) and civil society groups also play a significant role in promoting environmental awareness and education through grassroots initiatives, community engagement, and advocacy campaigns (Salequzzaman & Davis, 2003). However, resource constraints, outdated teaching methods, and limited access to educational materials remain obstacles to widespread environmental education in the country (Salequzzaman & Stocker, 2001; Uddin, 2023). The current initiatives and development of EE in Bangladesh are discussed in the following section.

Limitations of EE Curricula and Instruction

Biswas et al. (2009) analyzed school-level textbooks and summarized key issues involved in teaching environment studies in Bangladesh before the major curriculum reformation took place in 2012. Before 2009, environmental education was integrated from both social and scientific perspectives on the environment at the school level. The socio-environmental topics covered in the school-level curricula include- family life, living organisms, cleanliness, the connection between health and environment, social environment, human lifestyles, and the effects of environmental changes. The scientific exploration of environmental issues includes basic knowledge about the earth, solar system, conservation of resources, food and nutrition, population, agricultural development, and natural resources. Courses related to environmental biology in advanced grades emphasize population education (Biswas et al., 2009). Thus, before a significant educational reform in 2010, the curriculum for environmental education lacked depth and details in discussion leading to a superficial understanding of complex environmental issues. Moreover, while the curriculum covers both social and scientific perspectives on the environment, there may be limited integration and alignment between those perspectives. Therefore, the previous curriculum failed to adopt a holistic approach and resulted in a fragmented understanding of environmental issues.

National Educational Policy of 2010

The curriculum at the school level in Bangladesh has undergone substantial changes, with significant reforms taking place from 2012-13 to 2023. The alterations, including both comprehensive and minor adjustments, have reshaped the content of school-level courses and brought about a significant transformation in the historical approach to teaching EE in schools (Abbas, 2023). The National Education Policy (EP) 2010 by the GoB is one of the notable policies in recent times that has taken the issue of environmental education into serious consideration. The National Education Policy 2010 of Bangladesh has encouraged the inclusion of EE awareness in the curricula from pre-primary to post-graduation levels. This policy also acknowledged that the promotion of EE is crucial in equipping students with the skills to address climate change and natural disasters (Islam et al., 2016). Following the EP of 2010, the 15th amendment to the Bangladeshi constitution in 2011(Article 18A) adopted the provision for the conservation of natural resources, environmental development, and biodiversity (Islam et al., 2016). In the following year, The National Curriculum (NC) 2012 was formulated to implement the EP of 2010. The introduction of climate change at the school level curricula was a new inclusion by the National Curriculum 2012 (Islam et al., 2016; Chowdhury, 2014).

National Curriculum of 2012 and Environmental Education

In 2012, details and substantial changes to the national school-level curriculum were introduced. Existing studies have analyzed the NC 2012 and pinpointed several positive sides as well as weaknesses of the curriculum (2012). By analyzing school textbooks from standard 1 to standard 5, Chowdhury (2014) concluded that the school textbooks primarily emphasize the physical geography of the environment, neglecting key environmental issues such as degradation of natural resources, biodiversity loss, energy concerns, urbanization, etc. Rather, the only environmental problems that are extensively covered are natural disasters, such as the storms throughout the southern parts of Bangladesh (Chowdhury, 2014). Chowdhury (2014) also expands his argument, contending that the National Curriculum failed to pay adequate emphasis on EE because the EE syllabus is brief, random, fragmented, incomplete, and lacks continuity and logical progression.

Additionally, there is a notable absence of clear goals for the learning outcomes of the courses offered, which leads to difficulties in assessing the learning outcomes of the students based on the established assessment criteria currently available for environmental literacy at the respective level of education (Chowdhury, 2014). Chowdhury (2014) further argued that the global environmental issues such as the climate change and global warming are inadequately represented in the current curriculum. Nevertheless, the inclusion of climate change content is a significant step forward. Chowdhury (2014) pointed out more limitations of the current EE curricula in Bangladesh as he argued that there is a duplication of the topic "Disaster Management and Bangladesh" under both secondary 'Social Science' and 'General Science' textbooks. This shows a failure by the National Curriculum and School Textbook Board to address redundancy in the environmental topic distribution in relevant textbooks for each standard.

Knowledge and Awareness Among Students About EE

Apart from the mentioned gaps and shortcomings in the policies and curriculum, the relatively poor level of knowledge and awareness compared to the expected level of knowledge and literacy of environmental education among the learners stands to be one of the most concerning challenges yet to be considered seriously to resolve through introducing an enriched and higher kind of curriculum and instructional design model. For instance, Islam et al. (2016) show that the younger generation in Bangladesh lacks expected levels of environmental awareness, knowledge, and practices. This is a clear indication of the inefficiency of the EE curricula and the teaching methods. Sultana et al. (2017) investigated the environmental knowledge and attitude factors among students and found that students who study sciences possessed greater environmental knowledge, whereas learners from arts and humanities exhibited higher environmental attitudes.

About fifty percent of the surveyed students responded that television was their most important source of getting information and environmental awareness for them (Sultana et al., 2017). In contrast, the course curriculum was the main source of environmental awareness for only 11.3% of their respondents. This study highlights a potential gap in the effectiveness of the educational system in fostering environmental awareness among students. Respondents of Sultana et al. (2017) have also exhibited a high level of environmental knowledge. However, the findings of Uddin (2014) revealed that many secondary-level students do not have a clear concept of the sources of air pollution and the natural hazards that are related to their surroundings and environment.

Increasing Environmental Courses and Majors at the University Level

In recent times, only a few studies have explored the current situation of EE at the tertiary level. Haque and Sharif (2021) explored the potential for environmental engineering education at the tertiary level. According to them, at present, environmental engineering programs at the undergraduate level do not solely focus on environment-related issues in Bangladesh. Instead, most of the EE programs are taught under the departments of civil engineering, and these programs offer a wide range of interdisciplinary content knowledge, experience, and practice. Recently, some universities have changed the names of the Department of Civil Engineering to Civil and Environmental Engineering (CEE) (Haque & Sharif, 2021). Existing studies also show that an increased number of universities have introduced more programs related to EE (departments) in recent years (Mohammad, 2010). While more and more universities are introducing programs on Environmental Sciences, there is still an urgent need to integrate emerging issues such as climate change into the core curriculum (MEF, 2016). Thus, the introduction, evolution, and development of environmental education have been a continuous process in the different education systems in Bangladesh. Historically and similarly, the topics covered in environmental engineering education in Bangladesh include pollution prevention, environmental sustainability, and ensuring the goals of sustainable development (Haque & Sharif, 2021).

The Role of NGOs and Civil Society

Along with the government agencies, NGOs and civil societies also play a crucial role in promoting environmental education in Bangladesh through various awareness-building activities, fostering community building, and advocating for policy change (Chowdhury, 2004). Non-governmental organizations (NGOs), community-based organizations (CBOs), and grassroots movements actively work to fill the gaps available in formal education systems by organizing workshops, seminars, and awareness campaigns on environmental issues (Sutdhibhasilp, 2000). These efforts often target vulnerable communities, empowering them with knowledge and skills to address local environmental challenges effectively (Thomas Muñoz, 2022). BAPA (Bangladesh Environment Movement) and BELA (Bangladesh Environment Lawyers Associaton) are two civil society organizations based in Bangladesh that are constantly working to promote environmental awareness and prevent environmental degradation in Bangladesh (Chowdhury, 2004).

In addition, NGOs have several projects promoting EE among both formal and informal learners throughout the country. EE activities conducted by NGOs include group discussions, non-formal schools, community mobilization, and mass media campaigns. NGOs have emerged as effective grassroots development organizations capable of influencing government policies and institutions. (Chowdhury, 2004). One recent example of an NGO's involvement in EE can be drawn from BRAC's recent initiatives. The BRAC Institute of Education Development (BIED) has partnered with 400 pre-primary government schools across Bangladesh to establish Green Play Labs. These innovative spaces cater to children aged four to five, allowing them to engage in play-based activities centered around nature. Within these labs, children immerse themselves in creative play, crafting stories and experiences that deepen their connection to the physical world. Through this initiative, BIED aims to foster environmental awareness and appreciation among young learners while nurturing their cognitive and imaginative development (Sabbab & Ahmed, 2024).

THE PREPARATION FOR THE FUTURE OF ENVIRONMENTAL EDUCATION

Bangladesh is one of the most vulnerable countries in the world due to climate change, greenhouse effects, and global warming, and even the vulnerability is expected to increase in the future due to the ongoing environmental deterioration (Haque & Sharif, 2021; Hoque et al., 2017; Kagawa, 2022). Experts predicted a rise in the frequency and intensity of extreme natural disasters such as floods, tropical cyclones, storm surges, and droughts in the country in the upcoming years (Chowdhury, 2014). To tackle the ever-increasing challenges posed by climate change and environmental degradation, environmental education emerges as a critical component of sustainable development in the country (Chowdhury, 2014; Hoque et al., 2017). In response, the educational paradigms worldwide are being reshaped, integrating current environmental concerns into the curricula to develop environmentally-conscious citizens and leaders (Hoque et al., 2017). Over the years, notable efforts have been made to integrate environmental issues into the existing curricula. However, there exist numerous shortcomings in the current state of environmental education in Bangladesh. Thus, the foundation of educating EE in the future in the country can be fortified by comprehensively understanding and addressing these deficiencies. The following section sheds light on the future direction of environmental education in Bangladesh.

Interdisciplinary, Activity and Practice-Oriented Curriculum Design

Environmental education possesses an interdisciplinary nature and requires the incorporation of knowledge from diverse scientific disciplines to fulfill its cumulative objectives (Rovňák et al., 2021). An interdisciplinary Environmental Studies program with hands-on, activity-based teaching can significantly enhance EE in Bangladesh (Chowdhury, 2014). Thus, EE is increasingly becoming interdisciplinary day by day (Chowdhury, 2014). However, previous studies have revealed that EE teaching in Bangladesh has traditionally been confined to specific subjects and has not been delivered in an interdisciplinary manner (Salequzzaman & Stocker, 2001; Chowdhury, 2014). Thus, researchers recommended the integration of multidisciplinary environmental courses and programs to promote environmentally aware attitudes, skills, and behavior patterns among the students and stakeholders (Haque & Sharif, 2021). Haque & Sharif (2021) emphasize the limited number of environment engineering departments in Bangladesh, and they also argue for including environmental dimensions in the existing civil engineering syllabi as well. Unfortunately, several factors influence the limiting the adoption of activity-oriented, practice-based, problem-based, and interdisciplinary EE in Bangladesh, including a shortage of qualified teachers, inadequate educational infrastructures, a lack of teaching aids, and inadequate training opportunities, et cetera (Chowdhury, 2014).

Need-based Curriculum

The limitations identified in the existing studies suggest that the future of the EE policies and curriculum design must prioritize the needs of the students, recruit qualified instructors for EE courses, and provide necessary training to them (Kagawa, 2022). This goal can be achieved by conducting curriculum audits, enhancing pedagogical approaches for educating about climate change, and building teachers' capacities in facilitating student engagement and promoting sustainable practices at school. Moreover, the content of future EE needs to be developed in alignment with both the local context and the global environmental agendas. In this regard, Chowdhury (2014) suggested that the current National Curriculum of Bangladesh should be revised and updated by an editorial panel of experts for curriculum development, and the textbooks of all levels need to be rewritten and organized cohesively to avoid repetition, adopting an integrated and objective approach. Moreover, Kagawa (2022) realized the importance of educating skills essential to maintaining life in an environment-friendly way to address various aspects of climate change crises and advocated for the integration of green skills into the secondary curriculum. Kagawa

(2022) believed that it could ensure clear progression in acquiring knowledge and skills related to climate change and foster cross-curricular links (Kagawa, 2022).

Scope for Change in Curriculum, Assessment, and Educational Policy

Evaluation of environmental literacy among students in Bangladesh is challenging due to the absence of well-established assessment criteria (Chowdhury, 2014). Kagawa (2022) highlighted that there is a lack of methods for assessing the learning outcomes of the courses taught for a specific period in a program regarding the EE (climate change, greenhouse gas, and effects, global warming, etc.). In addition, the school textbooks exhibit a significant organizational issue characterized by a lack of sequential arrangement (Chowdhury, 2014; Kagawa, 2022). Some of the topics persist in these textbooks from the primary to the secondary stage without adequate improvement in the level of, difficulty, learning, and complexity (Chowdhury, 2014). Moreover, current pedagogical approaches emphasize theoretical knowledge acquisition over practical skills and values in textbooks, leading to an imbalance in learning outcomes (Kagawa, 2022).

In addition, three fundamental deficiencies of EE in Bangladesh are exposed by Uddin (2023), which include institutional and policy deficiencies, inadequate methods for teaching and evaluation of learning outcomes, and cultural and traditional beliefs. Uddin (2023) emphasizes key institutional limitations such as deficiencies in outcome-based education, misalignment of curricula with student needs, unacceptable institutional performance, insufficient resources and technology, and shortcomings in EE content within textbooks. In this regard, it is urgent to integrate education and instruction regarding the risks associated with climate change, greenhouse effects, global warming, and sea level rise. Hence, the National Education Policy can be an effective tool to address the pointed limitations in building resilience and reducing the anticipated damages that might happen if unprepared in advance (Kagawa, 2022).

Increased Collaboration Between Industry and Academia for Improved Outcomes

The university curriculum in Bangladesh has traditionally delivered theoretical knowledge with very little room for practical implication and skill gain (Haque, 2022). In this regard, continuous and reciprocal collaboration between industry and academia can play a pivotal role in advancing environmental education by bridging the gap between theoretical knowledge and practical application and encouraging more students and institutions to engage in EE initiatives. Through collaborative efforts, academia can enjoy and gain access to real-world data, case studies, and practical experiences provided by industry partners, enriching the learning experience for both teachers and students. At the same time, industry professionals can contribute their expertise and insights to curriculum development, ensuring that educational programs align with current industry practices and emerging environmental challenges. In this way, teachers can provide the industry with innovation, emerging knowledge, techniques, etc., while the professionals from the industry can provide practical experiences leading to a double win-win collaboration. Thus, existing studies, i.e., Haque & Sharif (2021) recognized the industry-academia collaboration for the development of environmental protection skills in engineering students and recommended replacing the traditional teaching methods with different discussed methods, i.e., back-casting analysis, project-based learning, etc.

CONCLUSION AND RECOMMENDATIONS

Since the mid-80s, Bangladesh has made continuous efforts to introduce and establish environmental education into the curricula and practice with the support of UN agencies and other strategic partners in all levels of education. The EE curricula have evolved and improved over time, integrating emerging environmental concerns. However, evaluating the EE curricula, the experts and researchers argue that despite constant improvements and increasing attention, the EE in Bangladesh remains inadequate (Salequzzaman & Stocker, 2001). The textbooks at the school level lack comprehensive exploration of emerging global environmental issues, such as climate change, global warming, greenhouse effects, the rise of sea levels, and problems with the conservation of natural resources and biodiversity. In addition, the curriculum does not include local examples of hazards caused by climate change and activities (Kagawa, 2022). On top of that, the traditional teaching style and assessment techniques for the evaluation of learning outcomes have been criticized by scholars for not being interactive, participatory, and innovative. Thus, there is a significant potential for the growth of environmental education in Bangladesh. Hence, researchers have recommended the development of the content, appropriate resources, the structure of the curriculum, teaching methods, classroom design, and practical training to deliver the content knowledge of environmental issues and lead to solutions.

Exploration into the history of EE has revealed that earlier environmental issues were taught under the departments of Botany, Zoology, Geology, Ecology, and similar areas of studies. However, in recent years, the number of dedicated departments on EE at the university level has increased (Mohammad, 2010). Some studies have argued that the curriculum to teach EE at the university level is well crafted to build a strong foundation in principles to help sustainable development (Salequzzaman

& Stocker, 2001; Salequzzaman & Davis, 2003; MEF, 2016). In addition, recent investigations have indicated that the absence of content and resources standardized with sustainability features in higher education is limiting the capacity of university graduates to equip them with the necessary skills and competency to tackle environmental concerns effectively (Hoque et al., 2017). To minimize the aforementioned limitations, existing studies have recommended employing more diversified, handy, practical, and instructional approaches to motivate students to adopt constructive and responsible behaviors (Sultana et al., 2017).

Overall, future research on the issues revolving around EE can focus on innovative, integrative, well-equipped, and practically translatable teaching methods and the integration of technologies to enhance the learning experience. In this circumstance, researchers and experts can be encouraged to invest more time and effort into developing technically sound devices or tools to conduct the assessment process as a more reciprocal between theoretical knowledge and practically translatable skills in real circumstances for fighting against the dangers caused by climate issues. In addition, researchers can also explore the effectiveness of the programs currently available related to environmental studies, examining their role in knowledge construction, raising awareness, and improving behavioral habits among the students. Another potential area for further research is how the collaboration between industry and academia can facilitate environmental education at the tertiary level in the context of Bangladesh, being a country of the Global South. The existing studies show that there is a lack of evidence on the recent developments of EE at the university level.

Therefore, future research endeavors may delve into a critical analysis of the tertiary-level environmental education and curriculum in Bangladesh, aiming to identify strengths, weaknesses, and areas for enhancement and improvement. Finally, exploring current trends and the scope of employment opportunities in the job sector related to the environmental sciences seems another promising research avenue. By investigating these areas, researchers can contribute valuable insights to inform policymakers and other stakeholders to develop an improved curriculum and policy, ultimately fostering a more robust and responsive educational framework and workforce in the field of environmental studies.

Declarations

Conflict of Interest: There is no conflict of interest in this study.

Data availability: The current study does not require empirical data. Therefore, there is no conflict due to the unavailability of data or similar conflicts with data.

Funding information: This study was not supported by any funding opportunity.

REFERENCES

Abbas, S. M. (2023, December 19). Books of new curriculum in the press. *Dhaka Tribune*.https://www.dhakatribune.com/bangladesh/334300/books-of-new-curriculum-in-the-press

Anguish, D. (2015). Mass Media and the Evolution of the Environmental Movement: 1960-1979. [MA dissertation, University of Central Florida]. Retrieved from https://stars.library.ucf.edu/cgi/viewcontent.cgi?article=2350&context=etd

Arthington, A. H., Bhaduri, A., Bunn, S. E., Jackson, S. E., Tharme, R. E., Tickner, D., Young, B., Acreman, M., Baker, N., Capon, S., Horne, A. C., Kendy, E., McClain, M. E., Poff, N. L., Richter, B. D., & Ward, S. (2018). The Brisbane Declaration and Global Action Agenda on Environmental Flows (2018). *Frontiers in Environmental Science*, 6, 364344. 10.3389/fenvs.2018.00045

Baste, I. A., & Watson, R. T. (2022). Tackling the climate, biodiversity and pollution emergencies by making peace with nature 50 years after the Stockholm Conference. *Global Environmental Change*, 73, 102466. 10.1016/j.gloenvcha.2022.102466

Begum, J. A., Shikdar, A. H., & Das, K. M. (2007). *Secondary social science: For class IX-X [in Bengali]* (New ed.). Dhaka: National Curriculum and Textbook Board.

Biswas, T. K., Saha, S. K., & Nasrin, F. (2009). An Assessment of Formal Environmental Education (FEE) System at Tertiary level in Bangladesh. *Journal of Teacher's World*, 35-36, 155–168.

Bocken, N. M., & Short, S. W. (2021). Unsustainable business models – Recognising and resolving institutionalised social and environmental harm. *Journal of Cleaner Production*, 312, 127828. 10.1016/j.jclepro.2021.127828

Chadwick, M., Soussan, J., Mallick, D., & Alam, S. (1998). *Understanding indigenous knowledge: Its role and potential in water resource management in Bangladesh.* University of Leeds and Bangladesh Centre for Advanced Studies.

Chowdhury, M. H. (2004, February). Environmental education for sustainability in Bangladesh. In *Asia Pacific Environmental Education Seminar* (pp. 11-14).

Chowdhury, M. T. (2014). Nature of environmental education in Bangladesh: A school level assessment with reference to the national curriculum. *International Electronic Journal of Environmental Education*, 4(1), 53–60. 10.18497/iejee-green.67994

Clark, B. Y., & Whitford, A. B. (2011). Does more federal environmental funding increase or decrease states' efforts? *Journal of Policy Analysis and Management*, 30(1), 136–152. 10.1002/pam.20547

Dorn, C. (2020). A New Global Ethic: A History of the United Nations International Environmental Education Program, 1975-1995. *Foro de educación, 18*(2), 83-108.

Environmental Protection Agency (EPA). (2016). What is Environmental Education? Retrieved January 17, 2024, from https://www.epa.gov/education/what-environmental-education

Gough, A. (2016). Teacher education for sustainable development: Past, present and future. *Teaching education for sustainable development at university level*, 109-122.

Haque, A. (2022, December 3). Bridging the industry-academia gap through collaboration. *The Business Standard.* Retrieved from https://www.tbsnews.net/features/panorama/bridging-industry-academia-gap-through-collaboration-544090

Haque, F. (2013). Education for Sustainable Development: An evaluation of the new curriculum of the formal primary education in Bangladesh. *European Scientific Journal, 1*(Special Edition), 320-329.

Haque, M. A., Shafiullah, M., Uddin, M. N., Roy, S., & Das, K. M. (2007). *Secondary general science: For class IX-X* (Revised ed.). (Sarker, M. T. H., Islam, M. R., Rahman, M. S., Bhadra, S. K., & Begum, J. A., Trans.). National Curriculum and Textbook Board.

Haque, M. S., & Sharif, S. (2021). The need for an effective environmental engineering education to meet the growing environmental pollution in Bangladesh. *Cleaner Engineering and Technology*, 4, 100114. 10.1016/j.clet.2021.100114

Hoque, A., Clarke, A., & Sultana, T. (2017). Environmental sustainability practices in South Asian university campuses: An exploratory study on Bangladeshi universities. *Environment, Development and Sustainability*, 19(6), 2163–2180. 10.1007/s10668-016-9845-0

Islam, M. R. (2020). Environment and disaster education in the secondary school curriculum in Bangladesh. *SN Social Sciences*, 1(1), 23. 10.1007/s43545-020-00025-1

Islam, M. T., Abdullah, A. B., Shahir, S. A., Kalam, M. A., Masjuki, H. H., Shumon, R., & Rashid, M. H. (2016). A public survey on knowledge, awareness, attitude, and willingness to pay for WEEE management: Case study in Bangladesh. *Journal of Cleaner Production*, 137, 728–740. 10.1016/j.jclepro.2016.07.111

Islam, S. T. (2015, June). Climate Change Education for Sustainable Development in Bangladesh. UNESCO. Retrieved January 25, 2023 from https://unesdoc.unesco.org/ark:/48223/pf0000246779/PDF/246779eng.pdf.multi

Jahan, F., & Quddusi, K. S. K. A. (2013). Climate change and Bangladesh: Geographical and socio-economic impacts. *Dhaulagiri*, 7, 113–132. 10.3126/dsaj.v7i0.10439

Kagawa, F. (2022). *The Heat is On! Towards a Climate Resilient Education System in Bangladesh. Kathmandu:* UNICEF Regional Office for South Asia. Retrieved January 23, 2024 from https://www.unicef.org/rosa/media/17566/file/The%20Heat%20is%20On%20-%20Bangladesh.pdf

Kremen, C., Merenlender, A. M., & Murphy, D. D. (1994). Ecological monitoring: A vital need for integrated conservation and development programs in the tropics. *Conservation Biology*, 8(2), 388–397. 10.1046/j.1523-1739.1994.08020388.x

McCrea, E. J. (2006). The Roots of Environmental Education: How the Past Supports the Future. Environmental Education and Training Partnership (EETAP). Retrieved from https://files.eric.ed.gov/fulltext/ED491084.pdf

Ministry of Environment and Forest (MFE). (2005, November). National Adaptation Programme of Action (NAPA). Retrieved from http://nda.erd.gov.bd/en/c/publication/national-adaptation-programme-of-action-napa-2005

Ministry of Environment and Forest (MFE). (2009, September). Bangladesh Climate Change Strategy and Action Plan 2009. Retrieved from http://nda.erd.gov.bd/files/1/Publications/CC%20Policy%20Documents/BCCSAP2009.pdf

Ministry of Environment and Forests (MEF). (2016). Bangladesh National Conservation Strategy (NCS) (2016-2031). Government of the People's Republic of Bangladesh. Published in September 2016. Retrieved from https://bforest.portal.gov.bd/sites/default/files/files/bforest.portal.gov.bd/notices/c3379d22_ee62_4dec_9e29_75171074d885/Executive%20Summary(NCS).pdf

Mohammad, N. (2010). Development of the Environmental Education in Higher Studies in Bangladesh: An Approach for Environmental Sustainability. *Environmental Research and Technology,* 148-155.

Mojid, M. A. (2020). Climate change-induced challenges to sustainable development in Bangladesh. [). IOP Publishing.]. *IOP Conference Series. Earth and Environmental Science*, 423(1), 012001. 10.1088/1755-1315/423/1/012001

National Curriculum and Textbook Board [NCTB]. (2003). *National curriculum for primary level, 2002-03*. Dhaka: Ministry of Education, Government of Bangladesh.

Rithu, M., Akhter, N., Rahman, M. M., & Al-Maruf, A. (2016). The Status of Geography and Environmental Studies at Secondary Level Education in Bangladesh. *Archives of Current Research International*, 5(4), 1–9. 10.9734/ACRI/2016/29017

Rovňák, M., Novotny, R., Barlak, J., Bakon, M., & Chovanec, F. (2021). Implementation and funding of projects focused on non-formal environmental education. *Studia Ekonomiczne. Gospodarka, Społeczeństwo, Środowisko, (2/2021 (8))*, 136-144.

Sabbab, S., & Ahmed, S. (2024, January 21). Restructuring climate education in Bangladesh. *The Dhaka Tribune*. Retrieved from https://www.dhakatribune.com/opinion/op-ed/337332/restructuring-climate-education-in-bangladesh

Salequzzaman, Md., & Davis, J. K. (2003). Environmental education and environmental management in Bangladesh and their sustainability. *Environmental Informatics Archives*, 1, 70–82.

Salequzzaman, M., & Stocker, L. (2001). The context and prospects for environmental education and environmental careers in Bangladesh. *International Journal of Sustainability in Higher Education*, 2(2), 104–127. 10.1108/14676370110388309

Sangwan, S., & Vyas, R. (2020). Assessment of Green Consumerism on Environmental Sustainability. *International Journal of Current Microbiology and Applied Sciences*, 9(10), 389. 10.20546/ijcmas.2020.910.389

Sarkar, M. (2011). Secondary Student's Environmental Attitudes: The Case of Environmental Education in Bangladesh. *International Journal of Academic Research in Business & Social Sciences*, 1, 106.

Sultana, N., Hossen, M. S., & Khatun, R. (2017). Assessment of environmental knowledge and attitude of secondary level students of Tangail, Bangladesh. *Int. J. Res. Environ. Sci*, 3, 41–46.

Sutdhibhasilp, N. (2000). *Environmental adult education: two case studies of Thai non-governmental organizations promoting green consumerism*. University of Toronto.

Thomas Muñoz, R. (2022). Environmental education networks for social empowerment and global citizenship: a case of non-formal education from Mexico. In *Educational Response, Inclusion and Empowerment for SDGs in Emerging Economies: How do education systems contribute to raising global citizens?* (pp. 191–203). Springer International Publishing. 10.1007/978-3-030-98962-0_12

Uddin, M. K. (2023). Environmental education for sustainable development in Bangladesh and its challenges. *Sustainable Development*. Advance online publication. 10.1002/sd.2728

Uddin, M. M. M. (2014). An analysis of knowledge on sources of environmental pollution of students' at secondary level in Bangladesh. *International Journal of Academic Research in Business & Social Sciences*, 4(3), 2222–6990. 10.6007/IJARBSS/v4-i3/731

Ullah, M. O., Abu Hasan, M. D., & Uddin, T. (2013). Environmental Awareness and Disaster Factors in Bangladesh. *Journal of Applied Quantitative Methods*, 8(4).

UN. (n.d.). United Nations Conference on the Human Environment, 5-16 June 1972, Stockholm. Retrieved February 12, 2024, from https://www.un.org/en/conferences/environment/stockholm1972

UNESCO. (2014). Shaping the Future We Want: UN Decade of Education for Sustainable Development (2005-2014) Final Report. Retrieved from https://unesdoc.unesco.org/ark:/48223/pf0000230302/PDF/230302eng.pdf.multi

APPENDIX: QUESTIONS AND ANSWERS

1. What is the overall problem presented in the current study?

Answer: The overall problem presented in this case is the inadequacy of environmental education (EE) in Bangladesh. While EE has been recognized as a crucial tool in addressing environmental challenges and promoting sustainable development, its implementation in Bangladesh faces several limitations. These include fragmented and incomplete curricula, insufficient assessment criteria, a lack of focus on emerging environmental issues such as climate change, and outdated teaching methods. Consequently, there is a need for policy reform and curriculum revision to make EE more effective and responsive to the country's environmental needs.

2. What are the factors affecting the problem(s) related to this study?

Answer: Several factors contribute to the limitations of environmental education (EE) in Bangladesh, including inadequate policies and the lack of specific guidelines for implementation and assessment, fragmented and incomplete EE curriculum, and outdated teaching methods that lack a focus on interactive and activity-oriented approaches, absence of well-defined assessment criteria for environmental literacy, limited teacher training, and finally the resource constraints.

3. Discuss managerial, organizational, and technological issues and resources related to the current case.

Answer: Managerial issues highlighted in the case primarily center around the need for effective policy implementation, curriculum reform, and teacher training. While policies recognizing the importance of EE exist, they often lack specific guidelines for implementation and assessment, leading to fragmented and ineffective EE programs. This points to a need for stronger managerial oversight and coordination in translating policy into practice. Additionally, there is a need for updated teaching methods that foster active student engagement and critical thinking, moving away from outdated and passive approaches. This requires effective management of the transition to new pedagogical models and ensuring teacher buy-in.

Organizational issues relate to resource constraints and the capacity of educational institutions to deliver quality EE. Schools and universities often lack the funding, teaching materials, and technology to support effective EE programs. This includes a shortage of qualified EE teachers and inadequate professional development opportunities for existing educators. Addressing these issues requires organizational solutions, such as securing funding, developing or acquiring teaching resources, and providing teacher training programs. Moreover, the fragmented nature of EE across different subjects and levels of education suggests a need for better organizational integration and coordination of EE efforts.

Technological issues primarily relate to the lack of resources and integration of technology in teaching methods for environmental education in Bangladesh. Schools and universities often lack the necessary technology to support effective EE programs, hindering the adoption of modern teaching methods that could foster active student engagement and critical thinking. Limited internet access in some universities further restricts students' ability to access updated environmental information. Additionally, there is a lack of focus on integrating technology into the curriculum to enhance the learning experience and equip students with digital skills relevant to environmental monitoring and problem-solving.

4. What role do different players (decision-makers) play in planning and implementing policies and practices in developing countries, especially in Bangladesh?

Answer: The text does not explicitly discuss the role of decision-makers in the planning and implementing EE policies within the context of environmental education in Bangladesh. However, it does highlight the need for collaboration among various stakeholders, including government agencies, educational institutions, teachers, NGOs, and industry partners, to improve EE in the country. This suggests that decision-makers at all levels play a crucial role in creating and implementing policies that support the integration of EE policies, providing funding and resources for infrastructures and teacher training, and fostering partnerships that promote the development and use of educational technologies.

The following section discusses the key players that contribute to the planning and implementation of policies in environmental education in Bangladesh:

Government Agencies:

Ministry of Education (MoE): This body sets the national curriculum guidelines, which can include IT-based environmental education resources. It might also manage funding for the involved initiatives.

Ministry of Environment, Forest and Climate Change (MoEFCC): Develops environmental policies and strategies that could be integrated with IT tools for education. They might also be involved in funding and overseeing environmental education projects.

Department of Information and Communication Technology (ICT): Provides technical expertise and resources for developing and implementing IT solutions in education, including environmental education.

Educational Institutions:

Bangladesh Curriculum Development and Evaluation Directorate (BCPED): Works under the MoE and is responsible for developing and revising the national curriculum, which can include integrating IT into environmental education.

Schools, Colleges, and Universities: These institutions are responsible for implementing the curriculum set by the MoE. Teachers play a crucial role in utilizing and advocating for effective IT-based environmental education tools.

Non-Governmental Organizations (NGOs):

Many NGOs in Bangladesh work on environmental issues and education. They can develop IT-based educational tools, conduct training for teachers, and pilot innovative approaches in schools. Several NGOs in Bangladesh are at the forefront of promoting IT-based tools for environmental education. Here are some key examples:

BRAC: This large NGO offers E-education programs for interactive learning and conducts teacher training programs. These initiatives could be adapted to incorporate environmental education modules and specific IT-based approaches.

ASA: This organization focuses on developing online courses on environmental topics and creating interactive school websites with engaging environmental resources. ASA Education Program opens 15,750 learning centers in villages across the country to provide tuition aid to unprivileged children.

The Hunger Project: While their main focus is food security, they recognize the link to environmental sustainability. They could develop educational videos on sustainable agriculture and partner with radio stations for interactive environmental programs.

These are just a few examples, and many other NGOs are actively contributing to this space. Their specific contributions vary based on their expertise and resources, but all play a vital role in bringing innovative IT-based approaches to environmental learning in Bangladesh.

Private Sector:

Technology companies can collaborate with the government and educational institutions to develop policies and implement solutions for environmental education. This could involve creating educational apps, games, or online platforms.

International Organizations:

UNESCO, UNICEF, UNDP, and other international organizations may provide funding and technical assistance for developing and implementing IT-based environmental education programs in Bangladesh.

The successful planning and implementation of policies in environmental education requires collaboration between all these stakeholders. Each player brings their unique strengths to the table:

- Government: Provides policy direction, funding, and infrastructure.
- Educational Institutions: Implement the curriculum and reach students directly.
- NGOs: Offer innovative approaches and community outreach.
- Private Sector: Provides technological expertise and solutions.
- International Organizations: Offer financial and technical support.

5. What are the possible alternatives in dealing with the problem(s) related to this study?

Answer:

Based on the problems identified in the text, the following alternatives can be inferred:

Status Quo: Continuing with the current EE curriculum and teaching methods.

- **Pros:** Requires minimal immediate effort or resources.
- **Cons:** Perpetuates the identified shortcomings in EE, leading to ineffective learning outcomes and unpreparedness for environmental challenges.

Gradual Reform: Incremental improvements to the existing curriculum and teaching methods.

- **Pros:** Allows for adjustments based on feedback and resources, potentially less disruptive than a complete overhaul.
- **Cons:** This may not fully address the systemic issues and could prolong the process of achieving effective EE.

Complete Overhaul: Designing a new EE curriculum and adopting modern teaching methods.

- **Pros:** Opportunity to create a comprehensive, integrated, and responsive EE program aligned with current needs and best practices.
- **Cons:** Requires significant resources, time, and effort for curriculum development, teacher training, and implementation.

Focus on Teacher Training: Prioritizing professional development for teachers to enhance their EE knowledge and pedagogical skills.

- **Pros:** Empowers teachers to deliver EE more effectively within the existing curriculum, potentially leading to improved student outcomes.
- **Cons:** May not address the broader systemic issues related to curriculum content and assessment.

Technology Integration: Incorporating technology into EE through online platforms, educational apps, and multimedia resources.

- **Pros:** Enhances student engagement, facilitates access to updated information, and prepares students for the digital age.
- **Cons:** Requires investment in technology infrastructure and teacher training in educational technology.

Collaboration and Partnerships: Engaging with NGOs, government agencies, and international organizations to support EE initiatives.

- **Pros:** Leverages expertise, resources, and networks to strengthen EE programs and advocate for policy reform.
- **Cons:** Requires coordination and alignment of goals among diverse stakeholders.

6. What are some of the emerging technologies that should be considered through the process of policymaking related to the current study?

Answer:

Technologies that could be considered through the policymaking process and could make the classroom more interactive and activity-oriented approaches, such as online learning platforms, educational apps, and multimedia resources, could be beneficial for solving the problems related to environmental education in Bangladesh. Additionally, technologies that could help address the lack of access to updated environmental information, such as digital libraries and online databases, could also be considered.

7. What is the final solution that can be recommended to form, manage, and organize policies more efficiently in this case?

Answer:

A multifaceted approach to enhancing environmental education in Bangladesh can be a final solution to the problems presented in this chapter of the book. Based on the identified problems and limitations, the following multidimensional approach is recommended for the policymaking and implementation:

1. Policy and Curriculum Reform:
   - **Advocate for stronger policy implementation:** Collaborate with relevant government agencies and stakeholders to ensure effective implementation of existing policies on environmental education.
   - **Develop a comprehensive national EE curriculum:** This curriculum should be aligned with national and international standards, address emerging environmental issues, and promote active learning and critical thinking.
   - **Integrate EE across different subjects and educational levels:** Ensure that EE is not taught in isolation but is woven into the fabric of the entire curriculum.
2. Teacher Training and Development:
   - **Organize comprehensive EE training programs for teachers:** These programs should focus on modern pedagogical methods, environmental literacy, and the use of technology in teaching.
   - **Provide ongoing professional development opportunities for teachers:** This will help them stay updated on current environmental issues and best practices in EE.
   - **Establish a network of EE teacher trainers:** This network can provide support and mentorship to newly trained teachers.
3. Resource Development and Technology Integration:
   - **Secure funding for EE programs:** This can be done through public and private partnerships, grants, and fundraising initiatives.
   - **Develop or acquire interactive and engaging teaching materials:** This includes multimedia resources, digital tools, and hands-on activities that promote environmental awareness and action.

- **Invest in technology infrastructure:** Ensure that schools and universities have access to computers, internet connectivity, and other necessary technological resources.
- **Develop online learning platforms and educational apps:** These can provide students with access to updated environmental information and interactive learning experiences.

4. Collaboration and Partnerships:
   - **Collaborate with environmental NGOs and other stakeholders:** This will help to share resources, expertise, and best practices.
   - **Participate in international EE initiatives:** This will help to stay updated on global trends and best practices in EE.
   - **Promote community-based environmental education projects:** This will engage local communities in environmental action and stewardship.

Arguments in Support of the Recommended Solution:

- This multi-pronged approach addresses all of the key factors contributing to the limitations of environmental education in Bangladesh.
- It is based on best practices in EE and is aligned with national and international standards.
- It is comprehensive and sustainable, addressing both short-term and long-term needs.
- It is collaborative and involves the participation of a wide range of stakeholders.
- It is likely to be effective in improving the quality of environmental education in Bangladesh and in equipping students with the knowledge, skills, and values they need to become responsible environmental citizens.

By implementing this multi-pronged approach, the organization can make a significant contribution to improving the quality of environmental education in Bangladesh and to promoting sustainable development in the country.

EPILOGUE AND LESSONS LEARNED

The long-range effects of enhancing environmental education in Bangladesh are far-reaching and multifaceted. A well-implemented, comprehensive EE program can foster a generation of environmentally conscious citizens equipped with the knowledge, skills, and values to address pressing environmental challenges. This can lead to improved environmental stewardship, reduced pollution and resource depletion, and increased resilience to climate change impacts. Moreover, EE can contribute to sustainable economic development by promoting green skills and innovation, creating new job opportunities, and attracting environmentally responsible investments. In the long run, a strong EE foundation can enhance Bangladesh's overall well-being, ensuring a healthier, more sustainable, and prosperous future for its people and ecosystems.

Lessons learned

Five lessons can be learned from this study on environmental education (EE) in Bangladesh:

1. **EE is crucial for sustainable development:** The study highlights the importance of EE in addressing environmental challenges and promoting sustainable practices. This lesson underscores the need for continued investment in EE to equip individuals with the knowledge and skills to make informed decisions about the environment.
2. **EE curricula need continuous improvement:** The analysis of Bangladesh's EE curricula reveals gaps and limitations in addressing emerging environmental issues, integrating interdisciplinary approaches, and promoting active learning. This lesson emphasizes the need for regular curriculum revisions to ensure that EE remains relevant and effective.
3. **Effective EE requires diverse teaching methods:** The study critiques traditional teaching methods in EE and advocates for interactive, participatory, and innovative approaches. This lesson highlights the importance of diversifying teaching strategies to cater to different learning styles and promote a deeper understanding of environmental issues.
4. **Collaboration is key to successful EE:** The involvement of various stakeholders, including government agencies, NGOs, and civil society, is crucial for the successful implementation of EE programs. This lesson emphasizes the need for collaborative efforts to share resources, expertise, and best practices in EE.
5. **Technology can enhance EE:** The study suggests that integrating technology into EE can improve access to information, facilitate interactive learning experiences, and equip students with digital skills relevant to environmental monitoring and problem-solving. This lesson highlights the potential of technology to transform EE and prepare students for the digital age.

LIST OF ADDITIONAL SOURCES

1. Nature of environmental education in Bangladesh: A school-level assessment with reference to the national curriculum. Access link: https://dergipark.org.tr/tr/download/article-file/89025
2. The context and prospects for environmental education and environmental careers in Bangladesh. Access link: https://researchportal.murdoch.edu.au/esploro/outputs/journalArticle/The-context-and-prospects-for-environmental/991005545426807891
3. Environmental education and environmental management in Bangladesh and their sustainability. Access link: https://researchportal.murdoch.edu.au/esploro/outputs/journalArticle/Perspectives-of-environmental-education-and-environmental/991005543781507891
4. Report on Mapping of Education for Sustainable Development-climate change initiatives in Bangladesh. Access link: https://dshe.portal.gov.bd/sites/default/files/files/dshe.portal.gov.bd/moedu_office_order/1bdc724e_58d7_473c_9f74_d0056b6d959f/Report%20on%20ESD%20Net%202030%20Country%20Initiative%20Bangladesh.pdf
5. An evaluation of the new curriculum of the formal primary. https://eujournal.org/index.php/esj/article/view/2311/2184
6. Environmental Education in Bangladesh with Special Reference to Higher Studies. https://www.academia.edu/24026153/Environmental_Education_in_Bangladesh_with_Special_Reference_to_Higher_Studies
7. Environmental Education and Awareness. https://bforest.portal.gov.bd/sites/default/files/files/bforest.portal.gov.bd/notices/c3379d22_ee62_4dec_9e29_75171074d885/16.%20Environmental,%20Education%20and%20Awareness_NCS.pdf
8. https://www.bedsbd.org/
9. https://esdo.org/
10. Skills and Education for Greener Bangladesh. https://www.researchgate.net/publication/364357375_Skills_and_Education_for_Greener_Bangladesh
11. Environmental education for sustainable development in Bangladesh and its challenges. https://onlinelibrary.wiley.com/doi/abs/10.1002/sd.2728
12. Environmental Education at the School Level in Bangladesh: Observations with Reference to the National Curriculum. https://www.academia.edu/66565355/Environmental_Education_at_the_School_Level_in_Bangladesh_Observations_with_Reference_to_the_National_Curriculum
13. BANBEIS. https://banbeis.portal.gov.bd/sites/default/files/files/banbeis.portal.gov.bd/page/6d10c6e9_d26c_4b9b_9c7f_770f9c68df7c/Bangladesh%20Education%20Statistics%202022%20%281%29_compressed.pdf

Chapter 11
The Importance of Air Pollution as Part of the Environmental Literacy Studies to School Students

Lamia Galal Dawood
https://orcid.org/0009-0001-7588-5574
Indiana University of Pennsylvania, USA

EXECUTIVE SUMMARY

Education about environment, sustainability, and ethically sourced practices became mandatory in school curriculum. Pennsylvania State Board of Education adopted new academic standards for Science, Technology and Engineering, Environmental Literacy, and Sustainability (STEELS). The new academic standards will integrate new disciplines including science, technology, and engineering. Disciplines also include teaching the relationship between humans and the environment. Environmental literacy and sustainability discipline will focus on human social practices, ethical and non-ethical practices, and their impact on the environment. In this chapter, the authors introduce the addition of environmental literacy and sustainability to the Pennsylvania academic curriculum. The topic of air pollution was chosen as the effect of air pollution on the health of school students, absenteeism, and life expectancy are not covered by much scholarly research. Importance of education about air pollution and the knowledge of school students on air pollution sources, air quality regulations, air quality, and its substantial gains to public health are briefly included. In general, this chapter emphasizes the importance of including air quality science in school curriculum. To demonstrate what can be added to the atmospheric

DOI: 10.4018/979-8-3693-1710-5.ch011

science classroom, Clean Air Act (CAA) rule issued by the Environmental Protection Agency (EPA) and Corporate Social Responsibility (CSR) of the International Civil Aviation Organization (ICAO) are recommended as examples.

INTRODUCTION

Environmental justice, sustainable practices, climate adaptation planning for current and future generations will be included in the STEELS academic standards, which is planned to be implemented in 2025. The new academic standards aim at developing critical thinking, problem solving, decision making, and communication skills; and encouraging responsible citizenry and community skills among students (Martin, 2024). There will be a three-year implementation window for smooth implementation of the new integrated standards for science, technology & engineering, and environmental literacy & sustainability (Pennsylvania Department of Education, 2022). Air pollution is one of the newly added disciplines to the Pennsylvania new academic standards. Air pollution caused by Air Transport or ground transportation is rarely or briefly discussed in classrooms. Numerous factors cause air pollution including motor vehicles, industrial facilities, aviation industry, forest fires and others. Increase in population is also considered one of the contributors of air pollution because of the harmful practices conducted by humans. Malpractices like cutting down forests, burning fossil fuels and agriculture activities mount every day, which reduce the air quality of our planet (Fowler et al., 2020). School students do not have scientific knowledge about the causes of air pollution and its impact on public health. Schools in many major cities have a higher rate of air pollution, which affects the health of students and increases absenteeism. Climate change causes allergies and long-term exposure to air pollution has been associated with diseases of the heart and lungs, cancers, and other health problems (American Public Health Association, 2017). Although efforts to control air pollution are made by federal, State, regional, and municipal agencies, it is crucial that students learn about those efforts and their impact on the air quality levels.

School students of all levels have limited to no information about how to obtain information on air quality. School students also have minimal information on how to act to air quality hazards and the ethical practices taken to control their harmful effect. As human health is affected by the environment, air pollution is considered one of the main factors leading to severe diseases (Almetwally et al., 2020). Air pollution is defined by the World Health Organization (WHO, 2023) as the contamination of the environment whether indoor or outdoor by any chemical, physical or biological agent that modifies the natural characteristics of the atmosphere. The data of the WHO indicates that air pollution is mainly caused by combustion of fossil fuels, which affects 99% of the global population who breathe air that exceeds their guideline limits (WHO, 2023). The rapid growth of global aviation led to the increase of air pollutant emissions from aircraft engine exhausts. The main contribution to air pollution is made by civil aviation and military aviation. Aircraft emissions contribute to climate change, which is the depletion of the ozone layer and the greenhouse effect (Peeters, 1998). Ground Transport emissions resulted in an estimated 7.8 million years of life lost and approximately $1 trillion in health damages globally in 2015 (Anenberg et al., 2019).

Air pollution causes biological effects on students who could experience respiratory problems such as asthma and environmental effects represented by global warming, ozone layer depletion, and acid rain. Exposure to air pollution is one of the main causes of respiratory disease among children. Air pollution from industrial resources around public schools was found to be one of the major reasons for student's illness and school absence. This shows the importance of choosing locations that are safe from air pollution for schools. Many studies found an association between air pollutants and hospital admission due to respiratory diseases and lung functions changes in elementary students (Mohai et al., 2011 & Park et al., 2002). As air pollution can increase school illness-related absences, education about air pollution, its consequences and means to control it is essential for school students to mitigate their drastic effects on students' wellbeing. Accordingly, creating an intrinsic concern for the environment should start from school; as teaching air pollution should be part of the teaching curriculum.

As public concerns about air pollution caused by the aviation industry increases, awareness and education in a scientific manner about air pollution and its effects on public health effects became essential. School students should also be aware of the entities responsible for taking action to address it. It is important for students to learn about the initiatives made by the International Civil Aviation Organization (ICAO) to reduce the impacts of aviation emissions affecting global climate. As today's students will be tomorrow's leaders, managers, and employees in the air transport industry, being educated about air pollution, its impacts and efforts needed to address it should be part of the school curriculum.

Ground transportation also with its sub-sectors including diesels vehicles contributed most to the health burden and ozone mortality from transportation emissions. This highlights the importance of adopting more stringent vehicle emission regulations and policies especially that expected growth in the transportation sector will increase public health burdens (Anenberg, 2019). Knowledge about the efforts made by Clean Air Act (CAA) to control transport emissions, waste management, and land-use planning are crucial parts of the air pollution learning that school students should have.

Air pollution is now the world's fourth-largest risk factor for early death. Future emission trends and its magnitude impact on public health require considering alternative resource allocations associated with environmental externalities in aviation. From this, air quality education programs on sources of air pollution and means to control them can help students in disseminating air quality information to promote pollutant avoidance. Air quality education can also safeguard public health by regulating the emissions of these harmful air pollutants.

DISCUSSION

The Effect of Air Pollution on School Children

Being exposed to environmental pollutants can cause long-lasting health problems in children especially in their important times of physiological development. Mohai et al. (2011) made a study on the extent of air pollution from industrial sources around public schools in Michigan and found that schools located in areas with the highest air pollution levels had the lowest attendance rates, which can be considered a potential indicator of poor health, and the highest proportions of students who failed to meet state educational testing standards.

In a study about the association of air pollution with school absenteeism due to illness, Park et al. (2002) conducted a time-series analysis on air pollution and school absenteeism data. To conduct this study, the number of daily illness-related absences was regressed against the daily levels of air pollution. The study concluded that air pollution is associated with illness-related absences among elementary students. This shows how health problems caused by air pollution can lead to a decline in daily activities and in school performance, which will generally affect students' progress in education.

Life expectancy is associated with air quality. Studies showed that children who live in a neighborhood with high air pollution have lower IQ and experience some memory loss issues (Teksoz, 2011). A lower level of education has been associated with bad health outcomes and higher rate of mortality (O'Neill et al., 2008). In a study conducted by Kan et al. (2008), they found that residents with low educational attainment have a higher number of deaths from air pollution. They recommended conducting studies on the implication of local environmental and social policies, to gain more knowledge about the effects of those policies on human health.

Education About Air Pollution for School Students

Modern societies suffer from diverse environmental aspects which affect the well-being of human beings, animals, and plants. Access to clean water, clean unpolluted air, reduced noise neighborhood became a challenge for everyone especially those living in big modern cities. The causes of these problems can be attributed to the way individuals interact with the environment and the developmental choices they made (Dimitriou & Christidou, 2011). Air pollution can be caused by chemical, ecological, or biological practices intrigued by human's interest to increase profit and modernization requirements.

As air pollution has a harmful impact on public health, improvement in air quality can lead to substantial gains in public health. In a study to explore the knowledge of pupils of air pollution, its sources and consequence, 132 Greek pupils aged 7-13 were interviewed. The data showed that pupils have a general conceptualization of air pollutants sources and their impact on human health. The outcomes of the study emphasized the importance of adding air pollution education in the science classroom (Dimitriou & Christidou, 2007).

Myers et al. (1999) explored the understanding of school students about air pollution through distributing a questionnaire to school students aged 11-16 years. Students identified sources of pollutions as those generated by transportation, industry, and energy generation, which revealed that they have an overall understanding of the air pollution sources but they did not have a scientific knowledge of the composition of air. In another study conducted on pupils in schools in England about their understanding of the nature of air pollution, its biological and environmental consequences and means to reduce it, the school students had little knowledge of the constituents of air pollution but were informed about the consequences of air pollution. Most pupils agreed that legal compulsion is more important than financial penalties, however they did not have a clear idea of actions needed to reduce air pollution (Myers et al., 2004).

In a questionnaire conducted by Thornber et al. (1999) on school students' knowledge about air pollution, and its biological and physical effects. They found that school students do not have scientific knowledge about air pollution but general knowledge about the effects of environmental problems caused by transport and industry. Most of the school students who answered the questionnaire identified air pollution as killing plants and animals, but fewer thought this is also true of humans.

In a study made on Hong Kong secondary school students about their understanding and knowledge of air pollution's composition and their effects on human health, several misconceptions were found. The school students believed that oxygen was more common in unpolluted air than nitrogen. This study recommended deeper scientific education about air pollution and encouraging students to reflect more on their impact on the environment and their role as consumers of goods and services (Boyes et al., 2004).

Introduction of Environmental Literacy and Sustainability to Pennsylvania K-12 Curriculum

In January 2022, Pennsylvania State Board of Education adopted new academic standards for Science, Technology and Engineering, Environmental Literacy and Sustainability (STEELS). The new academic standards aim at equipping all students with scientific, technological, environmental, and engineering knowledge in a mission to support Pennsylvania's economic vitality and its civic strength (Pennsylvania Department of Education, 2022). The new integrated standards for science, technology & engineering, and environmental literacy & sustainability help students to acquire scientific discourse and deeper understanding of environmental systems and resources, environmental literacy skills, sustainability, and stewardship (Pennsylvania Department of Education, 2022).

Implementation of the new academic standards is July 1, 2025 and is designed to give students access to education that helps them to become life-long learners (STEELS Hub, 2022). Environmental Literacy & Sustainability will teach students about human social practices and processes that have negative effects on the environment and the eco-system. Students will learn about sustainable practices which lead to the integrity of ecological system, environmental justice, and species diversity. Environmental literacy & sustainability curriculum and instruction will provide students to engage in with the necessary foundational knowledge in the field. (STEELS Hub, 2022). K-12 environmental literacy and sustainability curriculum will provide students with knowledge needed to solve many of humanity's most pressing current and future ecological challenges and to learn about air quality and ways to reduce the harmful effects of humans on our planet. K-2 students will learn about the physical characteristics of nature, students in 3-5 grade will develop a model to learn how local environmental issues are connected to larger environmental and human systems. Students in their 6-8 grade will learn to collect and analyze environmental data and how to manage natural resources. Students in their 6-8 grade will evaluate and interpret how trends, technologies and policies impact the environmental system and resources (STEELS Hub, 2022).

To contribute to the social, economic and environmental transformations that are crucial to achieve sustainable development goals, education institutions are uniquely positioned. Husic (2024) emphasized that the STEELS's new Science, Technology & Engineering, Environmental Literacy & Sustainability academic standards should be a requirement for all state school districts for the 2025–2026 school year. To become future social entrepreneurs, students need to have civic engagement to prepare them to act as environmental stewards and leaders. For this reason, sustainability and environmental literacy should be incorporated across the curriculum on school and university campuses, not just in certain majors or in certain states (Husic, 2024).

STEELS's new academic standards for teaching and learning in K-12 classrooms are modeled after the Next Generation Science Standards (nextgenscience.org). The retired standards adopted in 2002 focused on science content knowledge, while the new standards are three-dimensional models designed to engage students in science practices in investigation, manipulate models and discuss evidence (Stone, 2024). To fill the gap between the old and the new academic standards, teachers need to redesign curriculum to align to the new standards in three primary shifts. First shift is to focus on phenomena, second is to focus on three dimensions of science learning, and the third shift is to include a partnership between students, parents, and educators to support inclusion, equity and belonging in science (Onatlas.com, 2023). Pennsylvania is unique in adding environmental literacy and sustainability which equips PA students with plans for a sustainable future. The funding from the PA Department of Education provides professional training to teachers in PA and free K-12 library for science kits and equipment (Stone, 2024). PA instructional approach is to provide teachers with guidance, professional support, and development to give students confidence within science education (Onatlas.com, 2023).

RECOMMENDATIONS

As STEELS new academic standards includes environmental literacy to school students, this section provides recommendations to add efforts made by the Environmental Protection Agency (EPA) regulation through its Clean Air Act (CAA) Law and the International Civil Aviation Organization (ICAO) Corporate Social Responsibility (CSR) efforts.

The Clean Air Act (CAA) as a Regulator for Air Pollution From Ground Transport

Road or ground transport is considered one of the main sources of air pollution. Although many countries, especially developed countries stopped using coal, emissions from ground transport industry have participated drastically in ozone depletion and climate change. Urban road traffic has been associated with declining human health. A variety of policies to this problem have been adopted including the adoption of three-way catalysts technology and local traffic reduction measures regulated in western European cities (Colvile et al., 2001). Nevertheless, the great increase in car ownership and use, and the increase in ground transport networks makes the air pollution problems persist. Much research was made on the efforts to control or reduce air pollution which led to global climate change; however, most are restricted to a very narrow scope.

Acidification, noise, environmental and economic costs of gasoline and diesel emissions, chemical pollutants, industrial emissions are caused by the rapid rates of urbanization. To control those harmful effects, switching over to renewable energy sources and maximizing fuel efficiency in vehicles through replacing gasoline powered cars with electric vars can contribute to limiting air quality-related respiratory diseases. Consequently, ensuring environmental law remains intact and properly enforced is the only way to maintain and improve air quality.

In 1970 the United States passed the Clean Air Act, which has been a crucial tool for reducing air pollution. The Clean Air Act is the law regulated by the Environmental Protection Agency (EPA) that defines responsibilities for protecting and improving the nation's air quality and the stratospheric ozone layer (EPA, 2023). The Clean Air Act (CAA) includes regulations for noise pollution, ozone protection, acid deposition control, aircraft emission standards, and others. The motor vehicle emission and fuel standards, which is part of the Clean Air Act, regulates imports of vehicles, fuel used in vehicles, and devices installed on vehicles that can disable emission control which are generally called defeat devices (EPA, 2023).

The state of California has served for over fifty years as a national "laboratory" for the testing of technological solutions and regulatory approaches to improve air quality (Pappalardo, 2021). Those tests were conducted under the Clean Air Act (CAA), however in 2019, the Trump administration revoked California's authority to set more stringent pollution emission standards (Pappalardo, 2021). As the Clean Air Act mandates specific solutions and processes for non-compliance, it is known to have given rise to significant capital expenditures (Kramer et al., 2006). Accordingly, initiatives to reduce local air pollution and mitigate the effects of climate change from mobile sources such as motor vehicles have been challenged. Nevertheless, the Clean Air Act (CAA) clearly guarantees the protection of human populations from adverse health effects associated with air pollution exposure (Pappalardo, 2021). This magnifies the role of the EPA represented by its CAA rule to find new ways to cooperate with all states to build and reform the transportation sector through additional research strategy and policy documents that can improve regulatory health science and reduce human health risks.

ICAO's Efforts to Reduce Air Pollutants Generated by Aviation Industry

Air transport is the fastest and most used method of transportation. ICAO is the UN specialized agency responsible for administrating international civil aviation standards, practices, and policies with the aim of securing safe, efficient, economically sustainable and environmentally responsible civil aviation sector (International Civil Aviation Organization [ICAO], 2023, About). One of the initiatives adopted by ICAO is the environmental protection of climate change due to NOx and CO2 emissions, which is expected to increase from 2.4 to 4.4 times by 2045, compared with the forecasted 3.3 times growth in international air traffic (ICAO, 2023, Trends in local air quality). As Emissions from combustion of aviation fuel affect human health, degrade air quality resulting in climate change, technology change and new regulations to keep the quality of air became inevitable for the ICAO. Air Traffic is doubling every 15 years, this shows the need to modernize aviation systems, balance capacity, efficiency, safety and environmental requirements to address sustainability challenges. ICAA and the Federal Aviation Authority (FAA) mandates certification of aircraft and ground equipment and organizational approvals to ensure continued and enhanced safety (Batuwangala et al., 2018).

Part of the Corporate Social Responsibility (CSR) of the International Civil Aviation Organization (ICAO) is to improve annual fuel efficiency and carbon neutral growth. To achieve such goals, ICAO is adopting several measures including improvement of operational and aircraft technology and developing methods to increase sustainable aviation biofuels. Corporate Social Responsibility (CSR) is any firm's practice to maintain the principles of sustainable development. It encompasses the actions and procedures applied by a firm to economize consumption of raw material, preserve natural resources, and enhance the society and surrounding communities. In brief, it is when firms hold accountable for social change with their business beliefs, practices, and efforts. Such actions actually reflect positively on the firm's profitability and business growth as people tend to respect and appreciate more firms who are socially, environmentally, and ethically conscious.

To achieve its CSR goals of limiting climate change, ICAO is pursuing a number of measures that includes the development of technology to use sustainable aviation fuels through the usage of bioenergy feedstock instead of petroleum (ICAO, 2023, Sustainable aviation fuel). This shift in fuel would require allocating funds to research and development on how to expand agricultural sector to power planes with blends including biofuels made from renewable organic raw materials which are more efficient than kerosene. After conducting several research, ICAO were able to issue directions to airplanes manufacturers to design their new planes, engines, fans, and propulsion technology according to certain standards. For planes, ICAO recommended producing lighter planes as reducing aircraft empty mass through improved metals is vital in reducing fuel consumption. For engines, ICAO put regulations for manufacturers to redesign planes' fans, turbines, and compressors more efficiently to reduce fuel burn. ICAO also mandated remodeling fan pressure ratio and thermo-propulsive efficiencies to be reduced significantly to decrease fuel combustion (ICAO, Climate Change Technology Standards, 2023).

These new standards in designing aircraft will contribute to reducing greenhouse gas and CO_2 emissions from the air transport system through encouraging more fuel-efficient technologies into plane designs. As part of the new regulations, planes that do not meet these standards can no longer be produced beyond 2028 unless the designs are modified. In addition, a metric system to test fuel burn performance would be implemented inside each plane to allow ICAO to monitor planes' operations. ICAO has also reduced the shelf life of planes' components and operating life, as de-rated conditions and old planes tend to consume more fuel. As for planes operators, their CSR responsibilities include recycling magazines, refurbishing seats with recycled fabrics, and maintaining planes in an airworthy condition through consistent engine cleaning, efficient parts and components' quarantine management (Airline Trends, 2023).

CSR efforts exerted by any firm are beneficial to the firm itself and to the world in general. Still, it is preferable that firms engage in CSR initiatives that are related to their industry in order to have a better understanding of the problem. It is also recommended that firms do not consider CSR as a marketing opportunity, but rather as initiatives and actions emerging from ethical consciousness, responsibility towards public health conservation and awareness of threats of natural resources' depletion.

CONCLUSION

STEELS new academic standards have a unique discipline in Pennsylvania, which is environmental literacy, that aims to introduce means to mitigate the adverse impacts of human activity on the environment. The new academic standards that will be added to the Pennsylvania school curriculum in 2025 will fill the gap of knowledge students have in air pollution and its resources. It will not only provide theoretical understanding of the current and future environmental issues, but it also provides students with practical hands-on experience through investigating and developing models. Students will also learn about sustainability and stewardship, which will prepare them to behave ethically to nature and to our planet in general. They will be given a chance to evaluate solutions for pressing environmental issues and seek information on how to establish environmental justice.

Although there are numerous air pollutants, in this chapter we focused on only air pollution caused by air and ground transport. Air pollution is caused by many sources and has impacts on human health and global atmospheric change. Urban road traffic and aircraft emissions have been briefly introduced in this chapter, however the contribution of Sulphur emissions from ships to acid deposition need to be included as one of the main air pollutants. Further studies on other sources of air pollution and its impact on school students are recommended.

The main goal of this chapter is to emphasis the importance of adding education about air pollution, its impacts on human health and means to control pollution resources to school curriculum. Scientific education about air pollution can contribute significantly to addressing future threats caused by air pollution and will raise a generation equipped with knowledge and skills to adequately reduce the harmful impacts of air pollution. Adding atmospheric science to the school curriculum became a necessity to spread awareness about it and to improve air quality. The science of air pollution, assessment of air quality, regional atmospheric chemistry and global atmospheric change will add to the knowledge of students. This in turn will participate in inventing new technologies that can abate air pollution emissions from any type of source. As future legislation, future sustainable integrated transport systems, economic incentives to control emissions and action to protect the environment will be one of the responsibilities of the future generation, hence atmospheric science education is a necessity.

Students who plan to work in the air transport industry should know their responsibility to spread awareness about the importance of recycling electronics. They should also be educated about the efforts made by aviation companies to replace plastic items used for in-flight catering services with recycled ones, which should prevail in most of the manufacturing processes. Students should learn about the ethical obligations of corporates as each company has a code of ethics that guides employees to perform their duties according to relevant laws, regulations, and technical standards. As students are the future decision makers, it is crucial to be well informed of air pollution challenges and means to control it. For this, efforts made by the ICAO and the EPA Are introduced in this chapter, efforts exerted by other government or international agencies to control air pollution are recommended to be included in future studies.

Reducing air pollution caused by rising traffic volumes and airborne emissions requires the development of alternative solutions that are eco-friendly and sustainable. As confronting those environmental issues especially air pollution is inevitable, education about the causes, consequences, and means to control them is crucial. Students who will be equipped with skills and knowledge about air pollution will be able to take decisions to develop more future sustainable solutions. Incorporating awareness about air pollution, efforts, and regulations to control it into school curriculum, became a necessity especially because of the environmental cost caused by human beings' practices.

EPILOGUE AND LESSONS LEARNED

Epilogue:

The adoption of new academic standards, such as those by the Pennsylvania State Board of Education for Science, Technology, Engineering, Environmental Literacy, and Sustainability (STEELS), can have significant long-range effects on students, educators, communities, and the broader society. Potential long-term impacts include equipping students with critical skills needed for the modern workforce, such as problem-solving, critical thinking, and technological literacy. It also increases students' preparedness to work in high-demand fields such as engineering, environmental science, and technology. Emphasis on environmental literacy and sustainability fosters a mindset of lifelong learning and adaptability, essential for navigating future challenges. Curricula will be updated to include current scientific knowledge, technological advances, and contemporary environmental issues, making education more relevant. Students educated under these standards will be more informed and responsible citizens, capable of making decisions that positively impact society and the environment. Emphasizing sustainability can lead to a more environmentally conscious society that values and practices sustainable living.

Lessons:

My Study can help highlight the importance of the following aspects when applying STEELS new academic standards.

- Application of the new academic standards will increase the learning curve not for only students but also for teachers. It will give teachers a chance for ongoing learning opportunities as they will engage in continuous professional development to stay current with new teaching methods and content knowledge. Educators may adopt more innovative, interdisciplinary, and student-centered teaching approaches, enhancing overall teaching quality. Students too will be equipped with 21st Century Skills as the STEELS initiative supports interactive learning through increasing use of hands-on, experiential learning and technology in the classroom, which can make education more engaging for students. Accordingly, students will be encouraged to think critically about environmental issues and sustainability practices, which help them to understand the complexity of these challenges and to develop solutions.

- Learning about the Clean Air Act (CAA) law in the United States and how it regulates air emissions from stationary and mobile sources to protect public health and the environment, is an essential knowledge for school students. Students will learn how this law helps control pollutants that are harmful to human health including sensitive populations such as children and the elderly. Students who learn about this law, as part of their environmental literacy curriculum will know means adopted by the CAA to enforce emission limitations, regulate emissions of hazardous air pollutants and technology used to achieve control, monitor and reduce emissions. Students will also learn about requirements for vehicle inspection, maintenance programs, and cost-effective reductions practices. They will learn how to address emissions of greenhouse gases that contribute to climate change, as well as initiatives to improve energy efficiency and promote clean energy technologies.
- The inclusion of the Corporate Social Responsibility (CSR) within the context of the International Civil Aviation Organization (ICAO) encompasses a wide range of initiatives and policies aimed at promoting sustainable development, environmental protection, and social responsibility in the global aviation industry. By including the CSR of the ICAO to the environmental literacy new academic standards, students will learn about the ICAO's pivotal role in setting international standards and regulations for aviation safety, security, efficiency, and environmental protection. This includes ICAO's initiatives to Carbon Offsetting and Reduction Scheme for International Aviation (CORSIA) and Sustainable Development Goals (SDGs). This will give students knowledge about the importance of CSR as an essential practice for any company or organization to maintain the principles of sustainable development.

REFERENCES

About, I. C. A. O. (2023). Retrieved from Icao.int website: https://www.icao.int/abouticao/Pages/default.aspx

Airline Trends. (2023). Retrieved December 1, 2023, from Airlinetrends.com website: http://www.airlinetrends.com/category/csr-sustainability/

Almetwally, A. A., Bin-Jumah, M., & Allam, A. A. (2020). Ambient air pollution and its influence on human health and welfare: An overview. *Environmental Science and Pollution Research International*, 27(20), 24815–24830. 10.1007/s11356-020-09042-232363462

Anenberg, S. C., Miller, J., Henze, D. K., Minjares, R., & Achakulwisut, P. (2019). The global burden of transportation tailpipe emissions on air pollution-related mortality in 2010 and 2015. *Environmental Research Letters*, 14(9), 094012–094012. 10.1088/1748-9326/ab35fc

Aviation and the Environment. (n.d.). Retrieved from https://www.icao.int/environmental-protection/Documents/EnvironmentalReports/2023/ENVReport2023_pg24-38.pdf

Batuwangala, E., Kistan, T., Gardi, A., & Sabatini, R. (2018). Certification challenges for next-generation avionics and air traffic management systems. *IEEE Aerospace and Electronic Systems Magazine*, 33(9), 44–53. 10.1109/MAES.2018.160164

Boyes, E., Stanisstreet, M., & Pui-ming Yeung, S. (2004). Air pollution: The knowledge and attitudes of secondary school students in Hong Kong. *International Research in Geographical and Environmental Education*, 13(1), 21–37. 10.1080/10382040408668790

Climate Change Technology Standards. (2023). Retrieved 1 December 2023, from https://www.icao.int/environmental-protection/Pages/ClimateChange_TechnologyStandards.aspx

CSR (Corporate Social Responsibility) Definition and Examples. (2023). Retrieved 1 December 2023, from https://e-csr.net/definitions/csr-definition/#a-simple-definition-of-csr

Dimitriou, A., & Christidou, V. (2007). Pupils' understanding of air pollution. *Journal of Biological Education*, 42(1), 24–29. 10.1080/00219266.2007.9656103

Dimitriou, A., & Christidou, V. (2011). Causes and consequences of air pollution and environmental injustice as critical issues for science and environmental education. *The Impact of Air Pollution on Health, Economy, Environment and Agricultural Sources*, 215-238.

Environmental Protection Agency. (n.d.). EPA. https://www.epa.gov/clean-air-act-overview/clean-air-act-text

Environmental Protection Agency. (n.d.-a). EPA. https://www.epa.gov/enforcement/air-enforcement#engines

Fowler, D., Brimblecombe, P., Burrows, J. P., Heal, M. R., Grennfelt, P., Stevenson, D. S., Jowett, A., Nemitz, E., Coyle, M., Lui, X., Chang, Y., Fuller, G. W., Sutton, M. A., Klimont, Z., Unsworth, M. H., & Vieno, M. (2020). A chronology of global air quality. *Philosophical Transactions - Royal Society. Mathematical, Physical and Engineering Sciences. Philosophical Transactions. Series A, Mathematical, Physical, and Engineering Sciences*, 378(2183), 20190314–20190314. 10.1098/rsta.2019.031432981430

Home Page. Next generation science standards. (n.d.). Retrieved from https://www.nextgenscience.org/

Husic, D. W. (2024). Reframing sustainability initiatives in higher education. *Sustainable Earth Reviews*, 7(1), 5. 10.1186/s42055-024-00076-9

Icao.int. (2023). Local Air Quality. Available at: https://www.icao.int/environmental-protection/Pages/local-air-quality.aspx

Icao.int. (2023). Trends in Emissions that affect Climate Change. Available at: https://www.icao.int/environmental-protection/Pages/ClimateChange_Trends.aspx

Kan, H., London, S. J., Chen, G., Zhang, Y., Song, G., Zhao, N., Jiang, L., & Chen, B. (2008). Season, sex, age, and education as modifiers of the effects of outdoor air pollution on daily mortality in Shanghai, China: The Public Health and Air Pollution in Asia (PAPA) Study. *Environmental Health Perspectives*, 116(9), 1183–1188. 10.1289/ehp.1085118795161

Kramer, C. B., Cullen, A. C., & Faustman, E. M. (2006). Policy Implications of Genetic Information on Regulation under the Clean Air Act: The Case of Particulate Matter and Asthmatics. *Environmental Health Perspectives*, 114(3), 313–319. 10.1289/ehp.829916507451

Martin, T. (2024). *Examining the Influence of a PLC-based Intervention on the Local Adoption of the New Pennsylvania Science Standards via the Concerns Based Adoption Model*. Arizona State University.

Mohai, P., Kweon, B. S., Lee, S., & Ard, K. (2011). Air pollution around schools is linked to poorer student health and academic performance. *Health Affairs*, 30(5), 852–862. 10.1377/hlthaff.2011.007721543420

Myers, G., Boyes, E., & Stanisstreet, M. (1999). Something in the air: School students' ideas about air pollution. *International Research in Geographical and Environmental Education*, 8(2), 108–119. 10.1080/10382049908667600

Myers, G., Boyes, E., & Stanisstreet, M. (2004). School students' ideas about air pollution: Knowledge and attitudes. *Research in Science & Technological Education*, 22(2), 133–152. 10.1080/0263514042000290868

New Pennsylvania Science, Technology & Engineering, Environmental Literacy & Sustainability (Steels) standards. (2023). Retrieved from https://www.onatlas.com/blog/new-pennsylvania-science-technology-engineering-environmental-literacy-sustainability-steels-standards

O'Neill, M. S., Bell, M. L., Ranjit, N., Cifuentes, L. A., Loomis, D., Gouveia, N., & Borja-Aburto, V. H. (2008). Air pollution and mortality in Latin America: The role of education. *Epidemiology (Cambridge, Mass.)*, 19(6), 810–819. 10.1097/EDE.0b013e3181816528l8633327

Pappalardo, C. (2021). What a Difference a State Makes: California's Authority to Regulate Motor Vehicle Emissions Under the Clean Air Act and the Future of State Autonomy. *Michigan Journal of Environmental & Administrative Law. Michigan Journal of Environmental & Administrative Law*, 10(1), 169–169. 10.36640/mjeal.10.1.what

Park, H., Lee, B., Ha, E. H., Lee, J. T., Kim, H., & Hong, Y. C. (2002). Association of air pollution with school absenteeism due to illness. *Archives of Pediatrics & Adolescent Medicine*, 156(12), 1235–1239. 10.1001/archpedi.156.12.123512444836

Peeters, J. H. A. M. (1998). Aviation and air pollution. In *Studies in environmental science* (Vol. 72, pp. 601–613). Elsevier.

Pennsylvania Department of Education. (n.d.). *Modernizing PA's science education standards*. https://www.education.pa.gov/Teachers%20-%20Administrators/Curriculum/Science/Pages/Science-Standards.aspx

STEELS Hub - SAS. (n.d.). Retrieved from https://www.pdesas.org/Page/Viewer/ViewPage/58/?fbclid=IwAR0x48gZxjpPHvkwu-B0rDGasOq9XfQrKHKF8GjQIFK5rSRW40TsGuma8D8_aem_AcCJ8EDhaHY5tMHE8erG4rJKOPWxgoOJ8YBg5fYWc3yYtDdnuV9WDHQhW4nq0I1bmntdSqvAuUZMtlGgfOyYcxAR

Stone, R. K. & V. (2024). New science standards will be in our classrooms soon. Retrieved from https://gettysburgconnection.org/new-science-standards-will-be-in-our-classrooms-soon/

Teksoz, G. T. (2011). Managing air pollution: How does education help. *The impact of air pollution on health, economy, environment and agricultural sources, 1*, 397-422.

Thornber, J., Stanisstreet, M., & Boyes, E. (1999). School students' ideas about air pollution: Hindrance or help for learning? *Journal of Science Education and Technology*, 8(1), 67–73. 10.1023/A:1009481521719

World Health Organization. (n.d.). *Air Pollution*. World Health Organization. https://www.who.int/health-topics/air-pollution#tab=tab_1

ADDITIONAL READING

Berkowitz, A. R., Ford, M. E., & Brewer, C. A. (2005). A framework for integrating ecological literacy, civics literacy, and environmental citizenship in environmental education. *Environmental education and advocacy: Changing perspectives of ecology and education, 227*, 66.

Bodzin, A. M., Fu, Q., Peffer, T. E., & Kulo, V. (2013). Developing energy literacy in US middle-level students using the geospatial curriculum approach. *International Journal of Science Education*, 35(9), 1561–1589. 10.1080/09500693.2013.769139

Climate changes health: Respiratory health. (n.d.). Retrieved 3 June 2024, from https://www.apha.org/topics-and-issues/climate-health-and-equity/air-quality

Faria, F., Klima, K., Posen, I. D., & Azevedo, I. M. (2015). A new approach of science, technology, engineering, and mathematics outreach in climate change, energy, and environmental decision making. *Sustainability (New Rochelle, N.Y.)*, 8(5), 261–271. 10.1089/SUS.2015.29023

Hartley, J., Pugh, C., Stevenson, K., & Daniel, J. (2019). Using EnviroAtlas to Build K-12 Environmental Literacy. *Solutions (Burlington, Vt.)*, 10(2), 70.36742016

Jurin, R. R., Roush, D., Danter, K. J., Jurin, R. R., Roush, D., & Danter, J. (2010). *Developing your environmental literacy. Environmental Communication* (2nd ed.). Skills and Principles for Natural Resource Managers, Scientists, and Engineers.

Klawinski, T. P. (2022). *Achieving Environmental Literacy: An Evidenced-Based Model for Public Schools*. Wilmington University.

Martin, T. (2024). *Examining the Influence of a PLC-based Intervention on the Local Adoption of the New Pennsylvania Science Standards via the Concerns Based Adoption Model.* Arizona State University.

Rowe, D. (2002). Environmental literacy and sustainability as core requirements: Success stories and models. *Teaching sustainability at universities*, 79-103.

Ruggiero, K. (2016). A criteria-based evaluation of environmental literacy plans in the United States. Academic Press.

KEY TERMS AND DEFINITIONS

Air Pollution: Air pollution is the contamination of the environment by any chemical, physical or biological agent. Emissions generated by air and ground transport are one of the causes of climate change which leads to threats to people's health, especially respiratory diseases and are important sources of morbidity and mortality.

Carbon Dioxide: The most commonly produced greenhouse gas. Carbon Dioxide is produced naturally by animals through their waste system, and by human activities such as burning oil, coal and natural gas.

Climate Change: The increasing changes in the measures of climate over a long period of time – including precipitation, temperature, and wind patterns. Climate change caused by air pollution, disease and extreme weather has harmful impact on human physical and mental health and caused increased global hunger.

Environmental Literacy: The knowledge, skills and understanding of environmental principles, moral and ethical dimensions of the environment. It also includes knowledge about sustainability, environmental challenges, and stewardship. It helps develop ethical planning, practices, and knowledgeable actions about our personal and collective impacts on the environment.

Global Warming: Global warming is just one aspect of climate change. It is a breach of global temperature limit due mainly to the increasing concentrations of greenhouse gases in the atmosphere. A shift in temperature and weather pattern caused by nature and human activities due to the burning of fossil fuels which produces heat-trapping gases.

Sustainability Literacy: Knowledge about the relationships between humanity, development, the environment, and natural resources. Learning of practices to use natural resources to meet the needs of the present without compromising the ability of future generations to meet their own needs.

APPENDIX

Questions and Answers

1. What is the overall problem presented in this case?

The importance of including air Pollution as part of the environmental literacy studies to school Students. The importance of school students to have environmental literacy and knowledge about sustainability discipline, ethical and non-ethical human social practices, and their impact on the environment.

2. What are the factors affecting the problem(s) related to this case?

The inclusion of education about the environment, sustainability, and ethically sourced practices in school curriculums is influenced by a variety of factors. These factors can be broadly categorized into policy and governance, societal influences, educational infrastructure, and resource availability. By addressing these factors, schools can help prepare students to become informed and responsible global citizens.

3. Discuss managerial, organizational, and technological issues and resources related to this case.

For managerial issues, school leaders need to have a clear vision and strategy for integrating environmental literacy into the curriculum. Training programs, workshops and seminars for teachers should be provided to enhance their knowledge on environmental and sustainable practices. Interactive learning tools like digital learning platforms and educational software and apps should be available to both students and teachers.

4. What role do different players (decision-makers) play in the overall planning, implementation and management of the information technology applications?

Stakeholders should involve teachers and staff to ensure educators understand the importance of environmental literacy. Decision makers should secure funding for materials, training, and activities related to environmental literacy. They should also designate staff or committees to oversee the new academic environmental literacy initiatives.

5. What are the possible alternatives and pros and cons of each alternative facing the organization in dealing with the problem(s) related to the case?

When an organization, such as a school district or educational institution, faces new academic standards like the ones adopted by the Pennsylvania State Board of Education for Science, Technology, Engineering, Environmental Literacy, and Sustainability (STEELS), it must navigate several potential challenges and decisions. Pros are a robust education in STEELS for school students prepares them for future academic and career opportunities and provides the opportunity to integrate innovative teaching practices and interdisciplinary approaches. Cons are that implementing new standards can require significant investment in training, materials, and curriculum development, also adjusting to new standards can be time-consuming, potentially disrupting existing instructional plans.

6. What are some of the emerging technologies that should be considered in solving the problem(s) related to the case?

Artificial intelligence can help suggest new methods of applying the new academic standards. Pilot Testing of the new academic standards also offers the opportunity to pilot test the new approach and materials before full-scale implementation, which reduces risks. Utilizing simulation tools to model environmental scenarios and engage students in problem-solving activities is helpful. Implementing VR and AR technologies to create immersive learning experiences about different ecosystems, climate change, and sustainability practices can make the learning practice more enjoyable.

7. What is the final solution that can be recommended to the management of the organization described in the case? Provide your arguments in support of the recommended solution.

Careful planning and coordination to ensure a smooth gradual transition makes it easier for teachers and students to adapt to the new academic standards. Pooling or sharing resources with other districts can reduce costs and effort associated with developing and implementing the new academic standards.

Compilation of References

Abbas, S. M. (2023, December 19). Books of new curriculum in the press. *Dhaka Tribune.*https://www.dhakatribune.com/bangladesh/334300/books-of-new-curriculum-in-the-press

About, I. C. A. O. (2023). Retrieved from Icao.int website: https://www.icao.int/abouticao/Pages/default.aspx

Adichie, C. N. (2009, July). The Danger of a single Story [Video]. TED Conferences. https://www.ted.com/talks/chimamanda_ngozi_adichie_the_danger_of_a_single-story

Aguayo, C., & Eames, C. (2023). Using mixed reality (XR) immersive learning to enhance environmental education. *The Journal of Environmental Education*, 54(1), 58–71. 10.1080/00958964.2022.2152410

Ahuja, A. (2016). *Essentials of Environmental Education*. Koros Press Limited.

Airline Trends. (2023). Retrieved December 1, 2023, from Airlinetrends.com website: http://www.airlinetrends.com/category/csr-sustainability/

Akçayır, M., & Akçayır, G. (2017). Advantages and challenges associated with augmented reality for education: A systematic review of the literature. *Educational Research Review*, 20, 1–11. 10.1016/j.edurev.2016.11.002

Akdağ, H. (2014). Sosyal Bilgilerin Tanımı, Amacı, Önemi ve Türkiye'deki Yeri. Sosyal Bilgiler Öğretiminde Yeni Yaklaşımlar I (2. Basım). (s.2-21). Ankara: Pegem Akademi.

Akınoğlu, O., & Sarı, A. (2009). Environmental education in primary school curriculum. *Atatürk Education Faculty Journal of Educational Sciences*, 1(30), 5–29.

Akkuzu, N. (2016). Towards a profound ecological understanding: Statistical attempts to measure our ecological intelligence. *International Journal of Social Sciences and Education*, 6(2), 198–216.

Al-Ansi, A. M., Jaboob, M., Garad, A., & Al-Ansi, A. (2023). Analyzing augmented reality (AR) and virtual reality (VR) recent development in education. *Social Sciences & Humanities Open*, 8(1), 100532. 10.1016/j.ssaho.2023.100532

Allison, E., & Goldston, M. J. (2018). Modern scientific literacy: A case study of multiliteracies and scientific practices in a fifth grade classroom. *Journal of Science Education and Technology*, 27(3), 270–283. 10.1007/s10956-017-9723-z

Almaki, S. H., Gunda, M. A., Idris, K., Hashim, A. T. M., & Ali, S. R. (2023). A systematic review of the use of simulation games in K-12 education. *Interactive Learning Environments*, 1–25. 10.1080/10494820.2023.2205894

Almetwally, A. A., Bin-Jumah, M., & Allam, A. A. (2020). Ambient air pollution and its influence on human health and welfare: An overview. *Environmental Science and Pollution Research International*, 27(20), 24815–24830. 10.1007/s11356-020-09042-232363462

Altundağ, C. K., Mustafaoğlu, F. T., & Yücel, A. S. (2022). Developing a scale: Perceptions of high school students about the relationship between chemistry and daily life. *Jurnal Ilmiah Peuradeun*, 10(3), 721–746. 10.26811/peuradeun.v10i3.718

Alzahrani, M. (2022). *Traditional Learning Compared to Online Learning During the COVID-19 Pandemic: Lessons Learned From Faculty's Perspectives*. Sage Journals. 10.1177/21582440221091720

Ambe, B. A., Agbor, C. E., Amalu, M. N., Ngban, A. N., Bekomson, A. N., Etan, M. O., Ephraim, I. E., Asuquo, E. E., Eyo, O. E., & Ogunjimi, J. O. (2024). Electronic media learning technologies and environmental education pedagogy in tertiary institutions in Nigeria. *Social Sciences & Humanities Open*, 9, 100760. 10.1016/j.ssaho.2023.100760

Anastasiou, E., Balafoutis, A. T., & Fountas, S. (2023). Applications of extended reality (XR) in agriculture, livestock farming, and aquaculture: A review. *Smart Agricultural Technology*, 3, 100105. 10.1016/j.atech.2022.100105

Anenberg, S. C., Miller, J., Henze, D. K., Minjares, R., & Achakulwisut, P. (2019). The global burden of transportation tailpipe emissions on air pollution-related mortality in 2010 and 2015. *Environmental Research Letters*, 14(9), 094012–094012. 10.1088/1748-9326/ab35fc

Anguish, D. (2015). Mass Media and the Evolution of the Environmental Movement: 1960-1979. [MA dissertation, University of Central Florida]. Retrieved from https://stars.library.ucf.edu/cgi/viewcontent.cgi?article=2350&context=etd

Anjos, F., Rocha, L., Silva, D., Pacheco, R., & Pinheiro, D. (2021). Impacts of the application of virtual and augmented reality on teaching-learning processes in engineering courses. *International Journal of Virtual and Personal Learning Environments*, 12(1), 1–19. 10.4018/IJVPLE.291541

Aristoteles, M. (2020, June). Book 8, section 1045a. (t.y.). Perseus Digital Library. *Access*, 13. https://www.perseus.tufts.edu/hopper/text?doc=urn:cts:greekLit:tlg0086.tlg025.perseus-eng1:8.1045a

Arthington, A. H., Bhaduri, A., Bunn, S. E., Jackson, S. E., Tharme, R. E., Tickner, D., Young, B., Acreman, M., Baker, N., Capon, S., Horne, A. C., Kendy, E., McClain, M. E., Poff, N. L., Richter, B. D., & Ward, S. (2018). The Brisbane Declaration and Global Action Agenda on Environmental Flows (2018). *Frontiers in Environmental Science*, 6, 364344. 10.3389/fenvs.2018.00045

Ashraf, S., Ali, M., Shrestha, S., Hafeez, M. A., Moiz, A., & Sheikh, Z. A. (2022). Impacts of climate and land-use change on groundwater recharge in the semi-arid lower Ravi River basin, Pakistan. *Groundwater for Sustainable Development*, 17, 100743. 10.1016/j.gsd.2022.100743

Asia Water Wire. (2006, January 23). *Pakistan, India join hands to clean canal.* River Basin Initiative. Retrieved April 1, 2024, from https://web.archive.org/web/20110724140814/http://www.riverbasin.org/newsmaster.cfm?&menuid=45&action=view&retrieveid=152

Aşıcı, M. (2009). Kişisel ve sosyal bir değer olarak okuryazarlik. *Değerler Eğitimi Dergisi*, 7(17), 9–26.

Assaraf, B. O., & Orion, N. (2005). Development of systems thinking skills in the context of Earth system education. *Journal of Research in Science Teaching*, 42(5), 518–560. 10.1002/tea.20061

Assaraf, O. B. Z., & Orion, N. (2010). System thinking skills at the elementary school level. *Journal of Research in Science Teaching*, 47(5), 540–563. 10.1002/tea.20351

Atabek-Yiğit, E., Köklükaya, N., Yavuz, M., & Demirhan, E. (2014). Development and validation of environmental literacy scale for adults (ELSA). *Journal of Baltic Science Education*, 13(3), 425–435. 10.33225/jbse/14.13.425

Atasoy, E., & Ertürk, H. (2008). A field study about environmental knowledge and attidudes of elementary school students, *Erzincan University Journal of Education Faculty, 10*(1), 105-122.

Ateskan, A., & Lane, J. F. (2018). Assessing teachers' systems thinking skills during a professional development program in Turkey. *Journal of Cleaner Production*, 172, 4348–4356. 10.1016/j.jclepro.2017.05.094

Athman, J. A., & Monroe, M. C. (2001). Elements of effective environmental education programs.

Attenborough, D. (Director). (2020). David Attenborough: A Life on Our Planet [Documentary]. Netflix.

Augmented Island Studios (AIS). (2024). Augmented Reality and Virtual Reality: A Game-Changer for Environmental Awareness. Retrieved from https://augmentedislandstudios.com/augmented-reality-and-virtual-reality-a-game-changer-for-environmental-awareness/

Aviation and the Environment. (n.d.). Retrieved from https://www.icao.int/environmental-protection/Documents/EnvironmentalReports/2023/ENVReport2023_pg24-38.pdf

Aydın, E., & Selvi, M. (2020). Development of ecosystem, biodiversity and environmental problems achievement test for secondary school students. *Journal of Research in Education and Society*, 7(2), 661–682.

Aydın, F., & Çepni, O. (2012). Investigation of primary education second grade students' attitudes towards environment in terms of various variables (Karabük City Case). *Dicle University Journal of Ziya Gokalp Education Faculty*, 18, 189–207.

Baker, A., Bramer, W. F., & Bobbitt, A. (2021). Agricultural literacy outcomes: A systematic literature review. *Journal of Agricultural Education*, 62(2), 214–229.

Baker, J. A. S. B. (2022). Impact Of Using Technology In Education Environment. *Journal of Pharmaceutical Negative Results*, 624–633.

Balgopal, M. M., & Wallace, A. M. (2009). Decisions and dilemmas: Using writing to learn activities to increase ecological literacy. *The Journal of Environmental Education*, 40(3), 13–26. 10.3200/JOEE.40.3.13-26

Ballew, M. T., Goldberg, M. H., Rosenthal, S. A., Gustafson, A., & Leiserowitz, A. (2019). Systems thinking as a pathway to global warming beliefs and attitudes through an ecological worldview. *Proceedings of the National Academy of Sciences of the United States of America*, 116(17), 8214–8219. 10.1073/pnas.181931011630962375

Barak, B., & Gönençgil, B. (2020). Dunyada ve Turkiye'de ortaokul ogretim programlarinin iklim degisikligi egitimi yaklasimina gore karsılastirilmasi. *Cografya Dergisi*, (40), 187–201. 10.26650/JGEOG2019-0039

Baratimehr, K., Moosavi, M. R., & Tahayori, H. (2023). Measures for evaluating IT2FSs constructed from data intervals. *Applied Soft Computing*, 136, 110084. 10.1016/j.asoc.2023.110084

Barberán, A., Hammer, T. J., Madden, A. A., & Fierer, N. (2016). Microbes should be central to ecological education and outreach. *Journal of Microbiology & Biology Education*, 17(1), 23–28. 10.1128/jmbe.v17i1.98427047584

Barth, M., Godemann, J., Rieckmann, M., & Stoltenberg, U. (2007). Developing key competencies for sustainable development in higher education. *International Journal of Sustainability in Higher Education*, 8(4), 416–430. 10.1108/14676370710823582

Bartolini, S., & Bonatti, L. (2008). The role of social capital in enhancing factor productivity: Does its erosion depress per capita GDP? *Journal of Socio-Economics*, 37(4), 1539–1553. 10.1016/j.socec.2007.03.005

Baste, I. A., & Watson, R. T. (2022). Tackling the climate, biodiversity and pollution emergencies by making peace with nature 50 years after the Stockholm Conference. *Global Environmental Change*, 73, 102466. 10.1016/j.gloenvcha.2022.102466

Batuwangala, E., Kistan, T., Gardi, A., & Sabatini, R. (2018). Certification challenges for next-generation avionics and air traffic management systems. *IEEE Aerospace and Electronic Systems Magazine*, 33(9), 44–53. 10.1109/MAES.2018.160164

Baumgartner, H., & Homburg, C. (1996). Applications of structural equation modeling in marketing and consumer research: A review. *International Journal of Research in Marketing*, 13(2), 139–161. 10.1016/0167-8116(95)00038-0

Beck, H. W. (2009). Lyra Virtual World Environment (Lyra VWE): Educational Applications in Agriculture and Natural Resources. In *7th World Congress on Computers in Agriculture Conference Proceedings*, 22-24 June 2009, Reno, Nevada (p. 1). American Society of Agricultural and Biological Engineers.

Bednarz, S. W. (2004). Geographic information systems: A tool to support geography and environmental education? *GeoJournal*, 60(2), 191–199. 10.1023/B:GEJO.0000033574.44345.c9

Begum, J. A., Shikdar, A. H., & Das, K. M. (2007). *Secondary social science: For class IX-X [in Bengali]* (New ed.). Dhaka: National Curriculum and Textbook Board.

Bekiaris, I., Leonidis, A., Korozi, M., Stratakis, C., Zidianakis, E., Doxastaki, M., & Stephanidis, C. (2021). GRETA: Pervasive and ar interfaces for controlling intelligent greenhouses. In Proceedings of the *17th International Conference on Intelligent Environments (IE)* (pp. 1–8). IEEE. 10.1109/IE51775.2021.9486584

Bellah, K. A., & Dyer, J. E. (2009). Attitudes and Stages of Concern of Elementary Teachers Toward Agriculture as a Context for Teaching across Grade Level Content Area Standards. *Journal of Agricultural Education*, 50(2), 12–25. Advance online publication. 10.5032/jae.2009.02012

Benedick, P. L., Robert, J., & Le Traon, Y. (2021). A systematic approach for evaluating artificial intelligence models in industrial settings. *Sensors (Basel)*, 21(18), 6195. Advance online publication. 10.3390/s2118619534577398

Benke, K., & Tomkins, B. (2017). Future food-production systems: vertical farming and controlled-environment agriculture. Sustainability: Science. *Practice and Policy*, 13(1), 13–26.

Bentler, P. M. (1980). Multivariate analysis with latent variables: Causal modeling. *Annual Review of Psychology*, 31(1), 419–456. 10.1146/annurev.ps.31.020180.002223

Bercasio, R. R. O., & Remolacio, L. K. Q. (2022). Mainstreaming environmental education in the teacher education curriculum in the Philippines. *International Journal of Evaluation and Research in Education*, 11(3), 1552–1564. 10.11591/ijere.v11i3.21748

Berenjkar, P., Li, Y. Y., & Yuan, Q. (2021). The application of system dynamics in different practices of a waste management system. *Environment, Development and Sustainability*, 23(11), 15695–15724. 10.1007/s10668-021-01362-8

Berkowitz, A. R., Ford, M. E., & Brewer, C. A. (2005). A Framework for Integrating Ecological Literacy, Civics Literacy and Environmental Citizenship in Environmental Education. In Johnson, E. A., & Mappin, M. J. (Eds.), *Environmental Education or Advocacy: Perspectives of Ecology and Education in Environmental Education* (pp. 227–265). Cambridge University Press.

Berkowitz, A. R., Ford, M. E., & Brewer, C. A. (2005). A framework for integrating ecological literacy, civicsliteracy, and environmental citizenship in environmental education. In Johnson, E. A., & Mappin, M. J. (Eds.), *Environmental education and advocacy: Changing perspectives of ecology and education* (pp. 227–266). Cambridge University Press.

Bhang, K. J., & Huh, J. R. (2023). Effectiveness of Fine Dust Environmental Education on Students' Awareness and Attitudes in Korea and Australia Using AR Technology. *Sustainability (Basel)*, 15(22), 16039. 10.3390/su152216039

Bielik, T., Jagemann, J., Krell, M., Krüger, D., & Ben-Zvi Assaraf, O. (2023). Using concept maps to evaluate preservice biology teachers' conceptualization of COVID-19 as a complex phenomenon. *Journal for STEM Education Research*, 6, 199–231. 10.1007/s41979-023-00087-9

Billinghurst, M. (2021). Grand challenges for augmented reality. *Frontiers in Virtual Reality*, 2, 12. 10.3389/frvir.2021.578080

Biswas, T. K., Saha, S. K., & Nasrin, F. (2009). An Assessment of Formal Environmental Education (FEE) System at Tertiary level in Bangladesh. *Journal of Teacher's World*, 35-36, 155–168.

Blatti, J. L., Garcia, J., Cave, D., Monge, F., Cuccinello, A., Portillo, J., Juarez, B., Chan, E., & Schwebel, F. (2019). Systems thinking in science education and outreach toward a sustainable future. *Journal of Chemical Education*, 96(12), 2852–2862. 10.1021/acs.jchemed.9b00318

BlockBirds. (2023, November 19). Revolutionizing Environmental Education: Exploring the Power of AR Technology. Retrieved from https://blockbirds.de/en/uncategorized/ar-in-environmental-education/

Bocken, N. M., & Short, S. W. (2021). Unsustainable business models – Recognising and resolving institutionalised social and environmental harm. *Journal of Cleaner Production*, 312, 127828. 10.1016/j.jclepro.2021.127828

Boehnert, J. (2013). *The visual communication of ecological literacy: Designing, learning and emergent ecological perception*. PhD, University of Brighton

Bolton, G. (2010). *Reflective practice: Writing and professional development* (3rd ed.). SAGE Publications.

Boucenna, M. R., Benzouai, M. C., & Adli, B. (2021). Effect of foreign direct investment on agriculture productivity: A PMG panel ARDL approach. *International Journal of Green Economics*, 15(1), 42. 10.1504/IJGE.2021.117676

Bowen, G. A. (2009). Document analysis as a qualitative research method. *Qualitative Research Journal*, 9(2), 27–40. 10.3316/QRJ0902027

Boyes, E., Stanisstreet, M., & Pui-ming Yeung, S. (2004). Air pollution: The knowledge and attitudes of secondary school students in Hong Kong. *International Research in Geographical and Environmental Education*, 13(1), 21–37. 10.1080/10382040408668790

Brauer, B. T. (2010). Twenty-first century skills: Learning for life in our times. *The Journal of General Education*, 59(1), 82–84. 10.5325/jgeneeduc.59.1.0082

Braun, V., & Clarke, V. (2006). Using thematic analysis in psychology. *Qualitative Research in Psychology*, 3(2), 77–101. 10.1191/1478088706qp063oa

Braus, J., & Wood, D. (1993). *Environmental education in the schools: Creating a program that works!* (Vol. 2). North American Association for Environmental Education.

Brečka, P., Valentová, M., & Tureková, I. (2022). Digital Technologies in Environmental Education. *TEM Journal*, 11(2), 726–730. 10.18421/TEM112-28

Brewer, R. (1994). *The science of ecology* (2nd ed.). Saunders College Publishing.

Brolin, D. E. (1992). *Life centered career education: Competency units for personal-social skills*. Council for Exceptional Children.

Brooke, R. (Ed.). (2003). *Rural voices: Place-conscious education and the teaching of writing*. Teachers College Press.

Bruyere, B. L. (2008). The effect of environmental education on the ecological literacy of first-year college students. *Journal of Natural Resources and Life Sciences Education*, 37(1), 20–26. 10.2134/jnrlse2008.37120x

Buasri, S. (1985). *Study and research in educational philosophy*. Srinakharinwirot University.

Buasri, S. (2001). *Education, philosophy of education, democracy, and ethics*. Suweeriyasart.

Buchanan, J., Pressick-Kilborn, K., & Maher, D. (2018). Promoting environmental education for primary school-aged students using digital technologies. *Eurasia Journal of Mathematics, Science and Technology Education*, 15(2), em1661.

Budak, U. S., & Ceyhan, G. D. (2023). Research trends on systems thinking approach in science education. *International Journal of Science Education*, 1–18. 10.1080/09500693.2023.2245106

Büyüköztürk, Ş. (2011). *Manual of data analysis for social sciences* [Sosyal bilimler için veri analizi el kitabı]. Ankara, Pegem Publishing.

Büyüköztürk, Ş., Kılıç Çakmak, E., Akgün, Ö. E., Karadeniz, Ş., & Demirel, F. (2014). *Scientific research methods* [Bilimsel araştırma yöntemleri]. Pegem Publishing.

Çabuk, B., & Karacaoğlu, C. (2003). Üniversite öğrencilerinin çevre duyarlılıklarının incelenmesi. Ankara Üniversitesi, Eğitim Bilimleri Fakültesi Dergisi, 36(1-2).

Çabuk, B., & Haktanır, G. (2017). *Okul öncesi öğretmenliği öğretmen adaylarının ekolojik okuryazarlıkla ilgili bilgi düzeylerinin incelenmesi* [Examining the knowledge levels of pre-school teacher candidates regarding ecological literacy]. Ankara Üniversitesi Çocuk Kültürü Araştırma Uygulama Merkezi.

Çakır Arıca, Ş., & Kağar, C. (2018). The key of leaving a life to the future generations: Ecological literacy. *Doğanın Sesi*, 2, 31–42.

Can, D., Üner, S., & Akkuş, H. (2016). Determination of the secondary school students' environmental literacy. *Ondokuz Mayis University Journal of Education Faculty*, 35(2), 23–35.

Cao, F., & Jian, Y. (2024). The Role of integrating AI and VR in fostering environmental awareness and enhancing activism among college students. *The Science of the Total Environment*, 908, 168200. 10.1016/j.scitotenv.2023.16820037918744

Capra, F. (1999). *Ecoliteracy. The challenge for education in the next century*. Liverpool Schumacher Lectures.

Caria, M., Todde, G., Sara, G., Piras, M., & Pazzona, A. (2020). Performance and usability of smartglasses for augmented reality in precision livestock farming operations. *Applied Sciences (Basel, Switzerland)*, 10(7), 2318. 10.3390/app10072318

Carruth, D. W., Hudson, C., Fox, A. A. A., & Deb, S. (2020). User interface for an immersive virtual reality greenhouse for training precision agriculture. In Chen, J. Y. C., & Fragomeni, G. (Eds.), *Virtual, Augmented and Mixed Reality. Industrial and Everyday Life Applications* (pp. 35–46). Springer International Publishing. 10.1007/978-3-030-49698-2_3

Cebrián, G., & Junyent, M. (2015). Competencies in education for sustainable development: Exploring the student teachers' views. *Sustainability (Basel)*, 7(3), 2768–2786. 10.3390/su7032768

Cervantes, B., Hemmer, L., & Kouzekanani, K. (2015). The impact of project-based learning on minority student achievement: Implications for school redesign. *Education Leadership Review of Doctoral Research*, 2(2), 50–66.

Chadwick, M., Soussan, J., Mallick, D., & Alam, S. (1998). *Understanding indigenous knowledge: Its role and potential in water resource management in Bangladesh*. University of Leeds and Bangladesh Centre for Advanced Studies.

Chatterjee, S., Rana, N. P., Dwivedi, Y. K., & Baabdullah, A. M. (2021). Understanding AI adoption in manufacturing and production firms using an integrated TAM-TOE model. *Technological Forecasting and Social Change*, 170, 120880. 10.1016/j.techfore.2021.120880

Cheng, C., Kang, L., Cai, S., Li, J., & Sun, R. (2018). Virtual display and interactive experience platform of farming culture based on unity3D. *IFAC-PapersOnLine*, 51(11), 637–642. 10.1016/j.ifacol.2018.08.125

Cheung, K. K., Jong, M. S., Lee, F. L., Lee, J. H., Luk, E. T., Shang, J., & Wong, M. K. (2008). FARMTASIA: An online game-based learning environment based on the VISOLE pedagogy. *Virtual Reality (Waltham Cross)*, 12(1), 17–25. 10.1007/s10055-008-0084-z

Chowdhury, M. H. (2004, February). Environmental education for sustainability in Bangladesh. In *Asia Pacific Environmental Education Seminar* (pp. 11-14).

Chowdhury, M. T. (2014). Nature of environmental education in Bangladesh: A school level assessment with reference to the national curriculum. *International Electronic Journal of Environmental Education*, 4(1), 53–60. 10.18497/iejee-green.67994

Cho, Y., & Park, K. S. (2023). Designing Immersive Virtual Reality Simulation for Environmental Science Education. *Electronics (Basel)*, 12(2), 315. 10.3390/electronics12020315

Christensen, L. (2008). Teaching for joy and justice: Re-imagining the language arts classroom. *Rethinking Schools*.

Clark, B. Y., & Whitford, A. B. (2011). Does more federal environmental funding increase or decrease states' efforts? *Journal of Policy Analysis and Management*, 30(1), 136–152. 10.1002/pam.20547

Clark, W. C., van Kerkhoff, L., Lebel, L., & Gallopin, G. C. (2016). Crafting usable knowledge for sustainable development. *Proceedings of the National Academy of Sciences of the United States of America*, 113(17), 4570–4578. 10.1073/pnas.1601266113270919 79

Clifton, L. (1993). The earth is a living thing. *The American Poetry Review*, 22(1), 20–20.

Climate Change Technology Standards. (2023). Retrieved 1 December 2023, from https://www.icao.int/environmental-protection/Pages/ClimateChange_TechnologyStandards.aspx

Clune, S., Crossin, E., & Verghese, K. 2017. Systematic review of greenhouse gas emissions for different fresh food categories. *Journal of Cleaner Production, 140*, 766-783.

Colvin, J., Blackmore, C., Chimbuya, S., Collins, K., Dent, M., Goss, J., Ison, R., Roggero, P. P., & Seddaiu, G. (2014). In search of systemic innovation for sustainable development: A design praxis emerging from a decade of social learning inquiry. *Research Policy*, 43(4), 760–771. 10.1016/j.respol.2013.12.010

Computeam. (2023, December 19). How Technology in Education Can Help Tackle Environmental Challenges. Retrieved from https://www.computeam.co.uk/videos-and-blog/article/how-technology-in-education-can-help-tackle-environmental-challenges

Comyaoun, C., Yawirach, P., Tunkeaw, S., & Rattanachuwang, P. (2019). Strategies for management of collaborative networks to enhance the quality of education management in small primary schools. *The Journal of Academic Social Science*, 12(2), 30–36.

Corpuz, A. M., San Andres, T. C., & Lagasca, J. M. (2022). Integration of environmental education (EE) in teacher education programs: Toward sustainable curriculum greening. Problems of Education in the 21st Century, 80(1), 119.

Cox, M., Elen, J., & Steegen, A. (2019). The use of causal diagrams to foster systems thinking in geography education: Results of an intervention study. *The Journal of Geography*, 118(6), 238–251. 10.1080/00221341.2019.1608460

Crane, P., & Richardson, L. (2000). *Reconnect action research kit*. Australia Department of Family and Community Services.

Creswell, J. W., & Poth, C. N. (2016). *Qualitative inquiry and research design: Choosing among five approaches*. Sage publications.

CSR (Corporate Social Responsibility) Definition and Examples. (2023). Retrieved 1 December 2023, from https://e-csr.net/definitions/csr-definition/#a-simple-definition-of-csr

Cutter-Mackenzie, A., & Smith, R. (2003). Ecological literacy: The 'missing paradigm' in environmental education (part one). *Environmental Education Research*, 9(4), 497–524. 10.1080/1350462032000126131

Darling-Hammond, L. (2020). Accountability in teacher education. *Action in Teacher Education*, 42(1), 60–71. 10.1080/01626620.2019.1704464

Datta, R. (2024). Relationality in indigenous climate change education research: A learning journey from indigenous communities in Bangladesh. *Australian Journal of Environmental Education*, 1–15. 10.1017/aee.2024.13

de Albuquerque Júnior, A. X., Andrade, E. P., de Morais, M. A. C., & Batista, W. M. N. (2021). Environmental Education and English Language: In search of a sustainable and liberating school. *Research. Social Development*, 10(13), e552101321465–e552101321465. 10.33448/rsd-v10i13.21465

De Haan, G. (2006). The BLK '21'programme in Germany: A 'Gestaltungskompetenz'-based model for education for sustainable development. *Environmental Education Research*, 12(1), 19–32. 10.1080/13504620500526362

De Haan, G. (2010). The development of ESD-related competencies in supportive institutional frameworks. *International Review of Education*, 56(2), 315–328. 10.1007/s11159-010-9157-9

Dewey, J. (1976). *The middle works, 1899-1924* (Vol. 13). SIU press.

Dhammakosachan, P. (2009). *Buddhism and science create a culture of wisdom.* Pithak Printing.

Díaz-López, C., Serrano-Jiménez, A., Chacartegui, R., Becerra-Villanueva, J. A., Molina-Huelva, M., & Barrios-Padura, Á. (2023). Sensitivity analysis of trends in environmental education in schools and its implications in the built environment. *Environmental Development*, 45, 100795. 10.1016/j.envdev.2022.100795

Dimitriou, A., & Christidou, V. (2011). Causes and consequences of air pollution and environmental injustice as critical issues for science and environmental education. *The Impact of Air Pollution on Health, Economy, Environment and Agricultural Sources*, 215-238.

Dimitriou, A., & Christidou, V. (2007). Pupils' understanding of air pollution. *Journal of Biological Education*, 42(1), 24–29. 10.1080/00219266.2007.9656103

DoForms. (2023, September 13). 5 Technologies That Help the Environment & How Electronic Forms Preserve The Planet's Wellbeing. Retrieved from https://www.doforms.com/technologies-that-help-environment/

Doganca Kucuk, Z., & Saysel, A. K. (2018). Developing seventh-grade students' understanding of complex environmental problems with systems tools and representations: A quasi-experimental study. *Research in Science Education*, 48(2), 491–514. 10.1007/s11165-017-9620-8

Dorn, C. (2020). A New Global Ethic: A History of the United Nations International Environmental Education Program, 1975-1995. *Foro de educación, 18*(2), 83-108.

Đurđević, S., Golubović, G., Maričić, K., & Vladić, G. (2022). Development of augmented reality video application. 10.24867/GRID-2022-p32

Eckert, M., Volmerg, J. S., & Friedrich, C. M. (2019). Augmented Reality in medicine: Systematic and bibliographic review. *JMIR mHealth and uHealth*, 7(4), e10967. 10.2196/1096731025950

Ecosystems and Human Well-Being. Synthesis. (2005). In *Millennium Ecosystem Assessment*. Island Press, Washington, DC. Retrieved April 5, 2024, from https://www.millenniumassessment.org/documents/document.356.aspx.pdf

Education Improvement Commission. (2000). *School improvement planning for principal, teachers and school councils*. Education Improvement Commission.

Eliot, T. S. (1964). *The Waste Land 1922*. Grolier Club.

Environmental Protection Agency (EPA). (2016). What is Environmental Education? Retrieved January 17, 2024, from https://www.epa.gov/education/what-environmental-education

Environmental Protection Agency. (n.d.). EPA. https://www.epa.gov/clean-air-act-overview/clean-air-act-text

Environmental Protection Agency. (n.d.-a). EPA. https://www.epa.gov/enforcement/air-enforcement#engines

Erbasan, Ö., & Erkol, M. (2020). Investigation of environmental knowledge, attitudes and behaviors of primary school teachers. *OPUS International Journal of Society Researches*, 15(24), 2443–2471. 10.26466/opus.619973

Erten, S. (2005). Okul öncesi öğretmen adaylarinda çevre dostu davranişlarin araştirilmasi. *Hacettepe Üniversitesi Eğitim Fakültesi Dergisi*, 28, 91–100.

Eruz, E. (1989). Kentleşmede ekolojik yaklaşım. *İstanbul Üniversitesi Deniz Bilimleri ve Coğrafya Enstitüsü Bülteni,* 79-83

Evagorou, M., Korfiatis, K., Nicolaou, C., & Constantinou, C. (2009). An investigation of the potential of interactive simulations for developing systems thinking skills in elementary school: A case study with fifth-graders and sixth-graders. *International Journal of Science Education*, 31(5), 655–674. 10.1080/09500690701749313

Faham, E., Rezvanfar, A., Mohammadi, S. H. M., & Nohooji, M. R. (2017). Using system dynamics to develop education for sustainable development in higher education with the emphasis on the sustainability competencies of students. *Technological Forecasting and Social Change*, 123, 307–326. 10.1016/j.techfore.2016.03.023

Faridi, H., Tuli, N., Mantri, A., Singh, G., & Gargrish, S. (2020, October 4). A framework utilizing augmented reality to improve critical thinking ability and learning gain of the students in Physics. *Computer Applications in Engineering Education*. https://scite.ai/reports/10.1002/cae.22342

Farrell, T. S. C. (2012). Reflecting on Reflective Practice: (Re)visiting Dewey and Schön. *TESOL Journal*, 3(1), 7–16. 10.1002/tesj.10

Fauville, G., Lantz-Andersson, A., & Säljö, R. (2014). ICT tools in environmental education: Reviewing two newcomers to schools. *Environmental Education Research*, 20(2), 248–283. 10.1080/13504622.2013.775220

Fernández, A. H., Camargo, C. D. B., & Nascimento, M. S. L. D. (2019). Technologies and Environmental Education: A Beneficial Relationship. *Research in Social Sciences and Technology*, 4(2), 13–30. 10.46303/ressat.04.02.2

Ferreira, J. A., Ryan, L., & Tilbury, D. (2007). Mainstreaming education for sustainable development in initial teacher education in Australia: A review of existing professional development models. *Journal of Education for Teaching*, 33(2), 225–239. 10.1080/02607470701259515

Field, A. (2013). *Discovering Statistics Using IBM SPSS Statistics* (4th ed.). SAGE.

Fien, J., Scott, W. A. H., & Tilbury, D. (2001). Education and conservation: Lessons from an evaluation. *Environmental Education Research*, 7(4), 379–395. 10.1080/13504620120081269

Fink-Hafner, D., Dagen, T., Doušak, M., Novak, M., & Hafner-Fink, M. (2019). Delphi method: Strengths and weaknesses. *Advances in Methodology and Statistics*, 16(2), 1–19. 10.51936/fcfm6982

Finney, C. (2010). This land is Your Land, This Land is My Land: People and public lands redux. *The George Wright Forum*, 27(3), 247–254.

Firdausi, A., & Wulandari, F. E. (2021). Development of web-based science learning module in improving the students' understanding of eco-literacy. *Jurnal Penelitian Dan Pengkajian Ilmu Pendidikan: E-Saintika*, 5(3), 241–252. 10.36312/esaintika.v5i3.426

Flowers, A. A., Carroll, J. P., Green, G. T., & Larson, L. R. (2015). Using art to assess environmental education outcomes. *Environmental Education Research*, 21(6), 846–864. 10.1080/13504622.2014.959473

Forrester, J. W. (1993). System dynamics and the lessons of 35 years. In *A systems-based approach to policymaking* (pp. 199–240). Springer. 10.1007/978-1-4615-3226-2_7

Fowler, D., Brimblecombe, P., Burrows, J. P., Heal, M. R., Grennfelt, P., Stevenson, D. S., Jowett, A., Nemitz, E., Coyle, M., Lui, X., Chang, Y., Fuller, G. W., Sutton, M. A., Klimont, Z., Unsworth, M. H., & Vieno, M. (2020). A chronology of global air quality. *Philosophical Transactions - Royal Society. Mathematical, Physical and Engineering Sciences. Philosophical Transactions. Series A, Mathematical, Physical, and Engineering Sciences*, 378(2183), 20190314–20190314. 10.1098/rsta.2019.031432981430

Fraenkel, J. R., Wallen, N. E., & Hyun, H. H. (2012). *How to design and evaluate research in education* (8th ed.). McGraw Hill.

Fretwell, S. (2017). Cultivating ecological literacy and rethinking our connections to nature (Master's thesis). *Georgia Southern University, Jack N. Averitt College of Graduate Studies*. Retrieved from Digital Commons@Georgia Southern.

Gallwey, S. (2016). Capturing transformative change in education: The challenge of tracking progress towards SDG Target 4.7. *Policy & Practice: A Development. Education Review*, 23, 124–138.

Gardner, H. (2008). The five minds for the future. *Schools*, 5(1/2), 17–24. 10.1086/591814

Garrard, G. (2007). Ecocriticism and education for sustainability. *Pedagogy*, 7(3), 359–383. 10.1215/15314200-2007-005

Garzón, J., & Acevedo, J. (2019). Meta-analysis of the impact of Augmented Reality on students' learning gains. *Educational Research Review*, 27, 244–260. 10.1016/j.edurev.2019.04.001

Garzon, J., Acevedo, J., Pavon, J., & Baldiris, S. (2020). Promoting eco-agritourism using an augmented reality-based educational resource: A case study of aquaponics. *Interactive Learning Environments*, 1–15. 10.1080/10494820.2020.1712429

Garzon, J., Baldiris, S., Acevedo, J., & Pavon, J. (2020). Augmented reality-based application to foster sustainable agriculture in the context of aquaponics. In *Proceedings of the 2020 IEEE 20th International Conference on Advanced Learning Technologies (ICALT)* (pp. 316–318). IEEE. 10.1109/ICALT49669.2020.00101

Giangrande, N., White, R. M., East, M., Jackson, R., Clarke, T., Saloff Coste, M., & Penha-Lopes, G. (2019). A competency framework to assess and activate education for sustainable development: Addressing the UN sustainable development goals 4.7 challenge. *Sustainability (Basel)*, 11(10), 2832. 10.3390/su11102832

Giannarou, L., & Zervas, E. (2014). Using Delphi technique to build consensus in practice. *International Journal of Business Science and Applied Management*, 9(2), 65–82.

Glackin, M., & King, H. (2020). Taking stock of environmental education policy in England – the what, the where, and the why. *Environmental Education Research*, 26(3), 305–323. 10.1080/13504622.2019.1707513

Goleman, D., Bennett, L., & Barlow, Z. (2012). *Ecoliterate: How educators are cultivating emotional, social, and ecological intelligence*. Jossey-Bass.

Gough, A. (2016). Teacher education for sustainable development: Past, present and future. *Teaching education for sustainable development at university level*, 109-122.

Griffiths, S., Wong, M. S., Kwok, C. Y. T., Kam, R., Lam, S. C., Yip, T. L., Heo, J. Y., Chan, B. S. B., Xiong, G., & Lu, K. (2019, July 24). Exploring Bluetooth Beacon Use Cases in Teaching and Learning: Increasing the Sustainability of Physical Learning Spaces. https://scite.ai/reports/10.3390/su11154005

Grimm, L. G., & Yarnold, P. R. (1995). *Reading and understanding multivariate statistics*. American Psychological Association.

Gruenewald, D. A. (2003). The best of both worlds: A critical pedagogy of place. *Educational Researcher*, 32(4), 3–12. 10.3102/0013189X032004003

Gull, R., & Gul, R. (2017). In the surge of healing-the case of Indus basin.

Güngör Cabbar, B., Gültekin, S., Güneş, E., Aytaç, E., & Daşgın, F. (2020). Analysis of environmental achievements in 2018 science and biology courses curriculum according to the revised bloom taxonomy. *Necatibey Faculty of Education Electronic Journal of Science and Mathematics Education*, 14(1), 504–527. 10.17522/balikesirnef.702537

Guthrie, S., & Fisher, D. (1999). Systems thinking and system dynamics in the CC-STADUS high school project (How high school students become system thinkers). Retrieved 07/05/06, 2006, from http://www. clexchange. org/ftp/documents/Implementation/IM1997-07SDLeveragePoints. Pdf.

Güven, E., & Aydoğdu, M. (2012). Development of environmental problems behaviour scale and determination of teacher candidates' behaviour levels. *Journal of Uludag University Faculty of Education*, 25(2), 573–589.

Ha, C., Huang, G., Zhang, J., & Dong, S. (2022). Assessing ecological literacy and its application based on linguistic ecology: A case study of Guiyang City, China. *Environmental Science and Pollution Research International*, 29(13), 18741–18754. 10.1007/s11356-021-16753-734704227

Häggströmi, M., & Schmidt, C. (2020). Enhancing children's literacy and ecological literacy through critical placebased pedagogy. *Environmental Education Research*, 26(12), 1729–1745. 10.1080/13504622.2020.1812537

Hair, J. F.Jr, Black, W. C., Babin, B. J., & Anderson, R. E. (2009). *Multivariate Data Analysis* (7th ed.). Prentice Hall.

Hambleton, R. K., & Patsula, L. (1999). Increasing the validity of adapted tests: Myths to be avoided and guidelines for improving test adaptation practices. *Journal of Applied Testing Technology.*, 1(1), 1–30.

Haque, A. (2022, December 3). Bridging the industry-academia gap through collaboration. *The Business Standard*. Retrieved from https://www.tbsnews.net/features/panorama/bridging-industry-academia-gap-through-collaboration-544090

Haque, F. (2013). Education for Sustainable Development: An evaluation of the new curriculum of the formal primary education in Bangladesh. *European Scientific Journal, 1*(Special Edition), 320-329.

Haque, M. A., Shafiullah, M., Uddin, M. N., Roy, S., & Das, K. M. (2007). *Secondary general science: For class IX-X* (Revised ed.). (Sarker, M. T. H., Islam, M. R., Rahman, M. S., Bhadra, S. K., & Begum, J. A., Trans.). National Curriculum and Textbook Board.

Haque, M. S., & Sharif, S. (2021). The need for an effective environmental engineering education to meet the growing environmental pollution in Bangladesh. *Cleaner Engineering and Technology*, 4, 100114. 10.1016/j.clet.2021.100114

Hardiansyah, F., Ar, M. M., Hidayatillah, Y., & Astutik, C. (2023). Utilization of eco-literacy in science learning as a teaching media in elementary schools. *Pendas: Jurnal Ilmiah Pendidikan Dasar*, 8(3), 1384–1396.

Hassan, A., Abiddin, N. Z., & Yew, S. K. (2014). The Philosophy of Learning and Listening in Traditional Classroom and Online Learning Approaches. *Higher Education Studies*, 4(2), 19. 10.5539/hes.v4n2p19

Hay Group. (2003). *Using competencies to identify high performers: An overview of the basics*. Hay Group, Inc.

He, M. (2023). VR in Environmental Science. Retrieved February 2, 2024 from https://www.vrwiki.cs.brown.edu/applications-of-vr/vr-in-environmental-science

Hensley, N. (2020). Educating for sustainable development: Cultivating creativity through mindfulness. *Journal of Cleaner Production*, 243, 118542. 10.1016/j.jclepro.2019.118542

Heraclitus, . (2020, June). (Stanford encyclopedia of philosophy). (t.y.). Stanford Encyclopedia of Philosophy. *Access*, 13. https://plato.stanford.edu/entries/heraclitus/#Flu

Hillison, J. (1996). The Origins of Agriscience: Or Where Did All That Scientific Agriculture Come From? *Journal of Agricultural Education*, 37(4), 8–13. 10.5032/jae.1996.04008

Hokayem, H., & Gotwals, A. W. (2016). Early elementary students' understanding of complex ecosystems: A learning progression approach. *Journal of Research in Science Teaching*, 53(10), 1524–1545. 10.1002/tea.21336

Holden, C. (2007). Young people's concerns. In Hicks, D., & Holden, C. (Eds.), *Teaching the global dimension: Key principles and effective practice* (pp. 49–60). Routledge.

Hollweg, K. S., Taylor, J. R., Bybee, R. W., Marcinkowski, T. J., McBeth, W. C., & Zoido, P. (2011). Developing a framework for assessing environmental literacy. Washington, DC: North American Association for Environmental Education.

Holmes, A. J. (2021). Public Writing for Social Change. Writing Spaces. https://writingspaces.org/wp-content/uploads/2021/09/holmes.pdf

Home Page. Next generation science standards. (n.d.). Retrieved from https://www.nextgenscience.org/

hooks, b. (2009). Belonging: A culture of place. Routledge.

Hoque, A., Clarke, A., & Sultana, T. (2017). Environmental sustainability practices in South Asian university campuses: An exploratory study on Bangladeshi universities. *Environment, Development and Sustainability*, 19(6), 2163–2180. 10.1007/s10668-016-9845-0

Horn, J. L. (1965). A rationale and test for the number of factors in factor analysis. *Psychometrika*, 30(2), 179–185. 10.1007/BF0228944714306381

Hossain, N. U. I., Dayarathna, V. L., Nagahi, M., & Jaradat, R. (2020). Systems thinking: A review and bibliometric analysis. *Systems*, 8(3), 23. 10.3390/systems8030023

Hsu, S. J., & Roth, R. E. (1998). An assessment of environmental literacy and analysis of predictors of responsible environmental behavior held by secondary teachers in the Hualien Area of Taiwan. *Environmental Education Research*, 4(3), 229–249. 10.1080/1350462980040301

Huang, G. W., & Ha, C. C. (2021). The relationship between ecological literacy and ecolinguistics. *Foreign Language Education*, 1, 15–19.

Huang, G. W., & Zhao, R. H. (2019). *What is ecolinguistics*. Shanghai Foreign Language Education Press.

Huang, R. L., & Nejdl, W. (Eds.). (2011). *Advances in Web-Based Learning – ICWL* (pp. 34–43). Springer. 10.1007/978-3-642-22763-9_4

Hubert, D., Frank, A., & Igo, C. (2000). Environmental and agricultural literacy education. *Water, Air, and Soil Pollution*, 123(1/4), 525–532. 10.1023/A:1005260816483

Hu, L. T., & Bentler, P. M. (1999). Cutoff criteria for fit indexes in covariance structure analysis: Conventional criteria versus new alternatives. *Structural Equation Modeling*, 6(1), 1–55. 10.1080/10705519909540118

Humes, B. (2008). Moving toward a liberatory pedagogy for all species: Mapping the need for dialogue between humane and anti-oppressive education. Green Theory & Praxis. *The Journal of Ecopedagogy*, 4(1), 65–85. 10.3903/gtp.2008.1.4

Hurley, M. M. (2001). Reviewing Integrated Science and Mathematics: The Search for Evidence and Definitions From New Perspectives. *School Science and Mathematics*, 101(5), 259–268. 10.1111/j.1949-8594.2001.tb18028.x

Husic, D. W. (2024). Reframing sustainability initiatives in higher education. *Sustainable Earth Reviews*, 7(1), 5. 10.1186/s42055-024-00076-9

Iatsyshyn, A., Kovach, V., Lyubchak, V., Zuban, Y., Piven, A., Sokolyuk, O., & Shyshkina, M. (2020). Application of augmented reality technologies for education projects preparation. *Cte Workshop Proceedings, 7*, 134-160. 10.31812/123456789/3856

Icao.int. (2023). Local Air Quality. Available at: https://www.icao.int/environmental-protection/Pages/local-air-quality.aspx

Icao.int. (2023). Trends in Emissions that affect Climate Change. Available at: https://www.icao.int/environmental-protection/Pages/ClimateChange_Trends.aspx

Institute of Culture and Arts. (2010). *In remembrance of Professor Dr. Saroj Buasri's contributions*. Srinakharinwirot University.

IPCC. (2023). *Climate Change 2023: Synthesis Report* (Core Writing Team, H. Lee, & J. Romero, Eds.). https://www.ipcc.ch/report/ar6/syr/downloads/report/IPCC_AR6_SYR_LongerReport.pdf

Irianto, D. M., Yunansah, H., Mulyati, T., Herlambang, Y. T., & Setiawan, D. (2020). Multiliteracy: Alternative learning models to improve ecological literacy of primary school students. *PalArch's Journal of Archaeology of Egypt/Egyptology, 17*(9), 614-632.

Irwin, R. (2020). Climate change and education. *Educational Philosophy and Theory*, 52(5), 492–507. 10.1080/00131857.2019.1642196

Islam, S. T. (2015, June). Climate Change Education for Sustainable Development in Bangladesh. UNESCO. Retrieved January 25, 2023 from https://unesdoc.unesco.org/ark:/48223/pf0000246779/PDF/246779eng.pdf.multi

Islam, M. R. (2020). Environment and disaster education in the secondary school curriculum in Bangladesh. *SN Social Sciences*, 1(1), 23. 10.1007/s43545-020-00025-1

Islam, M. T., Abdullah, A. B., Shahir, S. A., Kalam, M. A., Masjuki, H. H., Shumon, R., & Rashid, M. H. (2016). A public survey on knowledge, awareness, attitude, and willingness to pay for WEEE management: Case study in Bangladesh. *Journal of Cleaner Production*, 137, 728–740. 10.1016/j.jclepro.2016.07.111

Jacobi, P. (2003). Educação Ambiental, Cidadania e Sustentabilidade. Cad. Pesquisa, n. 118. https://www.scielo.br/scielo.php?pid=S0100-

Jahan, F., & Quddusi, K. S. K. A. (2013). Climate change and Bangladesh: Geographical and socio-economic impacts. *Dhaulagiri*, 7, 113–132. 10.3126/dsaj.v7i0.10439

Jakab, I., Zigová, M., & Pucherová, Z. (2019). Modernization of Environmental Education with the Use of Project-Based Learning, Outdoor Education, and Mobile Learning Supported by Information and Communication Technology. *Universities in the Networked Society: Cultural Diversity and Digital Competences in Learning Communities*, 223-248

Jane, Castillo, & Mallillin. (2022). Environmental Education of Students Pursuing Higher Education: Probing on Climate Change Awareness. Journal of Climate Change. 10.3233/JCC220020

Jardin, E. J. (2023). Exploring the Use of Technology in Environmental Education: Teachers' and Students' Perspectives. *Excellencia: International Multi-disciplinary Journal of Education (2994-9521), 1*(1), 118-126.

Jerowsky, M., & Borda, A. (2022, July 27). Virtual reality can support and enhance outdoor environmental education. *The University of British Columbia Magazine.* Retrieved from https://magazine.alumni.ubc.ca/2022/environment-technology/virtual-reality-can-support-and-enhance-outdoor-environmental-education

Jiawiwatkul, U. (2010). *Participatory action research: Concepts, principles, and lessons*. P.A. Leaving.

Johnson, B., & Manoli, C. C. (2010). The 2-MEV scale in the United States: A measure of children's environmental attitudes based on the theory of ecological attitude. *The Journal of Environmental Education*, 42(2), 84–97. 10.1080/00958964.2010.503716

Jones, D., Ghasemi, S., Gračanin, D., & Azab, M. (2023). Privacy, Safety, and Security in Extended Reality: User Experience Challenges for Neurodiverse Users. In Moallem, A. (Ed.), Lecture Notes in Computer Science: Vol. 14045. *HCI for Cybersecurity, Privacy and Trust. HCII 2023*. Springer., 10.1007/978-3-031-35822-7_33

Jónsdóttir, Á. B. (2017). *Artistic actions for sustainability: Potential of art in education for sustainability*. Lapland University Press.

Jöreskog, K. G., & Sörbom, D. (1993). *LISREL 8: Structural equation modeling with the SIMPLIS command language. Scientific Software International*. Lawrence Erlbaum Associates, Inc.

Jumpatip, W., Asavapoom, S., & Pakothang, J. (2014). Strategies for successful education management based on the educational reform in basic education institutions. *Journal Buriram Rajabhat University*, 9(2), 163–169.

Junaini, S., Kamal, A., Hashim, A., Shaipullah, N., & Truna, L. (2022). Augmented and virtual reality games for occupational safety and health training: A systematic review and prospects for the post-pandemic era. *International Journal of Online and Biomedical Engineering (Ijoe)*, 18(10), 43–63. 10.3991/ijoe.v18i10.30879

Kagawa, F. (2022). *The Heat is On! Towards a Climate Resilient Education System in Bangladesh. Kathmandu:* UNICEF Regional Office for South Asia. Retrieved January 23, 2024 from https://www.unicef.org/rosa/media/17566/file/The%20Heat%20is%20On%20-%20Bangladesh.pdf

Kaiser, H. F. (1958). The varimax criterion for analytic rotation in factor analysis. *Psychometrika*, 23(3), 187–200. 10.1007/BF02289233

Kammanee, T. (2020). *Teaching science: Body of knowledge for an effective learning process* (24th ed.). Chulalongkorn University Press.

Kan, H., London, S. J., Chen, G., Zhang, Y., Song, G., Zhao, N., Jiang, L., & Chen, B. (2008). Season, sex, age, and education as modifiers of the effects of outdoor air pollution on daily mortality in Shanghai, China: The Public Health and Air Pollution in Asia (PAPA) Study. *Environmental Health Perspectives*, 116(9), 1183–1188. 10.1289/ehp.1085118795161

Karaarslan Semiz, G., & Teksöz, G. (2019). Sistemsel dusunme becerilerinin tanimlanmasi, olculmesi ve degerlendirilmesi uzerine bir calisma: Kavram haritalari. *Baskent University Journal of Education*, 6(1), 111-126. https://buje.baskent.edu.tr/index.php/buje/article/view/177

Karakuş, G., & Çimen, O. (2020). The examination of ethical approaches of candidate teachers towards environmental problems. *Gazi University Journal of Gazi Educational Faculty*, 40(2), 411–435.

Kaza, S. (2008). *Mindfully green: A personal and spiritual guide to whole earth thinking*. Shambhala Publications.

Keahey, J. (2021). Sustainable development and participatory action research: A systematic review. *Systemic Practice and Action Research*, 34(3), 291–306. 10.1007/s11213-020-09535-8

Kesapat, P. (2000). *Strategic planning for educational management*. Suweeriyasat.

Khalifé, M., Chaker, R., & Gasparovic, S. (2022). Environmental education and digital solutions: An analysis of the Lebanese context's existing and possible digital actions. *Frontiers in Education*, 7, 958569. 10.3389/feduc.2022.958569

Khamkeaw, W. (2012). *Creating world peace based on the concept of 4 noble truths*. Faculty of humanities. Srinakharinwirote University.

Kılıçer, K., & Odabaşı, H. F. (2010). Bireysel yenilikçilik ölçeği (BYÖ): Türkçeye uyarlama, geçerlik ve güvenirlik çalışması. *Hacettepe Üniversitesi Eğitim Fakültesi Dergisi, 38*(38).

Kim, R., Kim, J., Lee, I., Yeo, U., Lee, S., & Decano-Valentin, C. (2021). Development of three-dimensional visualization technology of the aerodynamic environment in a greenhouse using CFD and VR technology, Part 2: Development of an educational VR simulator. *Biosystems Engineering*, 207, 12–32. 10.1016/j.biosystemseng.2021.02.018

Kıral, B. (2020). Document analysis as a qualitative data analysis method. *Siirt University Journal of Social Sciences Institute, 8*(15), 170-189.

Kline, R. B. (2005). *Principles and practice of structural equation modeling* (2nd ed.). Guilford Press.

Koçak Altundağ, C., & Yücel, A. S. (2022). Development of an anxiety scale for chemistry. *Journal for Educators. Teachers and Trainers*, 13(1), 1–8. 10.47750/jett.2022.13.01.001

Kramer, C. B., Cullen, A. C., & Faustman, E. M. (2006). Policy Implications of Genetic Information on Regulation under the Clean Air Act: The Case of Particulate Matter and Asthmatics. *Environmental Health Perspectives*, 114(3), 313–319. 10.1289/ehp.829916507451

Krein, A. T. (2023). *Model United Nations: A Practical Guide*. Springer Nature. 10.1007/978-3-031-13524-8

Kremen, C., Merenlender, A. M., & Murphy, D. D. (1994). Ecological monitoring: A vital need for integrated conservation and development programs in the tropics. *Conservation Biology*, 8(2), 388–397. 10.1046/j.1523-1739.1994.08020388.x

Krishnan, L., Poorani, S., & Sawmya, S. (2022, October 25). Emerging Trends in Training Knowledge Workers; *New Economy*. https://scite.ai/reports/10.5296/ijld.v12i4.20413

Krishnan, S., Robb, W., Manweiler, K., & Wise, M. (2020). Growing Beyond the Hype: Controlled Environment. *Agricultures (Montrouge)*, 2G.

Kumari, R., & Thakur, N. (2023). Role of youth in environmental preservation. *Role of Youth in Environmental Preservation, 10*(6).

Kumpulainen, K., Kajamaa, A., Erstad, O., Mäkitalo, Å., Drotner, K., & Jakobsdóttir, S. (2022). *Nordic childhoods in the digital age: Insights into contemporary research on communication, learning and education*. Taylor & Francis. 10.4324/9781003145257

Kunaporn, P. P. (2013). *Thailand's philosophy of education, Buddhadharma section: Educational leader.* Wat Nyanavesakavan.

Kutay, Y., & Güngör Cabbar, B. G. (2022). Investigation of high school students' attitudes of environmental problems. *Journal of Balıkesir University Institute of Science and Technology*, 24(1), 29–46. 10.25092/baunfbed.912537

Labadze, L., Grigolia, M., & Machaidze, L. (2023). Role of AI chatbots in education: Systematic literature review. *International Journal of Educational Technology in Higher Education*, 20(1), 56. 10.1186/s41239-023-00426-1

Lai, O. (2022, May 11). *Pakistan's Ravi is most polluted river in the world from pharmaceutical pollution*. Earth.Org. Retrieved April 1, 2024, from https://earth.org/most-polluted-river-in-the-world-pharmaceutical-pollution/

Larbaigt, J., & Lemercier, C. (2021). An evaluation of the acceptability of smart eyewear for plot diagnosis activity in agriculture. *Ergonomics in Design*. Advance online publication. 10.1177/10648046211018541

Lee, C., Chang, Y., & Wang, E. (2022). Crossing the rivers by feeling the stones: The effect of China's green credit policy on manufacturing firms' carbon emission intensity. *Energy Economics*, 116, 106413. 10.1016/j.eneco.2022.106413

Lee, E. B. (2008). Environmental attitudes and information sources among African American college students. *The Journal of Environmental Education*, 40(1), 29–42. 10.3200/JOEE.40.1.29-42

Lee, L., Kim, M., & Hwang, W. (2019). Potential of augmented reality and virtual reality technologies to promote wellbeing in older adults. *Applied Sciences (Basel, Switzerland)*, 9(17), 3556. 10.3390/app9173556

Lenoir, Y., & Hasni, A. (2010). Interdisciplinarity in Quebec schools: 40 years of problematic implementation. *Issues in Interdisciplinary Studies*, 28, 238–294.

Lewis, E., Mansfield, C., & Baudains, C. (2010). Going on a turtle egg hunt and other adventures: Education for sustainability in early childhood. *Australasian Journal of Early Childhood*, 35(4), 95–100. 10.1177/183693911003500412

Lightfoot-Wright, A. D. (2021). Environmental Education, Digital Technologies, and Confronting New Realities in the Time of COVID-19. [MA dissertation, California State University]. Retrieved from https://scholarworks.calstate.edu/downloads/9k41zk91t

Li, H., Chai, J., Qian, Z. F., & Chen, H. (2022). Cooperation strategies when leading firms compete with small and medium-sized enterprises in a potentially competitive market. *J Manag Sci Eng*, 7(3), 489–509. 10.1016/j.jmse.2022.02.003

Li, R., & Chen, Y. (2022). The influence of a green credit policy on the transformation and upgrading of heavily polluting enterprises: A diversification perspective. *Economic Analysis and Policy*, 74, 539–552. 10.1016/j.eap.2022.03.009

Liu, S. C. (2023). Examining undergraduate students' systems thinking competency through a problem scenario in the context of climate change education. *Environmental Education Research*, 29(12), 1–16. 10.1080/13504622.2022.2120187

Lombardi, D. (2022). Climate crisis mitigation and adaptation: Educational and developmental psychology's responsibility in helping face this threat. *Educational and Developmental Psychologist*, 39(1), 1–4. 10.1080/20590776.2021.2012834

Lopez, B. (1996). A literature of place. *US Society and Values*, 1(10), 10–12.

Lowan-Trudeau, G. (2023). Digital technologies and environmental education. *The Journal of Environmental Education*, 54(1), 1–7. 10.1080/00958964.2022.2152413

Lu, S. J., & Liu, Y. C. (2015). Integrating augmented reality technology to enhance children's learning in marine education. *Environmental Education Research*, 21(4), 525–541. 10.1080/13504622.2014.911247

Macy, J., & Brown, M. Y. (1998). *Coming back to life: Practices to reconnect our lives, our world*. New Society Publishers.

Maddison Project Database. (2020). Bolt, Jutta, and Jan Luiten van Zanden. Maddison style estimates of the evolution of the world economy. A new 2020 update.

Mader, C. (2013). Sustainability process assessment on transformative potentials: The Graz model for integrative development. *Journal of Cleaner Production*, 49, 54–63. 10.1016/j.jclepro.2012.08.028

Ma, L., Shahbaz, P., Haq, S. U., & Boz, I. (2023). Exploring the moderating role of environmental education in promoting a clean environment. *Sustainability (Basel)*, 15(10), 8127. 10.3390/su15108127

Mambrey, S., Schreiber, N., & Schmiemann, P. (2020). Young students' reasoning about ecosystems: The role of systems thinking, knowledge, conceptions, and representation. *Research in Science Education*, 1–20. 10.1007/s11165-020-09917-x

Maneelek, R. (2002). *Strategic planning in educational institutions*. Metheetips.

Marsh, H. W., Hau, K. T., Artelt, C., Baumert, J., & Peschar, J. L. (2006). OECD's brief self-report measure of educational psychology's most useful affective constructs: Cross-cultural, psychometric comparisons across 25 countries. *International Journal of Testing*, 6(4), 311–360. 10.1207/s15327574ijt0604_1

Martinez-Moyano, I. J. (2024). History of the Beer Game. *System Dynamics Review*, 40(2), 1–32. 10.1002/sdr.1767

Martin, T. (2024). *Examining the Influence of a PLC-based Intervention on the Local Adoption of the New Pennsylvania Science Standards via the Concerns Based Adoption Model*. Arizona State University.

Masood, T., & Egger, J. (2020). Augmented reality in support of Industry 4.0—Implementation challenges and success factors. *Robotics and Computer-integrated Manufacturing*, 58, 181–195. 10.1016/j.rcim.2019.02.003

Mayring, P. (2002). *Einführung in die qualitative sozialforschung* (5th ed.). Beltz Verlag.

McBride, B. B., Brewer, C. A., Berkowitz, A. R., & Borrie, W. T. (2013). Environmental literacy, ecological literacy, ecoliteracy: What do we mean and how did we get here? *Ecosphere*, 4(5), 1–20. 10.1890/ES13-00075.1

McCrea, E. J. (2006). The Roots of Environmental Education: How the Past Supports the Future. Environmental Education and Training Partnership (EETAP). Retrieved from https://files.eric.ed.gov/fulltext/ED491084.pdf

McMenamin, M. M., & Zirkel, P. A. (1998). Section 504, the Americans with Disabilities Act, and Public School Employment: A Legal Primer. *Journal of Personnel Evaluation in Education*, 11(3), 243–253. 10.1023/A:1007929221417

Meadows, D. H. (2008). *Thinking in systems: A primer.* Chelsea Green Publishing.

Meadows, D., Sweeney, L. B., & Mehers, G. M. (2016). *The climate change playbook: 22 systems thinking games for more effective communication about climate change.* Chelsea Green Publishing.

Mehr, K. B., Tahayori, H., & Moosavi, M. R. (2024). Information preserving approach for encoding words into interval type-2 fuzzy sets. *Granular Computing*, 9(3), 55. 10.1007/s41066-024-00479-1

Mendel, J. M. (2018). The perceptual computer: The past, up to the present, and into the future. *Informatik-Spektrum*, 41(1), 15–26. 10.1007/s00287-018-1088-z

Mendel, J., & Wu, D. (2010). *Perceptual computing: Aiding people in making subjective judgments.* John Wiley & Sons. 10.1002/9780470599655

Merriam, S. B., & Tisdell, E. J. (2016). Designing your study and selecting a sample. *Qualitative research: A guide to design and implementation, 67*(1), 73-104.

Metwally, A. H. S., Nacke, L. E., Chang, M., Wang, Y., & Yousef, A. M. F. (2021). Revealing the hotspots of educational gamification: An umbrella review. *International Journal of Educational Research*, 109, 101832. 10.1016/j.ijer.2021.101832

Meydan, A. (2014). Sosyal Bilgiler Öğretimde Gezi-Gözlem ve Doğa Eğitimi. Sosyal Bilgiler Öğretiminde Yeni Yaklaşımlar I (s. 242-260). Ankara: Pegem Akademi.

Mills, G. (2003). *Action research: A guide for the teacher researcher.* Merrill Prentice Hall.

Ministry of Education Singapore. (2010). *Nurturing our young for the future: Competencies for the 21st Century.* Ministry of Education, Singapore.

Ministry of Environment and Forest (MFE). (2005, November). National Adaptation Programme of Action (NAPA). Retrieved from http://nda.erd.gov.bd/en/c/publication/national-adaptation-programme-of-action-napa-2005

Ministry of Environment and Forest (MFE). (2009, September). Bangladesh Climate Change Strategy and Action Plan 2009. Retrieved from http://nda.erd.gov.bd/files/1/Publications/CC%20Policy%20Documents/BCCSAP2009.pdf

Ministry of Environment and Forests (MEF). (2016). Bangladesh National Conservation Strategy (NCS) (2016-2031). Government of the People's Republic of Bangladesh. Published in September 2016. Retrieved from https://bforest.portal.gov.bd/sites/default/files/files/bforest.portal.gov.bd/notices/c3379d22_ee62_4dec_9e29_75171074d885/Executive%20Summary(NCS).pdf

Mof, N. E. (2018). *Ortaöğretim Biyoloji Dersi (9, 10, 11 ve 12. Sınıflar).* Öğretim Programı. https://mufredat.meb.gov.tr/Dosyalar/20182215535566-Biyoloji%20d%C3%B6p.pdf

Mogensen, F., & Schnack, K. (2010). The action competence approach and the 'new' discourses of education for sustainable development, competence and quality criteria. *Environmental Education Research*, 16(1), 59–74. 10.1080/13504620903504032

Moghaddam, H., & Kunst, R. M. (2023). The Role of Natural Gas in Mitigating Greenhouse Gas Emissions: The Environmental Kuznets Curve Hypothesis for Major Gas-Producing Countries. *Sustainability (Basel)*, 2023(15), 4266. 10.3390/su15054266

Mohai, P., Kweon, B. S., Lee, S., & Ard, K. (2011). Air pollution around schools is linked to poorer student health and academic performance. *Health Affairs*, 30(5), 852–862. 10.1377/hlthaff.2011.007721543420

Mohammad, N. (2010). Development of the Environmental Education in Higher Studies in Bangladesh: An Approach for Environmental Sustainability. *Environmental Research and Technology,* 148-155.

Mohr, L. A., Ero lu, D., & Ellen, P. S. (1998). The development and testing of a measure of skepticism toward environmental claims in marketers' communications. *The Journal of Consumer Affairs*, 32(1), 30–55. 10.1111/j.1745-6606.1998.tb00399.x

Mojid, M. A. (2020). Climate change-induced challenges to sustainable development in Bangladesh. [). IOP Publishing.]. *IOP Conference Series. Earth and Environmental Science*, 423(1), 012001. 10.1088/1755-1315/423/1/012001

MoNE. (2022). Çevre Egitimi ve Iklim Degisikligi Dersi (6, 7 veya 8. Sınıflar) Ogretim Programi. Ankara: Millî Egitim Bakanlığı. http://mufredat.meb.gov.tr/ProgramDetay.aspx?PID=1143. Erişim Tarihi: 21. 01.2023.

Monroe, M. C., Andrew, E., & Biedenweg, G. (2007). A framework for environmental education strategies. *Applied Environmental Education and Communication*, 6(3-4), 205–216. 10.1080/15330150801944416

Monroe, M. C., & Richard, R. (2019, June 3). Plate, Annie Oxarart, Alison Bowers & Willandia A. Chaves (2017): Identifying effective climate change education strategies: a systematic review of the research. *Environmental Education Research*, 25(6), 791–812. Advance online publication. 10.1080/13504622.2017.1360842

Moon, J. A. (2006). *Learning journals: A handbook for reflective practice and professional development* (2nd ed.). Routledge. 10.4324/9780203969212

Moore, J. (2005). Is higher education ready for transformative learning? A question explored in the study of sustainability. *Journal of Transformative Education*, 3(1), 76–91. 10.1177/1541344604270862

Morales, M. P. E. (2016). Participatory action research (PAR) cum action research (AR) in teacher professional development: A literature review. [IJRES]. *International Journal of Research in Education and Science*, 2(1), 156–165. 10.21890/ijres.01395

Morgan, A. (2022, August 28). Information and Communication Technology impacts on Environmental Education Sustainable Developments of EU and USA. NEWS. Retrieved from https://www.uspa24.com/bericht-21509/information-and-communication-technology-impacts-on-environmental-education.html

Mostafa, M. M. (2009). Shades of green: A psychographic segmentation of the green consumer in Kuwait using self-organizing maps. *Expert Systems with Applications*, 36(8), 11030–11038. 10.1016/j.eswa.2009.02.088

Mulkam, S. (2002). *20 Methods of learning management to develop morality, ethics, values and autonomous learning management*. Printmaking.

Mustafaoğlu, F.M., & Yücel, A.S. (2022a). Developing context-based teaching competencies of chemistry teachers: designing and implementing context-based activities. *Jurnal Penelitian dan Pembelajaran IPA, 8*(2), 126-152. 10.30870/jppi.v8i2.16491

Mustafaoğlu, F. M., & Yücel, A. S. (2022b). Context-based teaching experiences of chemistry teachers: Expectations, gains and applicability conditions. *Journal of Turkish Science Education*, 19(3), 958–978. 10.36681/tused.2022.158

Myers, G., Boyes, E., & Stanisstreet, M. (1999). Something in the air: School students' ideas about air pollution. *International Research in Geographical and Environmental Education*, 8(2), 108–119. 10.1080/10382049908667600

Myers, G., Boyes, E., & Stanisstreet, M. (2004). School students' ideas about air pollution: Knowledge and attitudes. *Research in Science & Technological Education*, 22(2), 133–152. 10.1080/0263514042000290868

Nalçacı, A., & Beldağ, A. (2012). Identifying the environmental attitudes of 7th and 8th grade elementary students (Erzurum sample). *Eastern Geographical Review*, 17(28), 141–154.

National Curriculum and Textbook Board [NCTB]. (2003). *National curriculum for primary level, 2002-03*. Dhaka: Ministry of Education, Government of Bangladesh.

National Research Council. (2013). *Next generation science standards: For states, by states.*

National Strategy Secretariat Office, Office of the National Economic and Social Development Board. (2018). *National strategy (2018–2037).* Office of the National Economic and Social Development Board, Office of the Prime Minister.

Naugle, A., Verzi, S., Lakkaraju, K., Swiler, L., Warrender, C., Bernard, M., & Romero, V. (2023). Feedback density and causal complexity of simulation model structure. *Journal of Simulation*, 17(3), 229–239. 10.1080/17477778.2021.1982653

New Pennsylvania Science, Technology & Engineering, Environmental Literacy & Sustainability (Steels) standards. (2023). Retrieved from https://www.onatlas.com/blog/new-pennsylvania-science-technology-engineering-environmental-literacy-sustainability-steels-standards

Nguyen, H., & Santagata, R. (2021). Impact of computer modeling on learning and teaching systems thinking. *Journal of Research in Science Teaching*, 58(5), 661–688. 10.1002/tea.21674

Nigam, A., Kabra, P., & Doke, P. (2011, October). Augmented Reality in agriculture. In 2011 IEEE 7th International Conference on Wireless and Mobile Computing, Networking and Communications (WiMob) (pp. 445-448). IEEE.

Niklanović, M., Miljanović, T., & Pribićević, T. (2014). A model of interdisciplinary teaching of ecology in the high school. *Archives of Biological Sciences*, 66(3), 1291–1297. 10.2298/ABS1403291N

Nuhoglu, H. (2008). İlkogretim fen ve teknoloji dersinde sistem dinamigi yaklasiminin tutuma, basarıya ve farkli becerilere etkisinin arastirilmasi. [Doctoral dissertation, Gazi Üniversitesi].

Nuhoglu, H. (2020). The effect of deduction and induction methods used in modeling current environmental issues with system dynamics approach in science education. *Participatory Educational Research*, 7(1), 111–126. 10.17275/per.20.7.7.1

O'Donoghue, R., Henze, C., Shimray, C., Sarabhai, K. V., & Rivera, J. C. A. S. (2020). Hand-print CARE: Towards ethics-led action learning for ESD in school subject disciplines. *Journal of Education for Sustainable Development*, 14(1), 41–60. 10.1177/0973408220934647

O'Leary, Z. (2017). *The essential guide to doing your research project*. SAGE Publications Inc.

O'Neill, M. S., Bell, M. L., Ranjit, N., Cifuentes, L. A., Loomis, D., Gouveia, N., & Borja-Aburto, V. H. (2008). Air pollution and mortality in Latin America: The role of education. *Epidemiology (Cambridge, Mass.)*, 19(6), 810–819. 10.1097/EDE.0b013e3181816528 18633327

O'Rourke, N., Psych, R., & Hatcher, L. (2013). *A step-by-step approach to using SAS for factor analysis and structural equation modeling*. Sas Institute.

Öcal, E. E., & Önsüz, M. F. (2020). Reliability and validity of ecological literacy scale for primary school students. *European Journal of Public Health*, 30(Supplement_5), 5. 10.1093/eurpub/ckaa166.103

OECD. (2019). *OECD future of education and skills 2030: OECD Learning Compass 2030*. Organization for Economic Co-operation and Development. www.oecd.org

Office of the Education Council. (2009). *The situation of Thai education on the world stage*. Prikwan Graphic. (Original work published 2007)

Office of the Education Council. (2016). *Directions and research issues of Thai education to achieve the world's sustainable development goals* (2nd ed.). Office of the Education Council.

Office of the Education Council. (2017). *Directional framework of the national education plan 2017–2031* [Draft]. Ministry of Education.

Office of the Education Council. (2018). *Report on the research results and development of the competency framework for grade 1–3 students for basic education courses*. Office of the Education Council.

Office of the National Economic and Social Development Board, Office of the Prime Minister. (2016). *Strategies for national development according to the 12th national economic and social development plan (2017–2021)*. Office of the National Economic and Social Development Board, Office of the Prime Minister.

Office of the National Economic and Social Development Board, Office of the Prime Minister. (2019). *Brief master plan under the national strategy (2018–2037)*. (2nd Ed.). Office of the National Economic and Social Development Board, Office of the Prime Minister.

Office of the National Primary Education Commission. (2000). *9-year strategic planning for compulsory education management*. Department of Religious Affairs.

Oğuz, D., Çakci, I., & Kavas, S. (2010). Environmental awareness of university students in Ankara, Turkey. *African Journal of Agricultural Research*, 5(10), 2629–2636.

Okyay, Ö., Sayın, A., Demir, Z. G., & Özdemir, K. (2022). The impact of ecological literacy training on environmental awareness and attitudes of pre-school childrens. *Journal of National Education*, 51(236), 2851–2870.

Olapiriyakul, K., & Scher, J. M. (2011). A Methodology for Integrating Traditional Classroom Learning with Contemporary Online Learning.

Oliveira, W., Hamari, J., Shi, L., Toda, A. M., Rodrigues, L., Palomino, P. T., & Isotani, S. (2023). Tailored gamification in education: A literature review and future agenda. *Education and Information Technologies*, 28(1), 373–406. 10.1007/s10639-022-11122-4

Orr, D. (1992). *Ecological literacy*. State of New York Press.

Orr, D. (2005). Ecological literacy. In Pretty, J. N. (Ed.), *The Earthscan reader in sustainable agriculture* (pp. 21–29). Earthscan.

Orr, D. W. (1992). *Ecological literacy: Education and the transition to a postmodern world*. SUNY Press.

Orr, D. W. (2004). *Earth in mind: On education, environment, and the human prospect*. Island Press.

Oxfam Education. (2002). *Whole School*. Oxfam Education. https://www.oxfam.org.uk/education/whole-school

Özgün, B. B. (2018). *Öğretmen Adaylarının Eko-Okuryazarlık Düzeylerinin İncelenmesi. Yayımlanmamış doktora tezi*. İnönü Üniversitesi Eğitim Bilimleri Enstitüsü.

Pang, J., & Good, R. (2000). A review of the integration of science and mathematics: Implications for further research. *School Science and Mathematics*, 100(2), 73–82. 10.1111/j.1949-8594.2000.tb17239.x

Papavasileiou, V., Nikolaou, E., Andreadakis, N., Xanthacou, Y., & Kaila, M. (2021). The role of art in environmental education. *IJAEDU-International E-Journal of Advances in Education*, 6(18), 287–295.

Pappalardo, C. (2021). What a Difference a State Makes: California's Authority to Regulate Motor Vehicle Emissions Under the Clean Air Act and the Future of State Autonomy. *Michigan Journal of Environmental & Administrative Law. Michigan Journal of Environmental & Administrative Law*, 10(1), 169–169. 10.36640/mjeal.10.1.what

Park, H., Lee, B., Ha, E. H., Lee, J. T., Kim, H., & Hong, Y. C. (2002). Association of air pollution with school absenteeism due to illness. *Archives of Pediatrics & Adolescent Medicine*, 156(12), 1235–1239. 10.1001/archpedi.156.12.123512444836

Parry, S., & Metzger, E. (2023). Barriers to learning for sustainability: A teacher perspective. *Sustainable Earth Reviews*, 6(1), 2. 10.1186/s42055-022-00050-3

Past climate. (2021, September 22). Climate.gov. Retrieved April 5, 2024, from https://www.climate.gov/maps-data/climate-data-primer/past-climate

Paul, J., & Jefferson, F. (2019). A Comparative Analysis of Student Performance in an Online vs. Face-to-Face Environmental Science Course From 2009 to 2016. *Frontiers of Computer Science*, 1, 7. Advance online publication. 10.3389/fcomp.2019.00007

Pe'er, S., Goldman, D., & Yavetz, B. (2007). Environmental literacy in teacher training: Attitudes, knowledge, and environmental behavior of beginning students. *The Journal of Environmental Education*, 39(1), 45–59. 10.3200/JOEE.39.1.45-59

Peeters, J. H. A. M. (1998). Aviation and air pollution. In *Studies in environmental science* (Vol. 72, pp. 601–613). Elsevier.

Peker, R., & Ceylan, S., (2020). The investigation of primary school fourth grade student's attitudes and behaviors towards the environment, Bilecik Şeyh Edebali University Journal of Social Sciences, *5*(TBMM 100. Year Special Issue), 60-90. 10.33905/bseusbed.747762

Pennsylvania Department of Education. (n.d.). *Modernizing PA's science education standards*. https://www.education.pa.gov/Teachers%20-%20Administrators/Curriculum/Science/Pages/Science-Standards.aspx

Phillips, R. J. (1997). Can juniors read graphs? A review and analysis of some computer-based activities. *Journal of Information Technology for Teacher Education*, 6(1), 49–58. 10.1080/14759399700200005

Phochanathan, S., & Sudsawat, S. (2016). Local administrative organizations' strategies for educational management of Phuket Province. *Rajabhat University Academic Journal, 12*(1).

Phupattanasilp, P., & Tong, S.-R. (2019). Augmented reality in the integrative internet of things (AR-IoT): Application for precision farming. *Sustainability (Basel)*, 11(9), 2658. 10.3390/su11092658

Pitman, S. D., & Daniels, C. B. (2016). Quantifying ecological literacy in an adult western community: The development and application of a new assessment tool and community standard. *PLoS One 11*(3), e0150648, 1-18. 10.1371/journal.pone.0150648

Pitman, S. D., Daniels, C. B., & Sutton, P. C. (2018). Ecological literacy and psychographics: Lifestyle contributors to ecological knowledge and understanding. *International Journal of Sustainable Development and World Ecology*, 25(2), 117–130. 10.1080/13504509.2017.1333047

Pituch, K. A., & Stevens, J. P. (2016). *Applied multivariate statistics for the social sciences: Analyses with SAS and IBM's SPSS* (6th ed.). Routledge.

Planet Bee Foundation. (n.d.). The importance of eco-literacy in schools. Retrieved from https://www.planetbee.org/planet-bee-blog/the-importance-of-eco-literacy-in-schools

Ponnusamy, V., Natarajan, S., Ramasamy, N., Clement, C., Rajalingam, P., & Mitsunori, M. (2021). An IoT-enabled augmented reality framework for plant disease detection. *RIA*, 35(3), 185–192. 10.18280/ria.350301

Pope, A. (1798). *An Essay on Man*.

Porter, T., & Córdoba, J. (2009). Three views of systems theories and their implications for sustainability education. *Journal of Management Education*, 33(3), 323–347. 10.1177/1052562908323192

Pörtner, H. O., Roberts, D. C., Adams, H., Adler, C., Aldunce, P., Ali, E., ... Birkmann, J. (2022). Climate change 2022: Impacts, adaptation and vulnerability. *IPCC Sixth Assessment Report*, 37-118. 10.1111/padr.12497

Powell, D., Agnew, D., & Trexler, C. (2008). Agricultural literacy: Clarifying a vision for practical application. *Journal of Agricultural Education*, 49(1), 85–98. 10.5032/jae.2008.01085

Powers, A. L. (2004). Teacher preparation for environmental education: Faculty perspectives on the infusion of environmental education into preservice methods courses. *The Journal of Environmental Education*, 35(3), 3–11.

Prastiwi, L., Sigit, D. V., & Ristanto, R. H. (2019). Ecological literacy, environmental awareness, academic ability and environmental problem-solving skill at Adiwiyata school. *Indonesian Journal of Science and Education*, 3(2), 82–92. 10.31002/ijose.v3i2.1114

Radovanović, D., Holst, C., Belur, S. B., Srivastava, R., Houngbonon, G. V., Le Quentrec, E., Miliza, J., Winkler, A. S., & Noll, J. (2020). Digital literacy key performance indicators for sustainable development. *Social Inclusion (Lisboa)*, 8(2), 151–167. 10.17645/si.v8i2.2587

Rasmussen, W. D. (1989). *Taking the University to the People: Seventy-five Years of Cooperative Extension*. Iowa State University Press.

Ray, J. (2023). *Ecology of a cracker childhood*. Milkweed Editions.

Raykov, T., & Marcoulides, G. A. (2006). *A first course structural equation modeling*. Lawrence Erlbaum Assocation Inc. Publishers.

Reed, M. S., Fraser, E. D. G., & Dougill, A. J. (2006). An adaptive learning process for developing and applying sustainability indicators with local communities. *Ecological Economics*, 59(4), 406–418. 10.1016/j.ecolecon.2005.11.008

Reid, A., Dahlgren, M. A., Dahlgren, L. O., & Petocz, P. (2011). *From expert student to novice professional* (Vol. 99). Springer Science & Business Media. 10.1007/978-94-007-0250-9

Reilly, C., Stevenson, K., Warner, W., Park, T., Knollenberg, W., Lawson, D., Brune, S., & Barbieri, C. (2022). Agricultural and environmental education: A call for meaningful collaboration in a U.S. context. *Environmental Education Research*, 28(9), 1410–1422. 10.1080/13504622.2022.2040431

Rieckmann, M. (2012). Future-oriented higher education: Which key competencies should be fostered through university teaching and learning? *Futures*, 44(2), 127–135. 10.1016/j.futures.2011.09.005

Riess, W., & Mischo, C. (2010). Promoting systems thinking through biology lessons. *International Journal of Science Education*, 32(6), 705–725. 10.1080/09500690902769946

Rithu, M., Akhter, N., Rahman, M. M., & Al-Maruf, A. (2016). The Status of Geography and Environmental Studies at Secondary Level Education in Bangladesh. *Archives of Current Research International*, 5(4), 1–9. 10.9734/ACRI/2016/29017

Rivers of the World. (2021). *Rivers of the world.* British Council. Retrieved April 1, 2024, from https://england.britishcouncil.org/coventry2021/rivers-world

Rode, H., & Michelsen, G. (2008). Levels of indicator development for education for sustainable development. *Environmental Education Research*, 14(1), 19–33. 10.1080/13504620701843327

Rogers, Y., Connelly, K., Hazlewood, W. R., & Tedesco, L. P. (2009). Enhancing learning: A study of how mobile devices can facilitate sensemaking. *Personal and Ubiquitous Computing*, 14(2), 111–124. 10.1007/s00779-009-0250-7

Rong, Q., Lian, Q., & Tang, T. (2022). Research on the influence of AI and VR technology for students' concentration and creativity. *Frontiers in Psychology*, 13, 767689. Advance online publication. 10.3389/fpsyg.2022.76768935401322

Rooney-Varga, J. N., Hensel, M., McCarthy, C., McNeal, K., Norfles, N., Rath, K., Schnell, A. H., & Sterman, J. D. (2021). Building consensus for ambitious climate action through the World Climate simulation. *Earth's Future*, 9(12). 10.1029/2021EF002283

Rose, D. H.. (2018). Accurate and Informative for All: Universal Design for Learning (UDL) and the Future of Assessment. In Elliott, S., Kettler, R., Beddow, P., & Kurz, A. (Eds.), *Handbook of Accessible Instruction and Testing Practices*. Springer., 10.1007/978-3-319-71126-3_11

Rosenström, U., & Kyllönen, S. (2007). Impacts of a participatory approach to developing national level sustainable development indicators in Finland. *Journal of Environmental Management*, 84(3), 282–298. 10.1016/j.jenvman.2006.06.00816901619

Rosmansyah, Y., Achiruzaman, M., & Hardi, A. B. (2019). A 3D multiuser virtual learning environment for online training of agriculture surveyors. *Journal of Information Technology Education*, 18, 481–507. 10.28945/4455

Rossetto, A. G. de M., Martins, T. C., Silva, L. A., Leithardt, D. R. F., Bermejo-Gil, B. M., & Leithardt, V. R. Q. (2023). An analysis of the use of augmented reality and virtual reality as educational resources. *Computer Applications in Engineering Education*, 31(6), 1761–1775. Advance online publication. 10.1002/cae.22671

Rossini, M. (2022). Teaching children about climate change and sustainability. *The School Science Review*, 103(384), 76–79.

Roth, C. E. (1968). Curriculum Owerwiev for Developing Environmentally Literate Citizens. (ERIC Reproduction Service No. ED 032982).

Roth, C. E. (1992). Environmental literacy: its roots, evolution and directions in the 1990s.

Roth, C. E. (1992). Environmental Literacy: Its roots, evolution and directions in the 1990s. (ERIC Reproduction Service No. ED 348235).

Rovňák, M., Novotny, R., Barlak, J., Bakon, M., & Chovanec, F. (2021). Implementation and funding of projects focused on non-formal environmental education. *Studia Ekonomiczne. Gospodarka, Społeczeństwo, Środowisko, (2/2021 (8))*, 136-144.

Roychoudhury, A., Shepardson, D. P., Hirsch, A., Niyogi, D., Mehta, J., & Top, S. (2017). The need to introduce system thinking in teaching climate change. *Science Educator*, 25(2), 73–81.

Ruchter, M., Klar, B., & Geiger, W. (2010). Comparing the effects of mobile computers and traditional approaches in environmental education. *Computers & Education*, 54(4), 1054–1067. 10.1016/j.compedu.2009.10.010

Rychen, D. S., & Salganik, L. H. (Eds.). (2003). *Key competencies for a successful life and well-functioning society*. Hogrefe & Huber Publishers.

Sabbab, S., & Ahmed, S. (2024, January 21). Restructuring climate education in Bangladesh. *The Dhaka Tribune.* Retrieved from https://www.dhakatribune.com/opinion/op-ed/337332/restructuring-climate-education-in-bangladesh

Sajjadi, P., Bagher, M. M., Myrick, J. G., Guerriero, J. G., White, T. S., Klippel, A., & Swim, J. K. (2022). Promoting systems thinking and pro-environmental policy support through serious games. *Frontiers in Environmental Science*, 10, 957204. 10.3389/fenvs.2022.957204

Salequzzaman, M., & Stocker, L. (2001). The context and prospects for environmental education and environmental careers in Bangladesh. *International Journal of Sustainability in Higher Education*, 2(2), 104–127. 10.1108/14676370110388309

Salequzzaman, Md., & Davis, J. K. (2003). Environmental education and environmental management in Bangladesh and their sustainability. *Environmental Informatics Archives*, 1, 70–82.

Samaddar, K., & Mondal, S. (2024). AR and VR-based travel: A responsible practice towards sustainable tourism. *International Journal of Tourism Cities*, 10(1), 105–128. 10.1108/IJTC-05-2022-0135

Sangwan, S., & Vyas, R. (2020). Assessment of Green Consumerism on Environmental Sustainability. *International Journal of Current Microbiology and Applied Sciences*, 9(10), 389. 10.20546/ijcmas.2020.910.389

Saqib, Z. A., Zhang, Q., Ou, J., Saqib, K. A., Majeed, S., & Razzaq, A. (2020). Education for sustainable development in Pakistani higher education institutions: An exploratory study of students' and teachers' perceptions. *International Journal of Sustainability in Higher Education*, 21(6), 1249–1267. 10.1108/IJSHE-01-2020-0036

Saribas, D., Teksoz, G., & Ertepinar, H. (2014). The relationship between environmental literacy and self-efficacy beliefs toward environmental education. *Procedia: Social and Behavioral Sciences*, 116, 3664–3668. 10.1016/j.sbspro.2014.01.820

Sarkar, M. (2011). Secondary Student's Environmental Attitudes: The Case of Environmental Education in Bangladesh. *International Journal of Academic Research in Business & Social Sciences*, 1, 106.

Sayadi, N., & Gómez-Zará, D. (2024). Fostering Inclusive Virtual Reality Environments: Discussing Strategies for Promoting Group Dynamics and Mitigating Harassment. *University of Notre Dame, USA*. https://doi.org//arXiv.2405.0591710.48550

Scavarelli, A., Teather, R. J., & Arya, A. (2023, May). Exploring Selection and Search Usability Across Desktop, Tablet, and Head-Mounted Display WebXR Platforms. In *2023 9th International Conference on Virtual Reality (ICVR)* (pp. 505-514). IEEE.

Schermelleh-Engel, K., Moosbrugger, H., & Müller, H. (2003). Evaluating the Fit of Structural Equation Models: Tests of Significance and Descriptive Goodness-of-Fit Measures. *Methods of Psychological Research Online*, 8(2), 23–74.

Schreiber, J.-R., & Siege, H. (2016). *Curriculum framework: Education for sustainable development*. Cornelsen.

Scott, W., & Gough, S. (2003). Rethinking relationships between education and capacity—building: Remodelling the learning process. *Applied Environmental Education and Communication*, 2(4), 213–219. 10.1080/15330150390241409

Şeker, H., & Gençdoğan, B. (2006). *Measurement Tool Development in Psychology and Education* [Psikolojide ve Eğitimde Ölçme Aracı Geliştirme]. Nobel Press.

Sell, M., Schiera, R., Siddique, Z., White, S., & Babal, J. (2023). Decision Literacy, Multimodal Storytelling, and the Digital Storygame Project. *English Journal*, 112(3), 28–35. 10.58680/ej202332226

Senge, P. (2006). Systems citizenship. In *The Leader of The Future 2* (pp. 31–46). Visions, Strategies, and Practices for The New Era.

Senge, P. M. (1990). *The fifth discipline: The art and practice of the learning organization*. Crown Publishing.

Setianingrum, D. A., Matahari, D. B., Jumadi, J., & Wilujeng, I. (2023). Development of science e-book containing gamelan's local wisdom based on steam-poe to facilitate students' love of local culture. *Jurnal Penelitian Pendidikan IPA*, 9(6), 4791–4800. 10.29303/jppipa.v9i6.3760

Shahid, M. J., Arslan, M., Siddique, M., Ali, S., Tahseen, R., & Afzal, M. (2019). Potentialities of floating wetlands for the treatment of polluted water of river Ravi, Pakistan. *Ecological Engineering*, 133, 167–176. 10.1016/j.ecoleng.2019.04.022

Shakespeare, W. (2006). *The sonnets* (Vol. 70). Cambridge University Press. 10.1017/9781139836623

Shaleendra, T., Wishvamali, B. T., Gunarathne, N., Hareendran, S., & Abeygunawardhana, P. (2019). Simulation of the Influence of environmental factors related to Greenhouses using Augmented Reality. In *Proceedings of the 2019 International Seminar on Research of Information Technology and Intelligent Systems (ISRITI)* (pp. 462–467). IEEE. 10.1109/ISRITI48646.2019.9034572

Shapley, L. S. (1953). Stochastic Games. *Proceedings of the National Academy of Sciences of the United States of America*, 39(10), 1095–1100. 10.1073/pnas.39.10.109516589380

Shepardson, D. P., Niyogi, D., Choi, S., & Charusombat, U. (2009). Seventh-grade students' conceptions of global warming and climate change. *Environmental Education Research*, 15(5), 549–570. 10.1080/13504620903114592

Shepardson, D. P., Roychoudhury, A., & Hirsch, A. S. (2017). *Teaching and learning about climate change*. Routledge Publishing. 10.4324/9781315629841

Shepardson, D. P., Roychoudhury, A., Hirsch, A., Niyogi, D., & Top, S. M. (2014). When the atmosphere warms it rains and ice melts: Seventh-grade students' conceptions of a climate system. *Environmental Education Research*, 20(3), 333–353. 10.1080/13504622.2013.803037

Shiva, V. (2010). Earth democracy: Beyond dead democracy and killing economies. *Capitalism, Nature, Socialism*, 21(1), 83–95. 10.1080/10455751003655922

Shiva, V. (2013). Everything I Need to Know I Learned in the Forest. *NAMTA Journal*, 38(1), 273–276.

Siegel, L. S., Homer, J., Fiddaman, T., McCauley, S., Franck, T., Sawin, E., ... Interactive, C. (2018). *En-ROADS simulator reference guide. Technical Report.*

Silko, L. M. (1997). *Yellow woman and a beauty of the spirit: Essays on Native American life today*. Simon & Schuster.

Silva, C. H.Junior, Pessôa, A. C., Carvalho, N. S., Reis, J. B., Anderson, L. O., & Aragão, L. E. (2021). The Brazilian Amazon deforestation rate in 2020 is the greatest of the decade. *Nature Ecology & Evolution*, 5(2), 144–145. 10.1038/s41559-020-01368-x33349655

Silvers, P., Shorey, M., & Crafton, L. (2010). Critical literacy in a primary multiliteracies classroom: The hurricane group. *Journal of Early Childhood Literacy*, 10(4), 379–409. 10.1177/1468798410382354

Siribanpitak, P. (2010). *Educational management for sustainable development: Basic education in economic, social and environmental aspects* (2nd ed.). Thai Samphan.

Smith, G. A. (2002). Place-based education: Learning to be where we are. *Phi Delta Kappan*, 83(8), 584–594. 10.1177/003172170208300806

Smith, J., Johnson, A., & Thompson, D. (2018). The Impact of Augmented Reality on Student Engagement: A Case Study Analysis. *Journal of Educational Technology Systems*, 47(1), 122–139.

Sobel, D. (2004). *Place-based education: Connecting classrooms and communities*. Orion Society.

Soğukpınar, R., & Korucu, D. K. (2020). Investigation of elementary school students' attitudes towards environment. *Mehmet Akif Ersoy University Journal of Education Faculty*, 53, 583–606.

Sommer, C., & Lücken, M. (2010). System competence–Are elementary students able to deal with a biological system? *Nordic Studies in Science Education*, 6(2), 125–143. 10.5617/nordina.255

Sopegno, A., Calvo, A., Berruto, R., Busato, P., & Bocthis, D. (2016). A web mobile application for agricultural machinery cost analysis. *Computers and Electronics in Agriculture*, 130, 158–168. 10.1016/j.compag.2016.08.017

Sorensen, T. J., Lambert, M. D., & McKim, A. J. (2021). A statewide examination of secondary agricultural educators' ability and willingness to teach climate change topics. *Journal of Agricultural Education*, 62(1), 213–229.

Soujanya, N., Roopa, Y. M., & Babu, D. K. (2019). Virtual reality meets IoT Through telepresence. In *Proceedings of the 2019 3rd International Conference on Trends in Electronics and Informatics (ICOEI)* (pp. 972–976). IEEE. 10.1109/ICOEI.2019.8862746

Stapp, W. B., Bennett, D., Bryan, W., Fulton, J., MacGregor, J., Nowak, P., & Havlick, S. (1969). The concept of environmental education. *The Journal of Environmental Education*, 1(1), 30–31. 10.1080/00139254.1969.10801479

STEELS Hub - SAS. (n.d.). Retrieved from https://www.pdesas.org/Page/Viewer/ViewPage/58/?fbclid=IwAR0x48gZxjpPHvkwu-B0rDGasOq9XfQrKHKF8GjQIFK5rSRW40TsGuma8D8_aem_AcCJ8EDhaHY5tMHE8erG4rJKOPWxgoOJ8YBg5fYWc3yYtDdnuV9WDHQhW4nq0I1bmntdSqvAuUZMtlGgfOyYcxAR

Sterman, J. D. (2001). System dynamics modeling: Tools for learning in a complex world. *California Management Review*, 43(4), 8–25. 10.2307/41166098

Sterman, J. D. (2014). Interactive web-based simulations for strategy and sustainability: The MIT Sloan LearningEdge management flight simulators, Part I. *System Dynamics Review*, 30(1-2), 89–121. 10.1002/sdr.1513

Sterman, J. D., & Sweeney, L. B. (2002). Cloudy skies: Assessing public understanding of global warming. *System Dynamics Review*, 18(2), 207–240. 10.1002/sdr.242

Sterman, J., Franck, T., Fiddaman, T., Jones, A., McCauley, S., Rice, P., Sawin, E., Siegel, L., & Rooney-Varga, J. N. (2015). World Climate: A role-play simulation of climate negotiations. *Simulation & Gaming*, 46(3-4), 348–382. 10.1177/1046878113514935

Stibbe, A. (2009). *The handbook of sustainability literacy: Skill for a changing world*. Green Books, Dartington.

Stone, R. K. & V. (2024). New science standards will be in our classrooms soon. Retrieved from https://gettysburgconnection.org/new-science-standards-will-be-in-our-classrooms-soon/

Strapasson, A., Ferreira, M., Cruz-Cano, D., Woods, J., do Nascimento Maia Soares, M. P., & da Silva Filho, O. L. (2022). The use of system dynamics for energy and environmental education. *International Journal of Educational Technology in Higher Education*, 19(1), 1–31. 10.1186/s41239-021-00309-335106364

Sultana, N., Hossen, M. S., & Khatun, R. (2017). Assessment of environmental knowledge and attitude of secondary level students of Tangail, Bangladesh. *Int. J. Res. Environ. Sci*, 3, 41–46.

Sümer, N. (2000). Yapısal Eşitlik Modelleri: Temel Kavramlar ve Örnek Uygulamalar. *Türk Psikoloji Yazilari*, 3, 49–73.

Suratruangchai, W. (2003). Research for teaching and learning development. *Journal of Education Science*, 14(2), 31–42.

Sutdhibhasilp, N. (2000). *Environmental adult education: two case studies of Thai non-governmental organizations promoting green consumerism*. University of Toronto.

Sutthaso, W., Sutthaso, R., & Piantham, C. (2022). Strategies for the development of education management of Pak Kret Municipality Pak Kret District Nonthaburi Province under the 20-year national strategy. *Academic Journal Faculty of Humanities and Social Sciences Phranakhon Rajabhat University*, 6(1), 29–41.

Sutthirat, C. (2009). *80 Innovations: Learner-centered learning management*. P Balans Design and Printing.

Sutthirat, C. (2016). *Techniques for using questions to develop thinking* (4th ed.). Chulalongkorn University Book Center.

Su, X., Pan, C., Zhou, S., & Zhong, X. (2022). Threshold effect of green credit on firms' green technology innovation: Is environmental information disclosure important? *Journal of Cleaner Production*, 380, 134945. Advance online publication. 10.1016/j.jclepro.2022.134945

Sweeney, L. B., & Sterman, J. D. (2000). Bathtub dynamics: Initial results of a systems thinking inventory. *System Dynamics Review*, 16(4), 249–286. 10.1002/sdr.198

Sweeney, L. B., & Sterman, J. D. (2007). Thinking about systems: Student and teacher conceptions of natural and social systems. *System Dynamics Review*, 23(2–3), 285–312. 10.1002/sdr.366

Szajowski, K. (2019). Stochastic Games and Learning. In Baillieul, J., & Samad, T. (Eds.), *Encyclopedia of Systems and Control* (pp. 1–7). Springer London. 10.1007/978-1-4471-5102-9_33-3

Tabachnick, B. G., & Fidell, L. S. (2007). *Using Multivariate Statistics* (5th ed.). Allyn and Bacon.

Tabachnick, B. G., & Fidell, L. S. (2013). *Using multivariate statistics* (6th ed.). Allyn and Bacon.

Tabuenca, B., Moreno-Sancho, J. L., Arquero-Gallego, J., Greller, W., & Hernández-Leo, D. (2023). Generating an environmental awareness system for learning using IoT technology. *Internet of Things : Engineering Cyber Physical Human Systems*, 22, 100756. 10.1016/j.iot.2023.100756

Tangprapha, T. (2013). Educational management for student competency development: A case study of elevating education management for student competency and outcome development in the 21st century in Singapore. *Journal of Education*, 41(3), 203–227.

Taylor-Powell, E., & Steele, S. (1996). Collecting evaluation data: Direct observation. *Program Development and Evaluation. University of Wisconsin-Extension.*

Teksoz, G. T. (2011). Managing air pollution: How does education help. *The impact of air pollution on health, economy, environment and agricultural sources, 1*, 397-422.

TEMA (2011-2015) Türkiye Erozyonla Mücadele Vakfı Genel Müdürlüğü Arşivi, İstanbul.

Tenney, A. W. (1977). The FFA at 50: A Golden Past—A Bright Future. *Phi Delta Kappan*, 58(9), 658–662.

Teyfur, E. (2008). The effect of academic achievement and enviroment club studies on primary school students towards the enviroment attitudes (Pattern of İzmir). *Ege Journal of Education*, 9(1), 131–149.

The 20-year national strategy (2018–2037). (2018, 13 October). Vol. 135 Chapter 82, Kor. 1–63.

The Goan. (2022, July 12). Virtual reality can support, enhance outdoor environmental education. Retrieved from https://www.thegoan.net/tg-life/%EF%BB%BFvirtual-reality-can-support-enhance-outdoor-environmental-education/86403.html

Thieman, E. B., Henry, A. L., & Kitchel, T. (2020). Resilient agricultural educators: Teacher motivations and adaptations under Covid-19. *Journal of Agricultural Education*, 61(4), 238–254.

Thomas Muñoz, R. (2022). Environmental education networks for social empowerment and global citizenship: a case of non-formal education from Mexico. In *Educational Response, Inclusion and Empowerment for SDGs in Emerging Economies: How do education systems contribute to raising global citizens?* (pp. 191–203). Springer International Publishing. 10.1007/978-3-030-98962-0_12

Thornber, J., Stanisstreet, M., & Boyes, E. (1999). School students' ideas about air pollution: Hindrance or help for learning? *Journal of Science Education and Technology*, 8(1), 67–73. 10.1023/A:1009481521719

True, A. C. (1929). *A history of agricultural education in the United States, 1785-1925* (No. 36). US Government Printing Office.

Tummons, J. D., Langley, G. C., Reed, N. A., & Paul, M. P. (2021). Contemporary practices in secondary agricultural education: A national study. *Journal of Agricultural Education*, 62(4), 88–103.

Tuncer, G., Tekkaya, C., Sungur, S., Cakiroglu, J., Ertepinar, H., & Kaplowitz, M. (2009). Assessing pre-service teachers' environmental literacy in Turkey as a mean to develop teacher education programs. *International Journal of Educational Development*, 29(4), 426–436. 10.1016/j.ijedudev.2008.10.003

Tynan, L. (2021). What is relationality? Indigenous knowledge, practices and responsibilities with kin. *Cultural Geographies*, 28(4), 597–610. 10.1177/14744740211029287

Uddin, M. K. (2023). Environmental education for sustainable development in Bangladesh and its challenges. *Sustainable Development*. Advance online publication. 10.1002/sd.2728

Uddin, M. M. M. (2014). An analysis of knowledge on sources of environmental pollution of students' at secondary level in Bangladesh. *International Journal of Academic Research in Business & Social Sciences*, 4(3), 2222–6990. 10.6007/IJARBSS/v4-i3/731

Ullah, M. O., Abu Hasan, M. D., & Uddin, T. (2013). Environmental Awareness and Disaster Factors in Bangladesh. *Journal of Applied Quantitative Methods*, 8(4).

UN. (n.d.). United Nations Conference on the Human Environment, 5-16 June 1972, Stockholm. Retrieved February 12, 2024, from https://www.un.org/en/conferences/environment/stockholm1972

UNESCO. (2010). *World data on education 7th edition 2010/11*.http://www.ibe.unesco.org

UNESCO. (2014). *Shaping the future we want. UN decade of education for sustainable development (2005–2014)*. UNESCO. https://unesdoc.unesco.org/images/0023/002301/230171e.pdf (accessed 17 May 2017).

UNESCO. (2014). Shaping the Future We Want: UN Decade of Education for Sustainable Development (2005-2014) Final Report. Retrieved from https://unesdoc.unesco.org/ark:/48223/pf0000230302/PDF/230302eng.pdf.multi

UNESCO. (2015a). *Getting climate-ready: A guide for schools on climate action*. UNESCO.

UNESCO. (2015b). *Global citizenship education: UNESCO's approach*. UNESCO.

UNESCO. (2016). *School in Action global citizens for sustainable development: A guide for students*. UNESCO.

UNESCO. (2017). *Education for sustainable development goals: Learning objectives*. UNESCO.

UNESCO. (2018a). *Integrating education for sustainable development (ESD) in teacher education in South-East Asia: A guide for teacher educators*. UNESCO.

UNESCO. (2018b). *Issue and trends in education for sustainable development*. UNESCO Publishing.

United Nations. (2016). Transforming our world: The 2030 agenda for sustainable development. *Civil Engineering (Johannesburg, South Africa)*, 2016(1), 26–30.

Vanichsenee, S. (2015). Factors and success conditions for basic education management. Lessons learnt from the past decades and future possibilities in looking backward and forward of basic education reforms. Office of the Basic Education Commission. Ministry of Education.

Vare, P., Arro, G., de Hamer, A., Del Gobbo, G., de Vries, G., Farioli, F., Kadji-Beltran, C., Kangur, M., Mayer, M., Millican, R., Nijdam, C., Reti, M., & Zachariou, A. (2019). Devising a competence-based training program for educators of sustainable development: Lessons learned. *Sustainability (Basel)*, 11(7), 1890. 10.3390/su11071890

Verhelst, D., Vanhoof, J., & Van Petegem, P. (2023). School effectiveness for education for sustainable development (ESD): What characterizes an ESD-effective school organization? *Educational Management Administration & Leadership*, 51(2), 502–525. 10.1177/1741143220985196

Vincent, S. K., & Villa, J. (2021). Agricultural communications undergraduate enrollment growth, curriculum development, and faculty perceptions. *Journal of Applied Communications*, 105(2).

Viphoouparakhot, V. (2021). Area-based education management in participatory administration of basic school administrators. *Panyapiwat Journal*, 13(3), 320–332.

Vlachopoulos, D., & Makri, A. (2017). The effect of games and simulations on higher education: A systematic literature review. *International Journal of Educational Technology in Higher Education*, 14(1), 1–33. 10.1186/s41239-017-0062-1

Wahyudin, D. (2018). Peace education curriculum in the context of education sustainable development (ESD). *Journal of Sustainable Development Education and Research*, 2(1), 21–32. 10.17509/jsder.v2i1.12354

Wals, A. E. J. (2015). *Beyond unreasonable doubt. Education and learning for socio-ecological sustainability in the anthropocene*. Wageningen University.

Wang, F., Du, L., & Tian, M. (2023). Does Agricultural Credit Input Promote Agricultural Green Total Factor Productivity? Evidence from Spatial Panel Data of 30 Provinces in China. *International Journal of Environmental Research and Public Health*, 20(1), 529. Advance online publication. 10.3390/ijerph2001052936612851

Wechayanan, P., Siribanpitak, P., Sriumpai, A., & Intharamaphan, W. (2016). Family's basic education management towards learners' desirable characteristics based on sufficiency economy. *Journal of Educational Administration Khon Kaen University*, 12(1), 145–155.

Weiss, C. L. A., & Mettrick, J. E. (2015). Individuals with Disabilities Education Act (IDEA). In Reynolds, C. R., & Fletcher-Janzen, E. (Eds.), *Encyclopedia of Cross-Cultural School Psychology* (pp. 542–545). Springer., 10.1007/978-0-387-71799-9_212

Welton, D. A., & Mallaro, A. A. (1915). *The Teaching of Agriculture in the High School*. Vocational Education Association of the Middle West.

Wen, F., Sun, Z., He, T., Shi, Q., Zhu, M., Zhang, Z., Li, L., Zhang, T., & Lee, C. (2020). Machine learning glove using self-powered conductive superhydrophobic triboelectric textile for gesture recognition in VR/AR applications. *Advancement of Science*, 7(14), 2000261. 10.1002/advs.20200026132714750

WESSA. (2016). *Stepping up to the sustainability goals: A practical guide to integrating the SDGs in our daily lives*. WESSA.

White, R. M., & Harder, M. K. (2013). The journey towards sustainability via community: Lessons from two UK universities. In Sterling, S., Maxey, L., & Luna, H. (Eds.), *The Sustainable University: Progress and Prospects* (pp. 132–150). Routledge.

Wiek, A., Bernstein, M., Foley, R. W., Cohen, M., Forrest, N., Kuzdas, C., & Keeler, L. W. (2015). Operationalising competencies in higher education for sustainable development. In *Routledge handbook of higher education for sustainable development* (pp. 241–260). Taylor and Francis.

Wiek, A., Withycombe, L., & Redman, C. L. (2011). Key competencies in sustainability: A reference framework for academic program development. *Sustainability Science*, 6(2), 203–218. 10.1007/s11625-011-0132-6

Williams, S. (2019). The appearance of voice: EAP and academic literacies approaches to teaching reflective writing.

Williams, D. L. (1991). Focusing Agricultural Education Research: Strategies for the Discipline. *Journal of Agricultural Education*, 32(3), 7–12. 10.5032/jae.1991.01007

Willis, J., Weiser, B., & Kirkwood, D. (2014). Bridging the gap: Meeting the needs of early childhood students by integrating technology and environmental education. *International Journal of Early Childhood Environmental Education*, 2(1), 140–155.

Wongwanich, S. (2006). *Action research in classrooms* (9th ed.). Chulalongkorn University Press.

Wordsworth, W. (1979). *The world is too much with us*. Dreadnaught.

Wordsworth, W. (1980). Expostulation and reply. *Journal for the Education of the Gifted*, 4(1), 49–49. 10.1177/016235328000400108

Wordsworth, W., & Hayden, J. O. (1981). *The poems* (Vol. 1). Yale University Press.

World Health Organization. (n.d.). *Air Pollution*. World Health Organization. https://www.who.int/health-topics/air-pollution#tab=tab_1

Worthington, R. L., & Navarro, R. L. (2003). Pathways to the future: Analyzing the contents of a content analysis. *The Counseling Psychologist*, 31(1), 85–92. 10.1177/0011000002239402

Wu, Y., Wan, J., & Yu, W. (2023). Impact of environmental education on environmental quality under the background of low-carbon economy. *Frontiers in Public Health, 11*, 1128791.

Wu, B., Yu, X., & Gu, X. (2020). Effectiveness of immersive virtual reality using head-mounted displays on learning performance: A meta-analysis. *British Journal of Educational Technology*, 51(6), 1991–2005. 10.1111/bjet.13023

Yang, X., Shu, L., Chen, J., Ferrag, M. A., Wu, J., Nurellari, E., ... & Huang, K. (2021). A survey on smart agriculture: development modes, technologies, and security and privacy challenges. *IEEE/CAA Journal of Automatica Sinica, 8*(2), 273-302. 10.1109/JAS.2020.1003536

Yeats, W. B., Hagstrom, F., & Gillespie, O. J. (1989). The second coming.

Yıldırım, A., & Şimşek, H. (2013). *Qualitative research methods in the social sciences* (9th ed.). Seçkin Press.

Yıldırım, A., & Şimşek, H. (2013). *Sosyal bilimlerde nitel araştırma yöntemleri* [Qualitative research methods in the social sciences]. Seckin Publishing Company.

Yıldız, K., Sipahioğlu, Ş., & Yılmaz, M. (2007). *Çevre Bilimi* (Genişletilmiş 2. Basım). Ankara: Gündüz Yayıncılık.

Yılmaz, V., & Çelik, E. H. (2009). *Structural equation modeling with Lisrel: basic concepts, applications, programming* [Lisrel ile yapısal eşitlik modellemesi: temel kavramlar, uygulamalar, programlama]. Pegem Publishing.

Yongyuth, P., Prada, R., Nakasone, A., Kawtrakul, A., & Prendinger, H. (2010, October). AgriVillage: 3D multi-language internet game for fostering agriculture environmental awareness. In *Proceedings of the International Conference on Management of Emergent Digital EcoSystems* (pp. 145-152). 10.1145/1936254.1936280

Young, M., & Commins, E. (2002). *Global citizenship: The handbook for primary teaching*. Oxfam.

Youngs, H. (2018). *Developing 3R8C skills towards Thailand 4.0 innovative leaders*. Thailand Tribune.

Yu, L., Liu, W., Yang, S., Kong, R., & He, X. (2022). Impact of environmental literacy on farmers' agricultural green production behavior: Evidence from rural China. *Frontiers in Environmental Science*, 10, 990981. Advance online publication. 10.3389/fenvs.2022.990981

Zafar, M. W., Qin, Q., & Zaidi, S. A. H. (2020). Foreign direct investment and education as determinants of environmental quality: The importance of post Paris Agreement (COP21). *Journal of Environment*

Zeydanlı, U., & Tuğ, S. (2008). Biyolojik Çeşitlilik ve Orman Ekosistemlerimdeki Önemi. Orman ve Biyolojik Çeşitlilik (s. 15-37). Ankara: Doğa Koruma Merkezi.

Zhao, X., Ren, Y., & Cheah, K. S. L. (2023). Leading Virtual Reality (VR) and Augmented Reality (AR) in Education: Bibliometric and Content Analysis From the Web of Science (2018–2022). *SAGE Open*, 13(3), 21582440231190821. Advance online publication. 10.1177/21582440231190821

Zheng, M., & Campbell, A. G. (2019). Location-based augmented reality in-situ visualization applied for agricultural fieldwork navigation. In *Proceedings of the 2019 IEEE International Symposium on Mixed and Augmented Reality Adjunct (ISMAR-Adjunct)* (pp. 93–97). IEEE. 10.1109/ISMAR-Adjunct.2019.00039

Zheng, S., Zhang, X., & Wang, H. (2023). Green credit policy and the stock price synchronicity of heavily polluting enterprises. *Economic Analysis and Policy*, 77, 251–264. 10.1016/j.eap.2022.11.011

Zweifach, S., & Triola, M. (2019). Extended reality in medical education: Driving adoption through provider-centered design. *Digital Biomarkers*, 3(1), 14–21. 10.1159/00049892332095765

Zwozdiak-Myers, P. N. (2023). *The Teacher's Reflective Practice Handbook: Becoming an Extended Professional Through Enacting Evidence-informed Practice*. Taylor & Francis. 10.4324/9781003307143

About the Contributors

Lizoon Nahar is a teaching associate and a doctoral program student in the Department of Professional Studies in Education at Indiana University of Pennsylvania (IUP). Her research interest focuses on innovative science teaching techniques and the integration of technology into science education in developing countries. She has 12 years of experience working as a secondary school biology teacher in Dhaka, Bangladesh. She has a Master of Science degree in Zoology from the University of Dhaka. Lizoon is the winner of the Global Teacher Grant 2018 and Alumni Engagement Innovation Fund (AEIF) 2019, Julius P. Filcik Fellowship Award 2021, PEO International Peace Scholarship 2021, 2022, and Alumni Ties Small Grant 2023.

Nada Tayem is a teaching associate and a doctoral candidate in English Literature and Criticism at Indiana University of Pennsylvania. Her scholarly work is distinguished by its interdisciplinary approach and global perspective, with a research focus on environmental feminism, place, and belonging in the writings of the Caribbean and Middle East. she began her academic journey in Jordan, where she earned her B.A. and M.A. in English Literature before pursuing her research in the US. With over a decade of teaching experience, Nada has taught in various educational settings in higher education. Her expertise spans teaching and translation, instructing undergraduates in English literature, writing composition, and language learning, including UN programs.Beyond her academic pursuits, she is deeply committed to community service. She has held leadership positions in various academic organizations and dedicated her time to support English language learners and refugees in the US.

* * *

Farzana Akter Munni is a graduate student and researcher at the Department of Information Science and Library Management at the University of Rajshahi in Bangladesh. Since her early university career, she has shown keen interest in scholarly exploration in the emerging and promising areas in the field of her university education. Simultaneously, she grew a robust and substantial interest and motivation in learning and practicing research in her field. She continues the collaboration with multiple roles in several research societies based on the university campus. In the meantime, she has presented two full papers with her valuable findings at two international conferences. Her research interests include audio-visual materials, digital library systems, digital divide, resources sharing and networking, metadata, information management, informatics, learning management systems, information quality and strategies, big data, information visualization, etc. She wants to continue her scholarly pursuit and build her career as a researcher in the future.

Saroash Ali is a professional entrepreneur and architect with a passion for researching sustainable environmental development and children's education. She has worked on various architectural projects mainly focusing on her area of interest. She has also worked on projects that promote awareness of climate change and its adverse effects on young children in Pakistan.

Fatma Alkan is works as professor in Hacettepe University Faculty of Education, Chemistry Teaching Program. She has published many articles in national and international peer-reviewed journals. She works as a field editor in peer-reviewed journals.

Zeynep Isilay Baskahya graduated from the Middle East Technical University, Department of Elementary Mathematics Education, in 2016. And in October 2021, she graduated from the Middle East Technical University Mathematics Education master's program. Between 2019 and 2023, she worked as a mathematics teacher at Darüşşafaka Private Middle School in İstanbul. She was introduced to systems thinking in education while teaching mathematics at Darussafaka Private Middle School in 2020, and she is still a member of the Systems Thinking Association. As of September 2023, she started her doctoral studies in Educational Sciences at Eötvös Loránd University in Hungary.

Fatima Zahra Bessedik is a lecturer of English Literature at the Department of English, Faculty of Foreign Languages, University of Oran 2 (Algeria). She is specilized in English Literature and critical theory.

A. N. Bushra is a dedicated young researcher who has excelled in her academic journey in Public Administration. With both Bachelor's and Master's degrees achieved with distinction, Bushra demonstrates outstanding analytical prowess and a strong dedication to addressing social disparities, particularly in gender equality, public policy, e-governance in developing nations, public healthcare, and social justice. Her current research focuses on pivotal areas shaping the future of public administration, including data-driven decision-making, the impact of Artificial Intelligence and automation on public services, smart governance initiatives, and government innovation. Published in prestigious journals, her work has ignited significant discourse and holds the potential to influence policy-making. As Bushra embarks on her career, her intellect, perseverance, and commitment position her as a promising scholar ready to contribute substantially to the social sciences.

Esra Cantürk is a non-Thesis Master's Student, Institute of Educational Sciences, Hacettepe University, Turkey.

Gaye Defne Ceyhan is an Assistant Professor of Science Education at Bogazici University. After working as a middle school science teacher for eight years, she received her Ph.D. in Science Education from Syracuse University. Her research focuses on STEM and climate change education, exploring innovative ways to teach STEM and emphasizing systems thinking, sustainable living, and climate change awareness. Dr. Ceyhan has been involved in several projects and has written academic papers to improve climate change education. Her academic endeavors aim to bridge the gap between scientific understanding and practical application in addressing environmental challenges.

Meltem Ceylan Alibeyoglu studied mathematics at Middle East Technical University. She has a Ph.D. in environmental sciences at Bogazici University, Istanbul. Alibeyoglu worked as a mathematics teacher, project coordinator, PYP Coordinator, and middle school academic coordinator at Darussafaka Educational Institutions. Since 2022, she has been working as a Research and Development Coordinator. She is a part-time instructor at Bogazici University and MEF University. She is a member of the Systems Thinking Association in Turkey and many non-governmental organizations where she has been involved in the establishment and work. She is actively working in her organization and the Association to develop the applications of systems thinking in education.

Saadet Etli is a science teacher at Enka Middle School. After completing her undergraduate degree in Science Education at Bogazici University in 2014, she graduated master's program in Environmental Sciences and Engineering at Istanbul Technical University in 2021. While working as a Science teacher at Darussafaka Middle School, she joined the Systems Thinking in Education Association. Since 2021, she has also been a member of the Systems Thinking Association. She creates enriched science lessons and projects using a systems thinking approach and system dynamics tools.

Gorkem Girgin graduated from Istanbul University, Department of Physics. He has been working as a science teacher at Darussafaka Middle School for eight years. He actively applies Systems Thinking in his Science course. He is an active member of the Systems Thinking Association.

Emre Goktepe graduated from Middle East Technical University with a degree in Industrial Engineering in 1988. He worked in various positions in the family company from 1990 to 2013 and as a production consultant from 2001 to 2013. His interest in systems thinking and system dynamics continued throughout his working life and has become a full-time pursuit since his retirement in 2013. He is a co-founder of the Systems Thinking Association. He has participated in national and international studies at teachers, schools, universities, and non-governmental organizations. He has aimed to promote the systems thinking approach in K-12 education.

Nihal Güneş Demir is a graduate of the Faculty of Science Education at Bogazici University and has been a science teacher for 15 years. While working at Darussafaka Educational Institutions, she met with the Systems Thinking Association of Turkey, received training on systems thinking, participated in the process of preparing and implementing lesson plans, and had the opportunity to present her work at various conferences. She is currently working at Hisar Foundation Educational Institutions. She is also a postgraduate student at MEF University Learning Sciences. She is a member of the Systems Thinking Association. She is in the process of writing her final project on systems thinking.

Fatemeh Jamshidi is currently a Doctoral Candidate in Computer Science and Software Engineering at Auburn University. Her research interests lie at the intersection of Music Mining, Artificial Intelligence, Machine Learning, and Information Retrieval. She has published several research papers at premier conferences including ACM SIGCSE, and HCII. She obtained her MS in Computer Science and Software Engineering and Music Education from Auburn University, respectively in 2023 and 2021. During her Ph.D., Fatemeh established Computing + Music programs, which have engaged hundreds of participants from underrepresented groups since 2018. Fatemeh has served as a Director of the Persian Music Ensemble at Auburn University from 2020 -2023 and as an invited reviewer for ACM SIGCSE in 2021, 2022, 2023, and 2024.

Busra Karga received her MS in Mathematics and Science Education from Bogazici University in 2023. With a background in science education, she started her professional journey as a science teacher in private schools in 2020. Recently, she joined a project as a project manager focusing on good practices of circular economy initiatives in the context of school education. Her interests include practical applications of sustainability education, systems thinking approach, and STEM education.

Farheen Mahmood is a teaching associate and a doctoral program student in the Department of Professional Studies in Education at Indiana University of Pennsylvania (IUP). Her research interests focus on children's literature and the integration of technology into education. She has over 21 years of teaching and administrative experience in Pakistan, Yemen, Qatar, UAE, and the US. She is the winner of the Carter G. Woodson Multicultural Award (2021), the Julius P. Filcik Fellowship Award (2023), and the American Reading Forum Graduate Scholarship (2023).

Daniela Marghitu received her B.S. in Automation and Computing from Polytechnic University of Bucharest, and her Ph.D. degree in Automation and Computing from University of Craiova. She is a faculty member in the Computer Science and Software Engineering Department at Auburn University, where she has worked since 1996. Her teaching experience includes a variety of Information Technology and Computing courses (e.g., Object-Oriented Programming for Engineers and Scientists, Introduction to Computing for Engineers and Scientists, Network Programming with HTML and Java, Web Development and Design Foundations with HTML 5.0, CSS3.0 and JavaScript, Personal Computer Applications, Spreadsheet-Based Applications with Visual BASIC, Web Application Development with XML amd JSP). Her research areas include STEM K12 Inclusive Computing Research and Outreach; Web Applications Design and Development; Education and Assistive Technology; Software Engineering; Web and Software Engineering Usability and Accessibility. Dr. Marghitu has received seven awards for research and education projects from National Science Foundation (e.g.Co-PI of NSF RET Site: Project-Based Learning for Rural Alabama STEM Middle School Teachers in Machine Learning and Robotics; Co-PI of NSF INCLUDES Alliance: The Alliance of Students with Disabilities for Inclusion, Networking, and Transition Opportunities in STEM (TAPDINTO-STEM); Co-PI of NSF EEC "RFE Design and Development: Framing Engineering as Community Activism for Values-Driven Engineering"; Co-PI of NSF CISE "EAGER: An Accessible Coding Curriculum for Engaging Underserved Students with Special Needs in Afterschool Programs"; co-PI of NSF INCLUDES: South East Alliance for Persons with Disabilities in STEM, Co-PI of NSF CE 21 Collaborative Research: Planning Grant: Computer Science for All (CS4ALL)), and Co-PI for NIFA "Reimagining Controlled Environment Agriculture in a Low Carbon World". Dr. Marghitu was also PI of grants from ACM Special Interest Group on Computer Science Education, Center for Woman in Information Technology, Daniel F. Breeden Endowment for Faculty Enhancement, AccessComputing Alliance, Computer Science Collaboration Project, Microsoft Fuse Research, Altova Co., and Pearson Education Publishing Co. Dr. Marghitu has mentored over fifteen hundreds k12, computing undergraduate, graduate students including representatives of underserved/underrepresented communities, women, and people with disabilities. Dr. Marghitu has participated in numerous administrative activities at Auburn University. Among these activities are the following: Auburn University Board of Trustee Faculty Representative; Auburn University representative for National Center for Women in Information Technology, AccessComputing, Access10K, and AccessEngineering Alliances; Auburn University Persons with Disabilities Committee chair; Founder and Director Auburn University Laboratory for Education and Assistive Technology; faculty representative Auburn University Core Curriculum Oversight committee and

Multicultural Diversity Commission. Dr. Marghitu also served as World Usability Day Web Site Committee Chair; Alabama STEM Education board chair, Panel member for the National Science Foundation; member of the congressionally mandated Committee on Equal Opportunities in Science and Engineering; member of the Committee on the Future of NSF EPSCoR; member of the Computer Science for All (CSforAll) Accessibility Board, College Board's Advanced Placement Program (AP) Computer Science A Development Committee (DC) and Lime Connect University Partner. Dr. Marghitu published seven Information Technology books at Pearson Publishing Co., articles at International Journal On Advances in Software, International Journal On Advances in Internet Technology, Journal of Women and Minorities in Science and Engineering, National Science Teaching Association Journals, Journal of Computer Science Education,International Journal on Advances in Internet Technology Transactions of the SDPS: Journal of Integrated Design and Process Science, User Experience Magazine, Journal of Computing Sciences in Colleges, International Journal for Virtual Reality, Journal of SMET Education and Research. Dr. Marghitu has published peer reviewed papers and gave presentations at numerous international conferences(e.g. ACM Special Interest Group on Computer Science Education Technical Symposium, International Technology and Persons with Disabilities Conference, International Conference on Software Engineering Advances, EDUCAUSE, Association for Advancement of Computing in Education, International Society for Technology in Education, Society for Design and Process Science, American Society for Engineering Education, Human Computer Interaction International Conference, and International Academy, Research, and Industry Association) in USA, Canada, England, France, Germany, Spain, Italy, Portugal and Romania. Her work was also presented by co-authors at conferences in Brazil, Taiwan and S. Korea. Among Dr. Marghitu's honors and awards are the following: 2011 AccessComputing Capacity Building Award, the 2012 Auburn University Access award, the 2012 Society for Design and Process Science Outstanding Achievement Award, the 2013 Microsoft Fuse Research award, the 2015 DO-IT Trailblazer award, the 2017 International Academy, Research, and Industry Association Fellowship, the 2017 Society for Design and Process Science Fellowship, and the 2019 Samuel Ginn College of Engineering 100+ Women Strong Leadership in Diversity Faculty Award.

Shorif Mollah, PhD, is a young enthusiast in applying cutting-edge technologies in the educational sector to make a positive change in the field, and he has achieved a Ph.D. in Educational Technology from the Department of Educational Technology at Beijing Normal University, Beijing, China. His research interests include the use of artificial intelligence in education, microlearning (ML), digital divide, critical pedagogies, micro-credentials (MC), blockchain in education, social & digital media in education, digital citizenship & discourse analysis, intelligent technologies & pedagogy, OER, VR, AR, LA, IoT, cloud storage for education & blended learning, design, etc. Dr. Mollah has been continuing his academic contribution through different conferences, symposia, academic publications, international discussions, dialogue, and knowledge sharing through professional roles, and he wants to stick to his academic mission in the future.

Ece Unsal graduated from Yeditepe University's Secondary Mathematics Teaching Department in 2017. The same year, she began working as a mathematics teacher at Darussafaka Schools. Since 2021, she has been the same institution's team leader for the secondary school mathematics department. She is also pursuing a master's degree in Educational Technologies at Bahcesehir University and is in the thesis stage. Since 2021, she has also been a member of the Systems Thinking Association. She creates enriched mathematics lessons and projects using a systems thinking approach and system dynamics tools.

Ulkem Yararbas, a professor of nuclear medicine, serves on the medical faculty at Ege University. She is one of the founders of the Systems Thinking Association in Turkey. In addition to the elective systems thinking courses she gives to medical school students, she voluntarily supports the activities for the infusion of ST/SD in the education system conducted by the System Thinking Association in Turkey.

Sena Yildiz Degirmenci graduated from the Department of Physics Education at Boğaziçi University, followed by a master's degree in Education, Learning, and Technology at the University of Oulu in Finland. While working as a science teacher at Darussafaka Middle School, she became acquainted with the Systems Thinking Association in Turkey and participated in projects focused on integrating systems thinking into science education. Currently, she works as a research assistant at the Technical University of Munich and remains an active member of the Systems Thinking Association in Turkey, contributing voluntarily to both national and international initiatives related to systems thinking in education.

Index

A

B

C

D

E

H

I

L

M

P

R

S

T

V

www.ingramcontent.com/pod-product-compliance
Lightning Source LLC
LaVergne TN
LVHW080341011026
849161LV00009B/2149